CAMBRIDGE TRACTS IN MATHEMATICS

General Editors

180 Synthetic Geometry of Manifolds

Synthetic Geometry of Manifolds

ANDERS KOCK
University of Aarhus

CAMBRIDGE
UNIVERSITY PRESS

CAMBRIDGE UNIVERSITY PRESS
Cambridge, New York, Melbourne, Madrid, Cape Town, Singapore, São Paulo, Delhi

Cambridge University Press
The Edinburgh Building, Cambridge CB2 8RU, UK

Published in the United States of America by Cambridge University Press, New York

www.cambridge.org
Information on this title: www.cambridge.org/9780521116732

First published 2010

Printed in the United Kingdom at the University Press, Cambridge

A catalogue record for this publication is available from the British Library

Library of Congress Cataloguing in Publication data
Kock, Anders.
Synthetic geometry of manifolds / Anders Kock.
p. cm. – (Cambridge tracts in mathematics ; 180)
Includes bibliographical references and index.
ISBN 978-0-521-11673-2 (hardback)
1. Geometry, Differential. 2. Manifolds (Mathematics) I. Title. II. Series.
QA641.K735 2009
516.3′62 – dc22 2009038164

ISBN 978-0-521-11673-2 Hardback

Contents

Preface

This book deals with a certain aspect of the theory of smooth manifolds, namely (for each k) the *kth neigbourhood of the diagonal.* A part of the theory presented here also applies in algebraic geometry (smooth schemes).

The neighbourhoods of the diagonal are classical mathematical objects. In the context of algebraic geometry, they were introduced by the Grothendieck school in the early 1960s; the Grothendieck ideas were imported into the context of smooth manifolds by Malgrange, Kumpera and Spencer, and others. Kumpera and Spencer call them "prolongation spaces of order k".

The study of these spaces has previously been forced to be rather technical, because the prolongation spaces are not themselves manifolds, but live in a wider category of "spaces", which has to be described. For the case of algebraic geometry, one passes from the category of varieties to the wider category of schemes; for the smooth case, Malgrange, Kumpera and Spencer, and others, described a category of "generalized differentiable manifolds with nilpotent elements" (Kumpera and Spencer, 1973, p. 54).

With the advent of topos theory, and of synthetic differential geometry, it has become possible to circumvent the construction of these various categories of generalized spaces, and instead to deal axiomatically with the notions. This is the approach we take; in my opinion, it makes the neighbourhood notion quite elementary and expressive, and in fact, provides a non-technical and geometric gateway to many aspects of differential geometry; I hope the book can be used as such a gateway, even with very little prior knowledge of differential geometry.

Therefore, the audience I have in mind with this book is anybody with a reasonable mathematical maturity, who wants to *learn* some differential geometry; but of course, I also invite the differential geometer to see aspects of his/her field from the synthetic angle, using the neighbourhood notion.

The specific requirements for reading the book are knowledge of multivariable calculus and linear algebra, and of basic commutative ring theory. Also, we require some basic category theory, in particular as it applies in the category of sets: the category of sets is a topos.

Concretely about the axiomatics: rather than specifying what (generalized) spaces *are*, we specify what a *category* $\mathscr{E}$ of generalized spaces should look like. And the simplest is to start this specification by saying: "the category $\mathscr{E}$ is a topos" (more precisely, a topos $\mathscr{E}$ in which there is given a commutative ring object† R). For, as we know now – through the work of Lawvere and the other topos theorists – toposes behave almost like the category of naive sets, so familiar to all mathematicians. In other words, whatever the objects (the "generalized spaces") are, we may reason about them *as if* they were sets – provided we only reason "constructively", e.g. avoid using the law of excluded middle. It is natural in differential geometry to avoid this law, since it is anyway by use of this law that non-smooth functions are constructed. An aspect of this "as-if" is that the words "set" and "space" are used synonymously: both mean just "an object of $\mathscr{E}$".

The reasoning in a topos *as if* it just were the topos of naive sets is the core in the synthetic method. The synthetic method opens the way to an *axiomatic* treatment of some aspects of differential geometry (as well as of analytic, algebraic, etc. geometry).

For many aspects of differential geometry, such axiomatic treatment is well documented in many publications; this particularly applies to the aspects deriving from the notion of *tangent vector* and *tangent bundle*, and their generalizations; see Kock (1981/2006) and the references therein (notably the references in the 2nd edition). We do not presuppose that the reader is familiar with Kock (1981/2006), nor with the other treatises on synthetic differential geometry, like Moerdijk and Reyes (1991) or Lavendhomme (1996) – provided he/she is willing to take the step of thinking in terms of naive set theory. We shall in the Appendix recapitulate the basic ingredients for the interpretation of naive set theory in toposes, but we shall not go into the documentation that the method is healthy. At the time of 1981 (1st edition of Kock, 1981/2006) or 1991 (Moerdijk and Reyes, 1991), this issue had to be dealt with more energetically: both for the question of *how* to interpret naive set theory in a topos, and for the question of actually *producing* toposes which were models for the various axioms.

The particular geometric notions and theorems that we expound in synthetic form are mainly paraphrased from the classical differential geometric literature; I have chosen such theories where the neigbourhood notions appeared to be natural and gave transparency. They all belong to *local* differential

† This ring object is intended to model the geometric line.

geometry; no global considerations enter. For this reason, the key kind of objects considered, *manifolds* M, may as well be thought of as open subsets of finite-dimensional vector spaces V; *locally*, any manifold is of course like this. Many proofs, and a few constructions, therefore begin with a phrase like "it suffices to consider the case where M is an open subset of a finite-dimensional vector space $V\ldots$"; and sometimes we just express this by saying "in a standard coordinatized situation…". However, it is important that the notions and constructions (but not necessarily the proofs) are from the outset coordinate free, i.e. are independent of choice of coordinatization of M by V. (The notion of *open* subset, and the derived notion of a space being *locally* something, we shall, for flexibility, take as axiomatically given; see Appendix Section A.6.)

I have not attempted (nor been able) to give historical credits to the classical notions and theories, since my sources (mainly textbooks) are anyway not the primary ones (like Riemann, Lie, Cartan, Ehresmann, …). Most of these topics expounded are discussed from the synthetic viewpoint in scattered articles (as referenced in the Bibliography). I shall not list these topics completely here, but shall just give a list of some "key words": affine connections, combinatorial differential forms, geometric distributions, jet bundles, (Lie) groupoids, connections in groupoids, holonomy and path connections, Lie derivative, principal bundles and principal connections, differential operators and their symbols, Riemannian manifolds, Laplace operator, harmonic maps.

For the reader with some previous experience in synthetic differential geometry, in the form as in Kock (1981/2006), Lavendhomme (1987, 1996) or Moerdijk and Reyes (1991), some comparison may be expedient.

Most of the theory which we develop here only depends on core axiomatics for synthetic differential geometry, and it is satisfied in all the standard models – both the well-adapted models for C^∞ manifolds (cf. Dubuc, 1979; Moerdijk and Reyes, 1991), and the topos models for algebraic geometry, as studied by the Grothendieck school, as in Demazure and Gabriel (1970).

For the most basic topics, like the "Kock–Lawvere" axiom scheme, and the multivariable calculus derived from it, we develop these issues from scratch in Chapter 1, and this chapter therefore has some overlap with Kock (1981/2006).

Otherwise, the overlap with Kock (1981/2006) is quite small; for, the synthetic part (Part I) of that book dealt with arbitrary "microlinear" spaces, and could therefore not go into the more specific geometric notions that exist only for finite-dimensional manifolds, and precisely such notions are the topic of the present book.

The reader should not take this book as anything like a complete survey of the present state of synthetic differential geometry; a wealth of important aspects are left out. This in particular applies to the applications of the synthetic method to the "infinite-dimensional" spaces that appear in functional analysis, say, in calculus of variations, continuum mechanics, distribution theory (in the sense of Schwartz)...; the theory of such spaces becomes more transparent by being seen in a cartesian closed category, and in fact, motivated the invention of, and interest in, cartesian closed categories in the mid-sixties, cf. Lawvere (1979). The bibliography in the 2nd edition (2006) of Kock (1981/2006) provides some references; notably Kock (1986), Kock and Reyes (1987, 2003, 2006).

The question of formulating integration axioms, and finding well-adapted topos models for them, is hardly touched upon in the present book, except that a possible formulation of the Frobenius integrability theorem is attempted in Section 2.6. Similarly for "infinitesimal-to-local" results. There are some deep investigations in this direction in Bunge and Dubuc (1987) and Penon (1985).

Neither do we touch on the role of "tininess/atomicity" of those infinitesimal objects that occur in synthetic differential geometry. To say that an object D in a cartesian closed category is *tiny* (or is an *atom*) is to say that the functor $(-)^D$ has a right adjoint. Except for the terminal object 1, naive reasoning is incompatible with tininess. On the other hand, tiny objects give rise to some amazing theory, cf. Lawvere (1998); e.g. to the construction of a category $\mathscr{E}_0$ of "discrete" spaces out of the category $\mathscr{E}$ of "all" spaces. Also, they give rise to construction of "spaces" classifying differential forms and de Rham cohomology, cf. Dubuc and Kock (1984); almost a kind of Eilenberg–Mac Lane space.

The neighbourhoods of the diagonal are, as said, invented in algebraic geometry, and make sense there even for spaces (schemes) which are not manifolds. Much of the theory developed here for manifolds therefore makes sense for more general schemes, as witnessed by the work of Breen and Messing (2001); I regret that I have not been able to include more of this theory. Some simple indication of the neighbourhoods of the diagonal, for affine schemes from a synthetic viewpoint, may be found in Dubuc and Kock (1984, §4) and in Kock (2004, §1).

We use the abbreviation "SDG" for synthetic differential geometry.

Acknowledgements

The mentors that I have had for this work are several, but three need to be mentioned in particular: Lawvere, Joyal, and C. Ehresmann. Lawvere opened up the perspective of synthetic/category-theoretic reasoning, with his study of categorical dynamics in 1967, leading ultimately, via Kock (1977), to the "Kock–Lawvere axiom" scheme for *R* ("KL axiom") as expounded in Chapter 1; Joyal pointed out that in this context, the neighbour relation could be used for a synthetic theory of differential forms and bundle-connections. Ehresmann formulated the jet notion, which is intimately related to the neighbourhoods of the diagonal (this relationship is the backbone in the book by Kumpera and Spencer, 1973, where also Ehresmann's use of differentiable groupoids is a main theme – as it also is in our Chapter 5).

I have already acknowledged my scientific debt to my mentors Lawvere, Joyal, and Ehresmann. (Unfortunately, I only met Ehresmann once, briefly.) I want to thank the two other mentors for many conversations on the topic of SDG (and on other subjects as well).

Also thanks for many crucial discussions with Gavin Wraith, Marta Bunge, Eduardo Dubuc, René Lavendhomme,† Ieke Moerdijk, Ronnie Brown, Larry Breen, Bill Messing, Paul Taylor, Hirokazu Nishimura, Joachim Kock, Marcelo Fiore, and in particular to my faithful collaborator over several decades, Gonzalo Reyes. Many others should be mentioned as well.

Most diagrams were made using Paul Taylor's package.

† Deceased.

Acknowledgements

1

Calculus and linear algebra

One does not get far in differential geometry without calculations. This also applies for that synthetic approach which we present. We develop in this chapter the basic calculus and algebra needed. The fundamental differentiation process (formation of directional derivatives) here actually becomes part of the algebra, since the classical use of limit processes is eliminated in favour of the use of infinitesimal subspaces of the number line R and of the coordinate vector spaces R^n. These infinitesimal spaces are defined in an algebraic, and ultimately coordinate-free, way, so that they may be defined as subspaces of arbitrary finite-dimensional vector spaces V. The combinatorial notion of "pairs of points in V which are k-neighbours" ($k = 0, 1, 2, \ldots$), written $x \sim_k y$, is introduced as an aspect of these infinitesimal spaces. The neighbour relations $\sim_k$ are invariant under all, even locally defined, maps. This opens up consideration of the neighbour relations in general manifolds in Chapter 2.

The content of this chapter has some overlap with the existing textbooks on SDG (notably with Part I of Kock, 1981/2006) and is, as these, based on the KL axiom scheme.

1.1 The number line R

The axiomatics and the theory to be presented involve a sufficiently nice category $\mathscr{E}$, equipped with a commutative ring object R, the "number line" or "affine line"; the symbol R is chosen because of its similarity with $\mathbb{R}$, the standard symbol for the ring of real numbers. The category $\mathscr{E}$ is typically a topos (although for most of the theory, less will do). Thus the axiomatics deals with a *ringed topos* $(\mathscr{E}, R)$. The objects of $\mathscr{E}$ are called "spaces", or "sets"; these words are used as synonyms, as explained in the Appendix. Therefore also "ring object" is synonymous with "ring". Also, "map" is synonymous with "smooth map", equivalently, the phrase "smooth" applies to all maps in $\mathscr{E}$, and therefore it is void, and will rarely be used.

Unlike $\mathbb{R}$, R is not assumed to be a *field*, because this would exclude the existence of the ample supply of nilpotent elements (elements $x \in R$ with $x^n = 0$ for some n) which are basic to the axiomatics presented here. We do, for simplicity, assume that R has characteristic 0, in the sense that the elements $1+1$, $1+1+1$, etc. are invertible; equivalently, we assume that R contains the field $\mathbb{Q}$ of rational numbers as a subring. (Part of the theory can be developed without this assumption, or with the assumption that $x+x=0$ implies $x=0$; in fact, as said in the Preface, part of the theory originates in algebraic geometry, where positive characteristic is taken seriously.) For some arguments, we need to assume that R is a local ring: "if a sum is invertible, then at least one of the terms in it is invertible". In Chapter 8, we shall furthermore assume that R is formally real, in the sense that if x_1 is invertible, then so is $\sum_{i=1}^n x_i^2$; or we shall even assume that R is Pythagorean, in the sense that a square root of such sum exists. No order relation is assumed on R.

Since R is not a field, and the logic does not admit the rule of excluded middle, the theory of R-modules is not quite so simple as the classical theory of vector spaces over a field. Therefore we have to make explicit some points and notions. A *linear* map is an R-linear map between R-modules. An R-module V is called a *finite-dimensional vector space* if *there exists* a linear isomorphism between V and some R^n, in which case we say that V *has dimension* n. The phrase (quantifier) "there exists" has to be interpreted according to sheaf semantics; in particular, it suffices that V is *locally* isomorphic to R^n. If U and V are finite-dimensional vector spaces, a linear inclusion $j : U \to V$ makes U into a *finite-dimensional subspace* of V if *there exists* a linear complement $U' \subseteq V$ with U' likewise finite dimensional.

An example of a linear subspace (submodule) of a finite-dimensional vector space, which is not itself a finite-dimensional vector space, is given in Exercise 1.3.4.

A manifold is a space which *locally* is diffeomorphic to a finite-dimensional vector space; to explain the phrase "locally", one needs a notion of *open* subset. This notion of "open", we shall present axiomatically as well (as in algebraic geometry), see the Appendix. A main requirement is that the set R^* of invertible elements in R is an open subset.

Note that R^* is stable under addition or subtraction of nilpotent elements: if x is nilpotent, say $x^{n+1} = 0$, and $a \in R^*$, then $a - x \in R^*$; for, an inverse for it is given by the geometric series which stops after the nth term, by the nilpotency assumption on x; thus, since $x^{n+1} = 0$,

$$(1-x)^{-1} = 1 + \sum_{k=1}^{n} x^n.$$

This relationship between "invertible" and "nilpotent", together with the stability properties of the property of being open, together imply that open subsets M of R^n are "formally open", meaning that if $a \in M$ and $x \in R^n$ is "infinitesimal" in the sense described in the next section, then $a + x \in M$. In most of the theory to be developed, the notion of open could be replaced by the weaker notion of formally open. In a few places, we write "(formally) open", to remind the reader of this fact. But we do not want to overload the exposition with too much esoteric terminology.

1.2 The basic infinitesimal spaces

We begin by describing some equationally defined subsets of R, of R^n (= the vector space of n-dimensional coordinate vectors), and of $R^{m \cdot n}$ (= the vector space of $m \times n$-matrices over R). The descriptions are then given in coordinate-free form, so that we can generalize them into descriptions of analogous subobjects, with R^n replaced by any finite-dimensional vector space V.

The fundamental one of these subsets is $D \subseteq R$,

$$D := \{x \in R \mid x^2 = 0\}.$$

More generally, for n a positive integer, we let $D(n) \subseteq R^n$ be the following set of n-dimensional coordinate vectors $\underline{x} = (x_1, \ldots, x_n) \in R^n$:

$$D(n) := \{(x_1, \ldots, x_n) \in R^n \mid x_j x_{j'} = 0 \text{ for all } j, j' = 1, \ldots, n\},$$

in particular (by taking $j = j'$), $x_j^2 = 0$, so that $D(n) \subseteq D^n \subseteq R^n$. The inclusion $D(n) \subseteq D^n$ will usually be a proper inclusion, except for $n = 1$. Note also that $D = D(1)$. Note that if $\underline{x}$ is in $D(n)$, then so is $\lambda \cdot \underline{x}$ for any $\lambda \in R$, in particular, $-\underline{x}$ is in $D(n)$ if $\underline{x}$ is. In general, $D(n)$ is not stable under addition. For instance, for d_1 and d_2 in D, $d_1 + d_2 \in D$ iff $(d_1, d_2) \in D(2)$ iff $d_1 \cdot d_2 = 0$.

The objects D and $D(n)$ may be called *first-order* infinitesimal objects. We also have *kth-order* infinitesimal objects: if k is any positive integer, $D_k \subseteq R$ is

$$D_k := \{x \in R \mid x^{k+1} = 0\}.$$

More generally,

$$D_k(n) := \{(x_1, \ldots, x_n) \in R^n \mid \text{ any product of } k+1 \text{ of the } x_i\text{s is } 0\}.$$

Note $D = D_1$, $D(n) = D_1(n)$; and that $D_k(n) \subseteq D_l(n)$ if $k \leq l$.

The notation for the spaces D, $D(n)$, D_k, and $D_k(n)$ is the standard one of SDG. The following space $\widetilde{D}(m,n)$ is less standard, and was first described in Kock (1981/2006, §I.16 and §I.18), with the aim of constructing a combinatorial notion of differential m-form; see Chapter 3.†

The subset $\widetilde{D}(m,n) \subseteq R^{m \cdot n}$ is the following set of $m \times n$ matrices $[x_{ij}]$ $(m,n \geq 2)$:

$$\widetilde{D}(m,n) := \{[x_{ij}] \in R^{m \cdot n} \mid x_{ij}x_{i'j'} + x_{i'j}x_{ij'} = 0 \\ \text{for all } i,i' = 1,\ldots,m \text{ and } j,j' = 1,\ldots,n\}.$$

We note that the equations defining $\widetilde{D}(m,n)$ are row–column symmetric; equivalently, the transpose of a matrix in $\widetilde{D}(m,n)$ belongs to $\widetilde{D}(n,m)$. Also, clearly any $p \times q$ submatrix of a matrix in $\widetilde{D}(m,n)$ belongs to $\widetilde{D}(p,q)$ (p and $q \geq 2$). For, if the defining equations

$$x_{ij}x_{i'j'} + x_{i'j}x_{ij'} = 0 \tag{1.2.1}$$

hold for all indices i,i',j,j', they hold for any subset of them. And since each of the equations in (1.2.1) only involve (at most) four indices i,i',j,j', we see that for an $m \times n$ matrix to belong to $\widetilde{D}(m,n)$, it suffices that all of its 2×2 submatrices belong to $\widetilde{D}(2,2)$.

If $[x_{ij}] \in \widetilde{D}(m,n)$, we get in particular, by putting $i = i'$ in the defining equations (1.2.1), that for any $j,j' = 1,\ldots,n$

$$x_{ij}x_{ij'} + x_{ij}x_{ij'} = 0.$$

Since 2 is assumed cancellable in R, we deduce from this equation that $x_{ij}x_{ij'} = 0$, which is to say that the ith row of $[x_{ij}]$ belongs to $D(n)$. Similarly, the jth column belongs to $D(m)$.

An $m \times n$ matrix is in $\widetilde{D}(m,n)$ iff all its $2 \times n$ submatrices are in $\widetilde{D}(2,n)$. We have a useful characterization of such $2 \times n$ matrices:

Proposition 1.2.1 *Consider a $2 \times n$ matrix as an element $(\underline{x},\underline{y})$ of $R^n \times R^n$. Then $(\underline{x},\underline{y}) \in \widetilde{D}(2,n)$ iff $\underline{x} \in D(n)$, $\underline{y} \in D(n)$ and for any symmetric bilinear $\phi : R^n \times R^n \to R$, $\phi(\underline{x},\underline{y}) = 0$.*

Proof. The left-hand sides of the defining equations (1.2.1) with $i = 1$ and $i' = 2$ generate the vector space of symmetric bilinear maps $R^n \times R^n \to R$, and for $i = i' = 1$, (1.2.1) means that $\underline{x} \in D(n)$, and similarly for $i = i' = 2$, (1.2.1) means that $\underline{y} \in D(n)$.

† The object $\widetilde{D}(m,n)$ is denoted $\Lambda^m D(n)$ in Kock (1996).

In Chapter 8, we shall have occasion to study an infinitesimal space $D_L(n) \subseteq R^n$ (the "L" is for "Laplace"); for $n \geq 2$,

$$D_L(n) := \{(x_1, \ldots, x_n) \in R^n \mid x_1^2 = \ldots = x_n^2 \text{ and } x_i \cdot x_j = 0 \text{ for } i \neq j\}.$$

It is easy to see that $D_1(n) \subseteq D_L(n) \subseteq D_2(n)$ (or, see the calculations after the proof of Proposition 8.3.2). (For $n = 1$, we put $D_L(n) := D_2(n)$)

Coordinate-free aspects of $D_k(n)$

We may characterize $D_k(n) \subseteq R^n$ in a coordinate-free way:

Proposition 1.2.2 *Let $\underline{x} \in R^n$. Then $\underline{x} \in D_k(n)$ if and only if for all $k+1$-linear $\phi : (R^n)^{k+1} \to R$, we have $\phi(\underline{x}, \ldots, \underline{x}) = 0$. Equivalently, $\underline{x} \in D_k(n)$ if and only if for all $k+1$-homogeneous $\Phi : R^n \to R$, we have $\Phi(\underline{x}) = 0$.*

Proof. This follows because the monomials of degree $k+1$ in n variables span the vector space of $k+1$-linear maps $(R^n)^{k+1} \to R$; and the $D_k(n)$ is by definition the common zero set of all these monomials.

In particular, $\underline{x} \in D(n)$ iff for all bilinear $\phi : R^n \times R^n \to R$, $\phi(\underline{x}, \underline{x}) = 0$.

Because of the proposition, we may define $D(V)$ and $D_k(V)$ for any finite-dimensional vector space (= R-module isomorphic to some R^n):

$$D(V) := \{v \in V \mid \phi(v, v) = 0 \text{ for any bilinear } \phi : V \times V \to R\}, \quad (1.2.2)$$

and similarly

$$D_k(V) := \{v \in V \mid \phi(v, v, \ldots, v) = 0 \text{ for any } (k+1)\text{-linear } \phi : V^{k+1} \to R\}, \quad (1.2.3)$$

or equivalently

$$D_k(V) = \{v \in V \mid \Phi(v) = 0 \text{ for any } (k+1)\text{-homogeneous } \Phi : V \to R\}.$$

(For the coordinate-free notion of "homogeneous map", see Section A.7.) For $V = R^n$, we recover the objects already defined, $D_k(R^n) = D_k(n)$.

It is clear from the coordinate-free presentation that if $\phi : V_1 \to V_2$ is a linear map between finite-dimensional vector spaces, then

$$\phi(D_k(V_1)) \subseteq D_k(V_2). \quad (1.2.4)$$

The construction D_k is actually a functor from the category of finite-dimensional vector spaces to the category of pointed sets (the "point" being $0 \in D_k(V) \subseteq V$).

Exercise 1.2.3 Prove that

$$D(V) = \{v \in V \mid \phi(v,v) = 0 \text{ for any symmetric bilinear } \phi : V \times V \to R\}.$$

(*Hint:* Use (1.2.2), and decompose ϕ into a symmetric bilinear map and a skew bilinear map.)

Proposition 1.2.4 *Let U be a finite-dimensional subspace of a finite-dimensional vector space V. Then $D(U) = D(V) \cap U$.*

Proof. The inclusion $\subseteq$ is trivial. For the converse, assume $x \in U \cap D(V)$. To prove $x \in D(U)$, it suffices, by the (coordinate-free version of) Proposition 1.2.2, to prove that $\phi(x,x) = 0$ for all bilinear $\phi : U \times U \to R$. But given such ϕ, it extends to a bilinear $\psi : V \times V \to R$, since U is a retract of V. Then $\psi(x,x) = 0$, since $x \in D(V)$, hence $\phi(x,x) = 0$.

Alternatively, prove the assertion for the special case where $i : R^m \to R^n$ is the canonical inclusion, and argue that the notions in question are invariant under linear isomorphisms.

A subset $S \subseteq D(V)$ is called a *linear subset* (we should really say: a finite-dimensional linear subset, to be consistent) if it is of the form $D(V) \cap U$ for a finite-dimensional linear subspace $U \subseteq V$. (Actually, under the axiomatics to be introduced in the next section, U is uniquely determined by S.) Then by Proposition 1.2.4, $S = D(U)$.

If $f : V_1 \to V_2$ is a linear isomorphism between finite-dimensional vector spaces, and $S \subseteq D(V_1)$ is a linear subset, then its image $f(S) \subseteq D(V_2)$ is a linear subset as well.

Proposition 1.2.5 *Let V be a finite-dimensional vector space. Then if $\underline{d} \in D_k(V)$ and $\underline{\delta} \in D_l(V)$, we have $\underline{d} + \underline{\delta} \in D_{k+l}(V)$.*

Proof. It suffices to consider the case where $V = R^n$, so $\underline{d} = (d_1, \ldots, d_n)$, $\underline{\delta} = (\delta_1, \ldots, \delta_n)$. The argument is now a standard binomial expansion: a product of $k+l+1$ of the coordinates of $(d_1 + \delta_1, \ldots, d_n + \delta_n)$ expands into a sum of products of $k+l+1$ d_is or δ_js; in each of the terms in this sum, there is either at least $k+1$ d-factors, or at least $l+1$ δ-factors; in either case, we get 0.

For any finite-dimensional vector space V, we define the *kth-order neighbour relation* $u \sim_k v$ by

$$u \sim_k v \text{ iff } u - v \in D_k(V).$$

If this holds, we say that u and v are *kth-order neighbours*. The relation $\sim_k$ is a reflexive relation, since $0 \in D_k(V)$, and it is symmetric since $d \in D_k(V)$ implies $-d \in D_k(V)$. It is not a transitive relation; but we have, as an immediate consequence of Proposition 1.2.5:

Proposition 1.2.6 *If $u \sim_k v$ and $v \sim_l w$ then $u \sim_{k+l} w$.*

We are in particular interested in the *first*-order neighbour relation, $u \sim_1 v$, which we sometimes therefore abbreviate into $u \sim v$; and we use the phrase u and v are *neighbours* when $u \sim_1 v$. The (first-order) neighbour relation is the main actor in the present treatise. The higher-order neighbour relation will be studied in Section 2.7. In Chapter 8 on metric notions, the second-order neighbour relation plays a major role.

It follows from (1.2.4) that any linear map between finite-dimensional vector spaces preserves the property of being kth-order neighbours. (In fact, under the axiomatics in force from the next section and onwards, *any* map preserves the kth-order neighbour relations.)

Remark. There are infinitesimal objects $\subseteq R^n$ which are not coordinate free, i.e. which cannot be defined for abstract finite-dimensional vector spaces V instead of R^n; an example is $D^n \subseteq R^n$, i.e. $\{(d_1,\ldots,d_n) \in R^n \mid d_i^2 = 0 \text{ for all } i\}$. Concretely, this can be seen by observing that D^n is not stable under the action on R^n of the group $GL(n,R)$ of invertible $n \times n$ matrices. The infinitesimal object $D_L(n)$ is not stable under $GL(n)$ either, but it is stable under $O(n)$, the group of orthogonal matrices, and is studied in Chapter 8 on metric notions.

Aspects of $\widetilde{D}$

The equations (1.2.1) defining $\widetilde{D}(m,n)$ can be reformulated in terms of a certain bilinear $\beta : R^n \times R^n \to R^{n^2}$, where $\beta(\underline{x},\underline{y})$ is the n^2-tuple whose jj' entry is $x_j y_{j'} + x_{j'} y_j$. Then an $m \times n$ matrix X ($m,n \geq 2$) is in $\widetilde{D}(m,n)$ if and only if $\beta(\underline{r}_i, \underline{r}_{i'}) = 0$ for all $i,i' = 1,\ldots,m$ ($\underline{r}_i$ denoting the ith row of X).

Note that this description is not row–column symmetric. But it has the advantage of making the following observation almost trivial:

Proposition 1.2.7 *If an $m \times n$ matrix X is in $\widetilde{D}(m,n)$, then the matrix X' formed by adjoining to X a row which is a linear combination of the rows of X, is in $\widetilde{D}(m+1,n)$.*

(There is, of course, a similar proposition for columns.) Combining this proposition with the observation that the rows of a matrix in $\widetilde{D}(p,n)$ are in $D(n)$, we therefore have

Proposition 1.2.8 *If X is a matrix in $\widetilde{D}(m,n)$, then any row in X is in $D(n)$, and also any linear combination of rows of X is in $D(n)$. Similarly for columns.*

We have a "geometric" characterization of matrices in $\widetilde{D}(m,n)$ in terms of the (first-order) neighbour relation $\sim$, namely the equivalence of (1) and (2) (or of (1) and (3)) in the following

Proposition 1.2.9 *Given an $m \times n$ matrix $X = [x_{ij}]$ $(m,n \geq 2)$. Then the following five conditions are equivalent: (1) the matrix belongs to $\widetilde{D}(m,n)$; (2) each of its rows is a neighbour of $0 \in R^n$, and any two rows are mutual neighbours; (3) each of its columns is a neighbour of $0 \in R^m$, and any two columns are mutual neighbours. (2') any linear combination of the rows of X is in $D(n)$; (3') any linear combination of the columns of X is in $D(m)$.*

Proof. We have already observed (Proposition 1.2.8) that (1) implies (2'), which in turn trivially implies (2).

Next, assume the condition (2). Let $\underline{r}_i$ denote the ith row of the matrix. Then the condition (2) in particular says that the $\underline{r}_i$ and $\underline{r}_{i'}$ are neighbours; this means that for any pair of column indices j, j',

$$(\underline{r}_i - \underline{r}_{i'})_j \cdot (\underline{r}_i - \underline{r}_{i'})_{j'} = 0$$

where for a vector $\underline{x} \in R^n$, $\underline{x}_j$ denotes its jth coordinate. So

$$(x_{ij} - x_{i'j}) \cdot (x_{ij'} - x_{i'j'}) = 0.$$

Multiplying out, we get

$$x_{ij}x_{ij'} - x_{ij}x_{i'j'} - x_{i'j}x_{ij'} + x_{i'j}x_{i'j'} = 0. \tag{1.2.5}$$

The first term vanishes because $\underline{r}_i \in D(n)$, and the last term vanishes because $\underline{r}_{i'} \in D(n)$. The two middle terms therefore vanish together, proving that the defining equations (1.2.1) for $\widetilde{D}(m,n)$ hold for the matrix; so (1) holds. This proves equivalence of (1), (2), and (2'). The equivalence of (1), (3), and (3') now follows because of the row–column symmetry of the equations defining $\widetilde{D}(m,n)$.

Remark 1.2.10 The condition (2) in this proposition was the motivation for the consideration of $\widetilde{D}(m,n)$, since the condition says that the m rows of the matrix, together with the zero row, form an *infinitesimal m-simplex*, i.e. an $m+1$-tuple of mutual neighbour points, in R^n; see Kock (1981/2006, §I.18; 2000), as well as Chapter 2 below. In the context of SDG, the theory of differential m-forms, in its combinatorial formulation, has for its basic input quantities such infinitesimal m-simplices. The notion of infinitesimal m-simplex, and of affine combinations of the vertices of such, make invariant sense in any manifold N, due to some of the algebraic stability properties (in the spirit of Proposition 1.2.11 below) which $\widetilde{D}(m,n)$ enjoys.

The set of matrices $\widetilde{D}(m,n)$ was defined for $m,n \geq 2$ only, but it will make statements easier if we extend the definition by putting $\widetilde{D}(1,n) = D(n)$, $\widetilde{D}(m,1) = D(m), \widetilde{D}(1,1) = D$ (here, of course, we identify R^p with the set of $1 \times p$ matrices, or $p \times 1$ matrices, as appropriate). By Proposition 1.2.8, the assertion that $p \times q$ submatrices of matrices in $\widetilde{D}(m,n)$ are in $\widetilde{D}(p,q)$ retains its validity, also for $p = 1$ or $q = 1$.

Proposition 1.2.11 *Let $X \in \widetilde{D}(m,n)$. Then for any $p \times m$ matrix P, $P \cdot X \in \widetilde{D}(p,n)$; and for any $n \times q$ matrix Q, $X \cdot Q \in \widetilde{D}(m,q)$.*

Proof. Because of the row–column symmetry of the property of being in $\widetilde{D}(k,l)$, it suffices to prove one of the two statements of the proposition, say, the first. So consider the $p \times n$ matrix $P \cdot X$. Each of its rows is a linear combination of rows from X, hence is in $D(n)$, by Proposition 1.2.8. But also any linear combination of rows in $P \cdot X$ is in $D(n)$, since a linear combination of linear combinations of some vectors is again a linear combination of these vectors. So the result follows from Proposition 1.2.9.

Since the neighbour relation $\sim$ applies in arbitrary finite-dimensional vector spaces V, it follows from the proposition that we may define $\widetilde{D}(m,V) \subseteq V^m$ as the set of m-tuples $v_1, \ldots, v_m$ of vectors in V such that $v_i \sim v_j$ for all $i, j = 1, \ldots, m$, and such that $v_i \sim 0$ for all $i = 1, \ldots, m$. Linear isomorphisms $V_1 \to V_2$ preserve this construction. With this definition, $\widetilde{D}(m,R^n) = \widetilde{D}(m,n)$. The notion of infinitesimal m-simplex in R^n (as in Remark 1.2.10) immediately carries over to arbitrary finite-dimensional vector spaces.

We leave it to the reader to derive the following coordinate-free corollary of Proposition 1.2.1:

Proposition 1.2.12 *Let V be a finite-dimensional vector space. Let x and y be elements of V. Then $(x,y) \in \widetilde{D}(2,V)$ iff $x \in D(V)$, $y \in D(V)$, and for any symmetric bilinear $\phi : V \times V \to R$, $\phi(x,y) = 0$.*

Let V be an R-module. Then there is a bilinear

$$R^{mn} \times V^n \to V^m$$

essentially given by matrix multiplication (viewing elements of R^{mn} as $m \times n$ matrices): the ith entry in $\underline{\underline{d}} \cdot \underline{v}$ is $\sum_j d_{ij} \cdot v_j$. For instance, a linear combination $\sum_{j=1}^{n} t_j \cdot \underline{v}_j$ is the matrix product $\underline{t} \cdot \underline{v}$, where $\underline{t}$ is the $1 \times n$ (row) matrix $(t_1, \ldots, t_n)$.

For any vector space (R-module) V, and any $m \times n$ matrix $\underline{\underline{t}}$, we therefore have a linear map $V^n \to V^m$ given by matrix multiplication $\underline{v} \mapsto \underline{\underline{t}} \cdot \underline{v}$, where $\underline{v} \in V^n$.

Proposition 1.2.12 has the following corollary:

Proposition 1.2.13 *Let* $(v_1, \ldots, v_k) \in \widetilde{D}(k, V)$, *i.e. the* v_i*s are mutual neighbours, and neighbours of* 0. *Then all linear combinations of these vectors are also mutual neighbours and are neighbours of* 0.

Proposition 1.2.11 has the following coordinate-free formulation:

Proposition 1.2.14 *If* $\underline{\underline{t}}$ *is an* $m \times n$ *matrix in* $\widetilde{D}(m,n)$, *then* $\underline{\underline{t}} \cdot \underline{v} \in \widetilde{D}(m, V)$, *for any* $\underline{v} \in V^n$.

It is clear that in a finite-dimensional vector space V, a $k+1$-tuple of points $(x_0, x_1, \ldots, x_k)$ in V are mutual neighbours iff

$$(x_1 - x_0, \ldots, x_k - x_0) \in \widetilde{D}(k, V).$$

An *affine combination* is a linear combination where the sum of the coefficients is 1. Since translations ($x \mapsto x - x_0$ for fixed x_0) preserve affine combinations, and also preserve the property of being neighbours, we immediately get from Proposition 1.2.13:

Proposition 1.2.15 *Let* $x_0, x_1, \ldots, x_k$ *be mutual neighbours in* V. *Then all affine combinations of them are also mutual neighbours.*

We leave to the reader to prove, in analogy with the proof of Proposition 1.2.4:

Proposition 1.2.16 *Let* U *be a finite-dimensional subspace of a finite-dimensional vector space* V. *Then* $\widetilde{D}(k, U) = \widetilde{D}(k, V) \cap (U \times \ldots \times U)$.

Exercise 1.2.17 Prove that $D(V \times V) \subseteq \widetilde{D}(2, V)$. (The "KL" axiomatics introduced in the next section will imply that the inclusion is a proper inclusion; see Exercise 1.3.5.) Prove that if V is 1-dimensional, $D(V \times V) = \widetilde{D}(2, V)$.

Proposition 1.2.18 *Let x and y be in $D(V)$, and let $B : V \times V \to V$ be bilinear. Then $x+y+B(x,y) \in D(V)$ iff $(x,y) \in \widetilde{D}(2,V)$.*

Proof. Assume $x+y+B(x,y) \in D(V)$. To prove that $(x,y) \in \widetilde{D}(2,V)$, it suffices by Proposition 1.2.12 to see that $\phi(x,y) = 0$, for any symmetric bilinear $\phi : V \times V \to R$. By the assumption and (1.2.2), we have

$$\phi(x+y+B(x,y), x+y+B(x,y)) = 0.$$

Use bilinearity of ϕ to expand this into nine terms; $\phi(x,x)$ and $\phi(y,y)$ vanish by (1.2.2); some others, like $\phi(y, B(x,y))$, vanish: it contains y in a bilinear position, so again (1.2.2) does the job. Only two terms remain, so we get $\phi(x,y) + \phi(y,x) = 0$. Since ϕ was assumed symmetric, we conclude from this that $\phi(x,y) = 0$, as desired.

Conversely, assume $(x,y) \in \widetilde{D}(2,V)$. To prove $x+y+B(x,y) \in D(V)$, we use (1.2.2): consider a bilinear $\phi : V \times V \to R$ and prove $\phi(x+y+B(x,y), x+y+B(x,y)) = 0$. Again, expand by bilinearity into nine terms; as before, only $\phi(x,y) + \phi(y,x)$ remains. But as a function of x and y, this is a *symmetric* bilinear function (even though ϕ itself was not assumed to be symmetric), and therefore it vanishes on the pair (x,y), by Proposition 1.2.12. This proves the proposition.

We shall make explicit a certain variant of the product rule for determinants.

Proposition 1.2.19 *Let $\underline{\underline{d}}$ be an $n \times n$-matrix, and let $F : V^n \to W$ be an n-linear alternating map into some vector space W. Then for any n-tuple $\underline{v}$ of vectors in V, $F(\underline{\underline{d}} \cdot \underline{v}) = \det(\underline{\underline{d}}) \cdot F(\underline{v})$.*

Exercise 1.2.20 Consider the function T which to a $k \times k$ matrix associates the product of its diagonal entries. As a function $(R^k)^k \to R$, T is clearly k-linear. Prove that its restriction to $\widetilde{D}(k,k)$ is alternating. Conclude that for $X \in \widetilde{D}(k,k)$, $k! \cdot T(X)$ equals the determinant of X. Conclude that if $k \geq 2$, then any diagonal matrix in $\widetilde{D}(k,k)$ has determinant 0.

1.3 The KL axiom scheme

The objects $D_k(n)$, $\widetilde{D}(m,n)$, etc. studied in the previous section are all *infinitesimal*, in a sense which can be made precise using the notion of "Weil-algebra", see e.g. Kock (1981/2006); the general KL axiom scheme† refers to all infinitesimal objects defined by such algebras. (We refer to Dubuc, 1979; Kock, 1981/2006; and Moerdijk and Reyes, 1991 for the question of models for the

† "KL" is short for "Kock–Lawvere".

axioms; a brief indication is given in Section A.3.) We shall henceforth only need some special cases, namely the ones corresponding to the particular infinitesimal objects introduced so far. These special cases are

• KL axiom for $D_k = D_k(1)$: Every map $D_k \to R$ is of the form

$$t \mapsto a_0 + a_1 \cdot t + \ldots + a_k \cdot t^k$$

for uniquely determined $a_0, a_1, \ldots, a_k \in R$.

This axiom clearly implies that R has the property: if $a_0, a_1, \ldots, a_k \in R$ are so that the function $t \mapsto a_0 + a_1 \cdot t + \ldots + a_k \cdot t^k$ is constant 0, then the a_is are 0. This property (R is "polynomially faithful") of R implies (cf. Section A.7 in the Appendix) that the notion of polynomial *map* between R-modules has good properties.

The KL axiom just stated is the case $n = 1$ of the following KL axiom, which we formulate in terms of the notion of polynomial map:

• KL axiom for $D_k(n)$: Every map $D_k(n) \to R$ extends uniquely to a polynomial map $R^n \to R$ of degree $\leq k$.

(The polynomials in question have coefficients from R.)

• KL axiom for $\widetilde{D}(m,n)$: Every map $\widetilde{D}(m,n) \to R$ can be written uniquely in the form $\underline{\underline{x}} \mapsto \sum_S \det(\underline{\underline{x}}_S) \cdot \alpha_S$, where S ranges over the set of submatrices of size $p \times p$ ($p \leq m$ and $p \leq n$) of $\underline{\underline{x}}$ (including the empty submatrix, whose determinant is taken to be 1), and where $\alpha_S \in R$.

There is also a KL axiom for $D_L(n)$, see Section 8.3.

Let us note some special cases and some immediate consequences:

(i) Every map $D(n) \to R$ extends uniquely to an affine map $R^n \to R$.
(ii) Every map $D(n) \to R$ taking 0 to 0 extends uniquely to a linear map $R^n \to R$.
(iii) Every map $D(n)^m \to R$ taking value 0 if one of the m input arguments is 0 extends uniquely to an m-linear map $(R^n)^m \to R$.
(iv) Every map $f : \widetilde{D}(m,n) \to R$ which has the property: "$f(\underline{\underline{x}}) = 0$ for all $m \times n$-matrices $\underline{\underline{x}}$ with a zero row" extends uniquely to an m-linear alternating map $(R^n)^m \to R$.

(Here, we identify the vector space of $m \times n$ matrices with the vector space $(R^n)^m$.)

The axioms and the consequences quoted imply immediately some "coordinate-free" versions for finite-dimensional vector spaces V, like "every map $D(V) \to R$ extends uniquely to an affine map $V \to R$"; we shall below state such coordinate-free versions, even with some further generality, namely we want to replace the codomain R by certain suitable R-modules W; the "suitable" ones are the *KL vector spaces* which we shall now define. (They are called *Euclidean modules* in Kock, 1981/2006.) The property of being a KL vector space is invariant under R-linear isomorphisms: if W and W' are isomorphic as R-modules, and W is KL, then so is W'. In particular, the axioms as they are stated above, may be expressed: "the R-module R is a KL vector space". It is easy to see that then also any R^n is a KL vector space, and hence any finite dimensional vector space is a KL vector space. But even some "infinite-dimensional" (whatever this means) R-modules, like R^M (= the set of functions from an arbitrary space M to R), are a KL vector space.

We need to make precise what we mean by a *polynomial* map $V \to W$, of degree $\leq k$, where V is a finite-dimensional vector space; we refer to the Appendix, Section A.7.

Then we say that the R-module W is a *KL vector space* if the following conditions hold:

• KL axiom for $D_k(V)$ relative to W: Every map $D_k(V) \to W$ extends uniquely to a polynomial map $V \to W$ of degree $\leq k$,

and

• KL axiom for $\widetilde{D}(m,n)$ relative to W: Every map $\widetilde{D}(m,n) \to W$ can be written uniquely in the form $\underline{\underline{x}} \mapsto \sum_S \det(\underline{\underline{x}}_S) \cdot \alpha_S$ where S ranges over the set of square submatrices of $\underline{\underline{x}}$ (including the empty submatrix, whose determinant is taken to be 1), and $\alpha_S \in W$.

From the first of the KL axioms here, we have in particular: if $\tau : D \to W$ is any map, there is a unique $w \in W$ such that for all $d \in D$, $\tau(d) = \tau(0) + d \cdot w$; this w is called the *principal part* of τ. (The reader may want to think of this w as $\tau'(0)$.)

Another immediate consequence of the first KL axiom here (again for $k = 1$) is the following:

• Let $x \in V$. Then every map $x + D(V) \to W$ extends uniquely to an affine map $V \to W$. (Affine = "a constant plus a linear map".)

For vectors w_1 and w_2 in a KL vector space W, we also deduce the following simple "cancellation principle": if $d \cdot w_1 = d \cdot w_2$ for all $d \in D$, then $w_1 = w_2$. Since the conclusion $w_1 = w_2$ depends on validity of $d \cdot w_1 = d \cdot w_2$ for *all* $d \in D$, we also express its use by saying that we are "cancelling *universally quantified* ds".

Consider a finite-dimensional vector space V and a KL vector space W; then

Proposition 1.3.1 *Let $K : V \to W$ be linear. Then K can be reconstructed from its restriction k to $D(V)$ as follows. For $v \in V$, $K(v)$ is the principal part of the map $D \to W$ given by $d \mapsto k(d \cdot v)$ (d ranging over D).*

Proof. This amounts to proving that for all $d \in D$, $d \cdot K(v) = k(d \cdot v)$; for, by the simple cancellation principle, this characterizes the vector $K(v)$. Since K is linear, $d \cdot K(v) = K(d \cdot v)$, and since K extends k and $d \cdot v \in D(V)$, $K(d \cdot v) = k(d \cdot v)$.

Let us note some further "cancellation principles" that come from the uniqueness assertions in the above KL axioms and their consequences:

- If $w \in W$ (a KL vector space) has $d \cdot w = 0$ for all $d \in D$, then $w = 0$.

By iteration of this, we get: If $d_1 \cdot d_2 \cdot w = 0$ for all $(d_1, d_2) \in D \times D$, then $w = 0$.

(We do not want to make the assumption that every $d \in D$ can be written in the form $d = d_1 \cdot d_2$ with d_1 and d_2 in D, i.e. we don't assert that the multiplication map $D \times D \to D$ is surjective. However, if f and g are maps $D \to R$ such that for all $(d_1, d_2) \in D \times D$, $f(d_1 \cdot d_2) = g(d_1 \cdot d_2)$, then $f(d) = g(d)$ for all $d \in D$; this follows immediately from the KL axiom for D. One sometimes expresses this by saying "R *perceives* the multiplication map $D \times D \to D$ to be surjective". There are several similar things that R (or, more generally, any microlinear object, cf. Section A.4) "perceive to be true"; cf. Kock, 1981/2006.)

Furthermore, for any positive natural number k, we have:

- If $\det(\underline{\underline{d}}) \cdot w = 0$ for all $\underline{\underline{d}} \in \widetilde{D}(k, k)$, then $w = 0$.

Another aspect of the cancellation principles concerns linear subspaces defined as zero sets of linear functions. For instance, consider a linear map $V \to W$ between R-modules, where W is KL; let $S \subseteq V$ be the zero set. Then in order for $x \in V$ to be in S, it suffices that $\det(\underline{\underline{d}}) \cdot x \in S$ for all $\underline{\underline{d}} \in \widetilde{D}(k, k)$.

It is possible to state the KL axiom for $\widetilde{D}(m, V)$ instead of $\widetilde{D}(m, n)$ (where V is a finite-dimensional vector space) but the uniqueness assertion cannot be stated in so elementary a way. Let us be content with observing some consequences; V denotes a finite-dimensional vector space, and W denotes a KL vector space:

- Every map $D(V)^m \to W$ taking value 0 if one of the m input arguments is 0 extends uniquely to an m-linear map $V^m \to W$.

• Every map $f : \widetilde{D}(m,V) \to W$ which has the property that it takes value 0 if one of the m input arguments is 0 extends uniquely to an m-linear alternating map $V^m \to W$.

In this sense, $D(V)^m$ is an *"m-linear map classifier"*, and $\widetilde{D}(m,V)$ is an *"m-linear-alternating map classifier"*. From the latter also follows that if $\phi : V^m \to W$ is m-linear, and $(v_1,\dots,v_m) \in \widetilde{D}(m,V)$, then ϕ "behaves *as if* it were alternating", in the sense that $\phi(v_{\sigma(1)},\dots,v_{\sigma(m)}) = \mathrm{sign}(\sigma)\phi(v_1,\dots,v_m)$.

(One can be more explicit: consider the object P defined by the pushout diagram

$$\begin{array}{ccc} \bigcup D(V)^{m-1} & \longrightarrow & D(V)^m \\ \downarrow & & \downarrow \\ 1 & \longrightarrow & P \end{array}$$

with the top map the inclusion, and the upper left corner the union of the coordinate hyperplanes. Then P is a pointed object, and base-point-preserving maps $P \to W$ are in bijective correspondence with m-linear maps $V^m \to W$. Similarly, the pushout object Q in

$$\begin{array}{ccc} \bigcup \widetilde{D}(m-1,V) & \longrightarrow & \widetilde{D}(m,V) \\ \downarrow & & \downarrow \\ 1 & \longrightarrow & Q \end{array}$$

will classify m-linear *alternating* maps $V^m \to W$, in the same sense.)

An immediate corollary of this, and of Proposition 1.2.16, is the following:

Proposition 1.3.2 *Let U be a finite-dimensional linear subspace of a finite-dimensional vector space V. If a bilinear alternating $\theta : V \times V \to W$ vanishes on $(U \times U) \cap \widetilde{D}(2,V)$, then it vanishes on $U \times U$.*

Proof. By assumption and by Proposition 1.2.16, θ vanishes on $\widetilde{D}(2,U)$. A bilinear alternating $U \times U \to W$ (like the restriction of θ to $U \times U$) which vanishes on $\widetilde{D}(2,U)$ is the zero map.

Proposition 1.3.3 *Let V be a finite-dimensional vector space and W a KL vector space. Given a bilinear $\phi : V \times V \to W$, then the following three conditions are equivalent: (1) ϕ is symmetric; (2) ϕ vanishes on $\widetilde{D}(2,V)$; (3) for any u and v in $D(V)$, $\phi(u,v)$ only depends on $u+v$.*

Proof. We first prove the equivalence of the first two conditions. Since the assertion is coordinate free (invariant under linear isomorphisms), we may assume that $V = R^n$. If ϕ is symmetric, it follows by coordinate calculations that it may be written in the form

$$(\underline{x}, \underline{y}) \mapsto \sum_{ij} (x_i y_j + x_j y_i) \cdot w_{ij},$$

and the coefficients $x_i y_j + x_j y_i$ vanish on $\widetilde{D}(2,n)$. Conversely, assume ϕ vanishes on $\widetilde{D}(2,n)$. Write $\phi = \phi_a + \phi_s$ with ϕ_a alternating and ϕ_s symmetric. We know already that ϕ_s vanishes on $\widetilde{D}(2,n)$, so we conclude that ϕ_a vanishes there as well. But bilinear alternating maps $R^n \times R^n \to W$ are determined by their restriction to $\widetilde{D}(2,n)$, by KL for W, so $\phi_a = 0$, meaning $\phi = \phi_s$, which is symmetric.

We next prove equivalence of (1) and (3); so assume symmetry of ϕ. Then for all $u, v \in V$, $\phi(u+v, u+v) = \phi(u,u) + \phi(v,v) + 2\phi(u,v)$. If u and v are in $D(V)$, the two first terms vanish, and we deduce that $\phi(u,v)$ is the half of $\phi(u+v, u+v)$, which only depends on $u+v$, so condition (3) holds. Conversely, if (3) holds, the restriction of ϕ to $D(V) \times D(V)$ is symmetric (since $u+v$ is symmetric in u and v), but by KL for W, a bilinear map $V \times V \to W$ is completely determined by its values on $D(V) \times D(V)$, so ϕ is symmetric.

Exercise 1.3.4 Let $D_\infty := \bigcup_k D_k \subseteq R$. (Equivalently, D_∞ is the set of nilpotent elements in R.) Then D_∞ is an R-module (a submodule of R); prove that it is not a KL vector space. Conclude that it is not a finite-dimensional vector space. More generally, $\bigcup_k D_k(n) \subseteq R^n$ is a submodule, for any n.

Exercise 1.3.5 Prove that the KL axioms imply that $R^{D(n)}$ is an $n+1$-dimensional vector space. In particular, $R^{D(V \times V)}$ is 5-dimensional if V is 2-dimensional. On the other hand, $R^{\widetilde{D}(2,V)}$ is 6-dimensional. The inclusion $D(R^2 \times R^2) \subseteq \widetilde{D}(2,2)$ induces the natural projection $R^6 \to R^5$. Hence it is a proper inclusion.

1.4 Calculus

Calculus means here the differential calculus of (smooth) maps between suitable vector spaces. More precisely, we consider (smooth) maps $f : M \to W$ where M is a suitable subset of a finite-dimensional vector space V, and where W is a KL vector space (so in particular, the theory applies if W itself is finite dimensional).

Which subsets $M \subseteq V$ are "suitable"? For the present exposition, the suitable ones should be stable under addition of elements from $D_k(V)$, in other words, $M \subseteq V$ should have the property

$$x \in M \text{ implies } x + D_k(V) \subseteq M$$

for all k. We call such subsets *formally open* (in some texts on SDG, one uses the terminology "formally étale inclusion").

Directional derivative, and differentials

Consider a map $f : M \to W$, where M is an open subset of a finite-dimensional vector space V, and where W is a KL vector space. Since M is formally open in V, we have that $x \in M$ and $v \in D(V)$ implies $x + v \in M$, so that $f(x+v) \in W$ makes sense. Let $x \in M$ be fixed. Applying KL to the map $D(V) \to W$ given by $v \mapsto f(x+v)$, we have immediately the existence of a unique linear map $df(x;-) : V \to W$ with

$$f(x+v) = f(x) + df(x;v) \text{ for all } v \in D(V).$$

This linear $df(x;-)$ is the *differential* of f at x. For any $v \in V$ (whether or not $v \in D(V)$), we call $df(x;v)$ the *directional derivative* of f at $x \in M$ in the *direction* of $v \in V$. For a $v \in V$ which is not necessarily in $D(V)$, $df(x;v) \in W$ is characterized by the validity, for all $d \in D$, of

$$f(x + d \cdot v) = f(x) + d \cdot df(x;v).$$

(Note that $x + d \cdot v$ is in M by openness of M in V.)

Let U and U' be vector spaces (= R-modules), and let $Lin(U,U')$ be the vector space of linear maps $U \to U'$. If U' is a KL vector space, then so is $W := Lin(U,U')$. Consider a map $f : M \to Lin(U,U')$, where $M \subseteq V$ is as above. Consider $x \in M$. Then $f(x)$ is a linear map $U \to U'$; we denote its value on $u \in U$ by $f(x;u)$,

$$f(x;u) := f(x)(u).$$

For each $v \in V$, we have the directional derivative $df(x;v) \in Lin(U,U')$; we denote its value on $u \in U$ by $df(x;v,u)$. It depends in a linear way on both v and u.

More generally, we may consider the KL vector space $W = k\text{-}Lin(U_1 \times \ldots \times U_k, U')$ of k-linear maps $U_1 \times \ldots \times U_k \to U'$, where the U_is are arbitrary vector spaces and U' is a KL vector space. For $f : M \to k\text{-}Lin(U_1 \times \ldots \times U_k, U')$ and $x \in M$, $f(x)(u_1, \ldots, u_k)$ is denoted $f(x;u_1,\ldots,u_k)$, and the directional derivative in the direction $v \in V$ is denoted $df(x;v,u_1,\ldots,u_k)$. This expression is

$k+1$-linear in the arguments after the semicolon. This is consistent with the following.

Convention: Arguments after the semicolon are linear; and, to the extent it applies, the *first* argument after the semicolon is the one indicating the direction in which the directional derivative has been taken.

Consider the basic situation $f: M \to W$. For $x \in M$ and $v \in V$, we thus have $df(x;v)$. Keep v fixed. For $u \in V$, we may take the directional derivative of $df(x;v)$, as a function of x, in the direction of u. This is denoted $d^2f(x;u,v) \in W$. It is characterized by

$$df(x+u;v) = df(x;v) + d^2f(x;u,v) \text{ for all } u \in D(V).$$

It is linear in u, but it is also linear in v, since $df(x;v)$ is so. Furthermore, $d^2f(x;u,v)$ is symmetric in u and v ("Clairaut's Theorem"); we give a proof in the present context below, Theorem 1.4.9.

Let V and W be R-modules. We call a map $f: V \to W$ *1-homogeneous* (in the sense of Euler) if $f(t \cdot v) = t \cdot f(v)$ for all $t \in R$ and $v \in V$.

Theorem 1.4.1 *Assume W is a KL vector space. Then if $f: V \to W$ is 1-homogeneous in the sense of Euler, it is linear.*

This is classical; in the context of SDG, it is proved in Kock (1981/2006, Proposition 10.2) or in Lavendhomme (1996, 1.2.4.) For completeness, we give the proof using the notions employed presently. To prove $f(x+y) = f(x) + f(y)$, it suffices, by one of the basic cancellation principles, to prove that for each $d \in D$, $d \cdot f(x+y) = d \cdot (f(x) + f(y))$. The left-hand side is, by the assumed homogeneity, $f(d \cdot x + d \cdot y)$; now taking the directional derivative, we get the first equality sign in

$$f(d \cdot x + d \cdot y) = f(d \cdot x) + df(d \cdot x; d \cdot y) = f(d \cdot x) + d \cdot df(d \cdot x; y). \quad (1.4.1)$$

For the last term, we may take the directional derivative of $df(-;y)$ from 0 in the direction of $d \cdot x$:

$$df(d \cdot x; y) = df(0;y) + d^2f(0; d \cdot x, y) = df(0;y) + d \cdot d^2f(0;x,y);$$

but in (1.4.1), there is already a factor d multiplied onto the term involving df, so this term vanishes, and the equation (1.4.1) therefore may be continued

$$= d \cdot f(x) + d \cdot df(0;y) = d \cdot f(x) + df(0; d \cdot y) = d \cdot f(x) + f(d \cdot y),$$

by taking the directional derivative of f from 0 in the direction $d \cdot y$ (note that $f(0) = 0$, likewise by homogeneity of f). Finally, we rewrite the last term as $d \cdot f(y)$, using the assumed homogeneity of f again, and so we get $d \cdot f(x) + d \cdot f(y)$, as desired.

A consequence of the existence of directional derivatives is what we shall call the "Taylor Principle", and which we use over and over. Let M and W be as above (M an open subset of a finite-dimensional vector space V, W a KL vector space). Recall that for $x, y \in M$, $x \sim y$ means $y - x \in D(V)$.

Taylor Principle: Let $f : M \times V \to W$ be linear in the second variable (f may depend on other variables as well). Then

$$\textbf{if } x \sim y \textbf{ in } M, \textbf{ then } f(x; y-x) = f(y; y-x), \tag{1.4.2}$$

i.e. occurrences of x in front of the semicolon may be replaced by y.

For, taking the directional derivative of $f(-; y-x)$ at x in the direction of $y - x$ yields

$$f(y; y-x) = f(x; y-x) + df(x; y-x, y-x),$$

and the second term here depends in a bilinear way on $y - x \in D(V)$, and so is 0 by Proposition 1.2.2.

Taylor polynomials

We consider a finite-dimensional vector space V and a KL vector space W. We have the notion of symmetric k-linear map $V^k \to W$; for $k = 1$, this just means "linear map $V \to W$"; for $k = 0$, it means "a constant map" (i.e. given by a single element in W). Recall from classical commutative algebra (cf. Theorem A.7.1 in the Appendix) that there is a bijective correspondence between k-linear symmetric maps $V^k \to W$, and k-homogeneous maps $V \to W$, given by diagonalization of k-linear maps. This correspondence takes care of the alternative formulation in the following theorem.

Theorem 1.4.2 (Taylor expansion) *For any $f : V \to W$, there exists a unique sequence of maps $f_0, f_1, f_2, \ldots$ with $f_k : V^k \to W$ a symmetric k-linear map, such that, for each $k = 0, 1, 2, \ldots$*

$$f(x) = f_0 + f_1(x) + f_2(x,x) + \ldots + f_k(x, \ldots, x) \textit{ for all } x \in D_k(V). \tag{1.4.3}$$

Equivalently, there exists a unique sequence of maps $f_{(0)}, f_{(1)}, f_{(2)}, \ldots$ with $f_{(k)} : V \to W$ k-homogeneous such that

$$f(x) = f_{(0)} + f_{(1)}(x) + f_{(2)}(x) + \ldots + f_{(k)}(x) \textit{ for all } x \in D_k(V). \tag{1.4.4}$$

Proof. We first prove

Lemma 1.4.3 *Let $g : D_k(V) \to W$ vanish on $D_{k-1}(V) \subseteq D_k(V)$. Then there exists a unique k-linear symmetric $G : V \times \ldots \times V \to W$ such that for all $v \in D_k(V)$,*

$$g(v) = G(v, \ldots, v).$$

Proof. Let $(v_1, \ldots, v_k) \in V \times \ldots \times V$. Consider the map $\tau : D^k \to W$ given by

$$(d_1, \ldots, d_k) \mapsto g(d_1 v_1 + \ldots + d_k v_k).$$

Note that $d_i v_i \in D_1(V)$ since $d_i \in D$; therefore (by Proposition 1.2.5) the sum $d_1 v_1 + \ldots + d_k v_k$ is in $D_k(V)$, and so it makes sense to evaluate g at it. If one of the d_is happens to be zero, the sum in question is even in $D_{k-1}(V)$, and so by assumption, g vanishes on it. In terms of $\tau : D^k \to W$, this means that τ vanishes on the k copies of $D^{k-1}(V)$ in $D^k(V)$, and so by the fact that W is a KL vector space, it follows that τ is of the form

$$\tau(d_1, \ldots, d_k) = d_1 \cdot \ldots \cdot d_k \cdot w$$

for a unique $w \in W$. This w we denote $G(v_1, \ldots, v_k)$, and we have thus defined a map $G : V^k \to W$; it is characterized by satisfying, for all $(d_1, \ldots, d_k) \in D^k$,

$$d_1 \cdot \ldots \cdot d_k \cdot G(v_1, \ldots, v_k) = g(d_1 v_1 + \ldots + d_k v_k).$$

Let us prove that G is multilinear. For simplicity, we prove that it is linear in the first variable, and by Theorem 1.4.1, it suffices to prove $G(t \cdot v_1, v_2, \ldots) = t \cdot G(v_1, v_2, \ldots)$. By definition, we have for all $(d_1, \ldots, d_k) \in D^k$ both equality signs in

$$\begin{aligned} d_1 \cdot \ldots \cdot d_k \cdot G(t \cdot v_1, v_2, \ldots) &= g(d_1 \cdot t \cdot v_1 + d_2 \cdot v_2 + \ldots) \\ &= (d_1 \cdot t) \cdot d_2 \cdot \ldots \cdot G(v_1, v_2, \ldots) \end{aligned}$$

(the last equality by applying the characterizing equation on the k-tuple $(d_1 \cdot t, d_2, \ldots) \in D^k$). Cancelling the universally quantified d_is one at a time, we get the desired $G(t \cdot v_1, v_2, \ldots) = t \cdot G(v_1, v_2, \ldots)$.

The proof of symmetry of G is much similar. Let us do the case $k = 2$. We have

$$\begin{aligned} d_1 \cdot d_2 \cdot G(v_2, v_1) &= d_2 \cdot d_1 \cdot G(v_2, v_1) \\ &= g(d_2 v_2 + d_1 v_1) = g(d_1 v_1 + d_2 v_2) = d_1 \cdot d_2 \cdot G(v_1, v_2), \end{aligned}$$

and cancelling d_1 and d_2 one at a time, we conclude $G(v_2, v_1) = G(v_1, v_2)$. This proves the lemma.

We return to the proof of the theorem.

We construct the f_k by induction. Clearly the constant $f_0 \in W$ has to be $f(0)$. Assume $f_0, \ldots, f_{k-1}$ have been constructed in such a way that on $D_{k-1}(V)$ we have

$$f(x) = f_0 + f_1(x) + f_2(x,x) + \ldots + f_{k-1}(x, \ldots, x).$$

Subtracting the two sides here, we therefore get a map $g : V \to W$ which vanishes on $D_{k-1}(V)$. The lemma gives a symmetric k-linear $G : V^k \to W$, and we put $f_k = G$. Then clearly (1.4.3) holds.

For the uniqueness: if $\tilde{f}_0, \ldots, \tilde{f}_k, \ldots$ is another sequence satisfying the conclusion of the theorem, we have $f_0 = \tilde{f}_0$, and we prove by induction that likewise $f_k = \tilde{f}_k : V^k \to W$. Let $H : V^k \to W$ be their difference; it is likewise a symmetric k-linear map, and it has the property that $H(x, \ldots, x) = 0$ for any $x \in D_k(V)$. Now the result follows from the uniqueness assertion in Lemma 1.4.3.

Corollary 1.4.4 *Let V and W be as in the theorem. Then for any map $f : D_k(V) \to W$, there exist unique symmetric r-linear functions $f_r : V^r \to W$ ($r = 0, 1, \ldots, k$) such that (1.4.3) holds. Equivalently, there exist unique r-homogeneous functions $f_{(r)} : V \to W$ such that (1.4.4) holds.*

Proof. By KL, the map $f : D_k(V) \to W$ extends to a (polynomial) map $\overline{f} : V \to W$; now apply the theorem to $\overline{f}$. Uniqueness is clear from the uniqueness assertion in the theorem.

The map $f_{(r)} : V \to W$ is called the *homogeneous component* of degree r of f.

Corollary 1.4.5 *Let V and V' be finite-dimensional vector spaces. Then if $f : D_k(V) \to V'$ takes 0 to 0, it factors through $D_k(V') \subseteq V'$.*

Proof. Write f in the form provided by the previous corollary. Then $f_0 = 0$, so

$$f(u) = f_1(u) + \ldots + f_k(u, \ldots, u);$$

to see that this is in $D_k(V')$, we apply the description (1.2.3) of $D_k(V')$; thus, consider a $k+1$-linear map $\phi : V' \times \ldots \times V' \to R$; we have to see that

$$\phi(f_1(u) + \ldots + f_k(u, \ldots, u), \ldots, f_1(u) + \ldots + f_k(u, \ldots, u)) = 0.$$

But this is clear: using the multilinearity of ϕ, this expression splits up into several terms, but each of these depends in an r-linear way ($r \geq k+1$) on u, and so vanishes because $u \in D_k(V)$.

Corollary 1.4.6 *Let V and V' be finite-dimensional vector spaces, and let $M \subseteq V$ be a (formally) open subset. Then any map $f : M \to V'$ preserves the relation $\sim_k$.*

Some rules of calculus

Theorem 1.4.7 (Chain Rule) *Let $f : M \to M'$ and $g : M' \to W$ where $M \subseteq V$ and $M' \subseteq V'$ are open subsets of finite-dimensional vector spaces V and V' and where W is a KL vector space. Then*

$$d(g \circ f)(x;v) = dg(f(x);df(x;v)).$$

Proof. Since both sides of this equation (for fixed x) are linear in $v \in V$, it suffices to see that they agree for $v \in D(V)$. So assume that $v \in D(V)$. Then

$$(g \circ f)(x+v) = (g \circ f)(x) + d(g \circ f)(x;v),$$

and

$$g(f(x+v)) = g(f(x) + df(x;v)) = g(f(x)) + dg(f(x);df(x;v)),$$

using for the last equality that $df(x;v) \in D(V')$ (by $v \in D(V)$ and $df(x;-)$ being linear). Since the two left-hand sides are equal, then so are the two right-hand sides, and this gives the desired equality.

Proposition 1.4.8 *If $f : V \to W$ is linear, $df(x;v) = f(v)$ for all $x \in V$.*

Proof. Since both sides of this equation (for fixed x) are linear in $v \in V$, it suffices to see that they agree for $v \in D(V)$. So assume that $v \in D(V)$. Then we have $f(x+v) = f(x) + df(x;v)$; we also have $f(x+v) = f(x) + f(v)$ since f is linear. This immediately gives the result.

Let $f : M \to W$ be a function defined on a (formally) open subset M of a finite-dimensional vector space V, and let W be a KL vector space.

Theorem 1.4.9 (Clairaut's Theorem) *The function $d^2f(x;u,v)$ is symmetric in the arguments u, v,*

$$d^2f(x;u,v) = d^2f(x;v,u).$$

Proof. Since both sides of this equation (for fixed x) are bilinear in u,v, it suffices to see that they agree for u and v in $D(V)$. So assume that u and v are in $D(V)$; we calculate $f(x+u+v) = f(x+v+u)$. We have

$$\begin{aligned} f(x+u+v) &= f(x+u) + df(x+u;v) \\ &= f(x) + df(x;u) + df(x;v) + d^2f(x;u,v). \end{aligned}$$

Similarly, we may calculate $f(x+v+u)$:

$$f(x+v+u) = f(x) + df(x;v) + df(x;u) + d^2f(x;v,u);$$

the three first terms in the two expressions cancel, and the resulting equality $d^2f(x;u,v) = d^2f(x;v,u)$ is the desired result.

By iteration, it follows that $d^kf(x;-,\dots,-) : V^k \to W$ is symmetric in the k arguments.

Using this, and making a parallel translation to $x \in V$, we get a more explicit and general form of Taylor expansion, namely

Theorem 1.4.10 *For $y \sim_k x$,*

$$\begin{aligned} f(y) = f(x) + df(x;y-x) + \frac{1}{2!}d^2f(x;y-x,y-x) + \cdots \\ + \frac{1}{k!}d^kf(x;y-x,\dots,y-x). \end{aligned}$$

We have the following variant of the Leibniz rule for a product:

Proposition 1.4.11 *Let $M \subseteq U$ be a (formally) open subset of a finite-dimensional vector space U. Let W_1, W_2 and W_3 be KL vector spaces; let $* : W_1 \times W_2 \to W_3$ be a bilinear map. Let $f_1 : M \to W_1$ and $f_2 : M \to W_2$. Then for $x \in M$, $u \in U$, we have*

$$d(f_1 * f_2)(x;u) = df_1(x;u) * f_2(x) + f_1(x) * df_2(x;u).$$

Proof. Calculate the expression

$$(f_1 * f_2)(x + d \cdot u) = f_1(x + d \cdot u) * f_2(x + d \cdot u)$$

in two ways, using the definition of directional derivative, and bilinearity of $*$.

Corollary 1.4.12 *Let V and V' be KL vector spaces, and let M be a (formally) open subset of some finite-dimensional vector space U. Let $b : M \to Lin(V,V')$,* *with each $b(x)$ an invertible linear map $V \to V'$ with inverse $\beta(x) : V' \to V$. Then, for $x \in M$, $u \in U$ and $v' \in V'$,*

$$db(x;u,\beta(x;v')) = -b(x;d\beta(x;u,v')).$$

Proof. We have that $b(x) \circ \beta(x)$ is independent of x, so directional derivatives of it are 0. Now apply Proposition 1.4.11 to the case where $W_1 = Lin(V,V')$, $W_2 = Lin(V',V)$ and where $*$ is the composition map

$$Lin(V,V') \times Lin(V',V) \to Lin(V',V')$$

(we compose from right to left here).

1.5 Affine combinations of mutual neighbour points

Recall that an infinitesimal k-simplex in a finite-dimensional vector space V is a $k+1$-tuple of mutual neighbour points in V.

We consider a finite-dimensional vector space V and a KL vector space W.

Proposition 1.5.1 *Let M be a (formally) open subset of V, and let $g : M \to W$ be an arbitrary (smooth) map. Then if $(x_0, \ldots, x_k)$ is an infinitesimal k-simplex in M, and $(t_0, \ldots, t_k)$ is a $k+1$-tuple of scalars with sum* 1,

$$g(\sum_{i=0}^{k} t_i \cdot x_i) = \sum_{i=0}^{k} t_i \cdot g(x_i).$$

Note that the affine combination appearing as input of g on the left-hand side of the equation is a neighbour of x_0 (by Proposition 1.2.15) and hence is in M, by openness.

Proof. Consider the restriction of g to the set $x_0 + D(V) \subseteq M$. By KL for W, this extends uniquely to an affine map $g_1 : V \to W$. Since $\sum_{i=0}^{k} t_i \cdot x_i \in x_0 + D(V)$, we have the first equality sign in

$$g(\sum_{i=0}^{k} t_i \cdot x_i) = g_1(\sum_{i=0}^{k} t_i \cdot x_i) = \sum_{i=0}^{k} t_i \cdot g_1(x_i),$$

because g_1 is an affine map and thus preserves affine combinations; but this sum equals term by term the sum $\sum_{i=0}^{k} t_i \cdot g(x_i)$ since g and g_1 agree on $x_0 + D(V)$.

Assume, as above, that V is a finite-dimensional vector space, and that W is a KL vector space.

Proposition 1.5.2 *Let* $\underline{v} = (v_1, \ldots, v_n) \in D(V) \times \ldots \times D(V)$ *and let* $\underline{\underline{d}} \in \widetilde{D}(m,n)$*; then* $\underline{\underline{d}} \cdot \underline{v} \in \widetilde{D}(m,V)$*. Furthermore, if* $\phi : V \to W$ *is a map with* $\phi(0) = 0$*, then*

$$\phi^m(\underline{\underline{d}} \cdot \underline{v}) = \underline{\underline{d}} \cdot \phi^n(\underline{v}).$$

Proof. The first assertion is a consequence of Proposition 1.2.14. So in particular all the vectors in the m-tuple $\underline{\underline{d}} \cdot \underline{v}$ are in $D(V)$. Also all the vectors in the n-tuple $\underline{v}$ are in $D(V)$, by assumption. Since $\phi(0) = 0$, it follows by KL for W that the restriction of ϕ to $D(V)$ extends to a *linear* map $\phi_1 : V \to W$. So in the equation to be proved, ϕ may be replaced by ϕ_1, and the result is then immediate.

Proposition 1.5.3 *Let* $f : D(V_1) \to D(V_2)$ *be a bijection which takes* 0 *to* 0 *(where* V_1 *and* V_2 *are finite-dimensional vector spaces). Then if* $S \subseteq D(V_1)$ *is a linear subset, its image* $f(S) \subseteq D(V_2)$ *is a linear subset as well.*

This follows because such a map f by KL extends to a linear isomorphism $V_1 \cong V_2$.

Exercise 1.5.4 Let $U \subseteq V$ be a finite-dimensional linear subspace. Prove the following: If a bilinear $V \times V \to V$ maps $D(U) \times D(U)$ into U, then it maps $U \times U$ into U. If an alternating bilinear $V \times V \to V$ maps $\widetilde{D}(2,U)$ into U, then it maps $U \times U$ into U. (Hint: Consider U as the kernel of some linear $V \to R^k$, and use the KL axiom.)

Degree calculus

Proposition 1.5.5 *Let* W_1*,* W_2 *and* W_3 *be KL vector spaces; let* $* : W_1 \times W_2 \to W_3$ *be a bilinear map. Let* V *be a finite-dimensional vector space. Let* k *and* l *be non-negative integers, and let* $n \geq k + l + 1$*. If a function* $f : D_n(V) \to W_1$ *vanishes on* $D_k(V)$ *and* $g : D_n(V) \to W_2$ *vanishes on* $D_l(V)$*, then the map* $f * g : D_n(V) \to W_3$ *vanishes on* $D_{k+l+1}(V)$*.*

Proof. We expand f as a sum of $n+1$ homogeneous maps $V \to W_1$, according to Proposition 1.4.4, $f = f_{(0)} + f_{(1)} + \ldots + f_{(n)}$. The assumption that f vanishes on $D_k(V)$ implies that the $k+1$ first terms here are actually 0, so

$$f = f_{(k+1)} + \ldots + f_{(n)}.$$

Similarly,

$$g = g_{(l+1)} + \ldots + g_{(n)}$$

as maps $V \to W_2$. Then the map $f * g : V \to W_3$ may by bilinearity of $*$ be written as a sum of functions of the form $f_{(p)} * g_{(q)}$ with $p \geq k+1$ and $q \geq l+1$. Clearly, such $f_{(p)} * g_{(q)} : V \to W_3$ is homogeneous of degree $p+q$. Also clearly $p+q \geq k+l+2$; and homogeneous functions of degree $k+l+2$ vanish on $D_{k+l+1}(V)$.

If a function $f : V \to W$ vanishes on $D_k(V)$, we say that it *vanishes to order* $k+1$.

2
Geometry of the neighbour relation

This chapter gives the basic configurations in the infinitesimal geometry: it describes and draws pictures of some of the figures which can be made out of the first-order neighbour relation $\sim_1$: infinitesimal simplices, infinitesimal parallelepipeda, geometric distributions, parallelograms given by affine connections. (Some further pictures, which derive from a symmetric affine connection, are found in Chapter 8.)

The section on jets deals with the kth-order neighbour relation $\sim_k$ for general k.

2.1 Manifolds

A *manifold* M of dimension n is a space such that there exists a family $\{U_i \mid i \in I\}$ of spaces equipped with open inclusions $U_i \to M$ and $U_i \to R^n$; the family $U_i \to M$ is supposed to be jointly surjective.

The meaning of this "definition" depends on the meaning of the term "open inclusion"(= "open subspace"), and on the meaning of "family".

For "open inclusion", we take the viewpoint that this is a primitive notion: we assume that among all the arrows in $\mathscr{E}$, there is singled out a subclass $\mathscr{R}$ of "open inclusions", with suitable stability properties, e.g. stability under pullback, as spelled out in the Appendix, Section A.6. Also, we require that the inclusion $Inv(R) \subseteq R$ of the set of invertible elements in the ring R should be open.

It will follow that all maps that are inclusions, which are open according to $\mathscr{R}$, are also *formally open*; this is the openness notion considered in Kock (1981/2006). For V a finite-dimensional vector space, $U \subseteq V$ is formally open if $x \in U$ and $y \sim_k x$ implies $y \in U$. The "formal open" notion has the virtue of being intrinsically *definable* in naive terms, in terms of R; but there are so many formal opens that some, notably integration statements, become rather shallow (amounting to "integration by formal power series"). For instance, the subspace $D_\infty = \cup D_k \subseteq R$ of nilpotent elements in R is formally open, and thus qualifies as a manifold if we take $\mathscr{R}$ to consist of the formal opens. Since D_∞ also is a group under addition, it would also qualify as a Lie group. And R, admitting an open subgroup, would not be connected.

Another openness notion which is intrinsically definable (even without using R) is due to Penon (1985), and studied by Bunge and Dubuc (1987). It is more sophisticated, and does not force R to be disconnected; it deserves more study.

In so far as the meaning of the phrase "family" of spaces: in the context of the definition of what a manifold is, we take this notion to be in the *external* sense: the index set I for a supposed atlas $\{U_i \mid i \in I\}$ for a manifold is an external (discrete) set, not an object in $\mathscr{E}$. This means that it is not subject to the naive reasoning, where "set = space". This is the price we have to pay for not having openness defined in naive (= intrinsic) terms.

Henceforth we assume that an openness notion $\mathscr{R}$ is given, at least as strong as formal étaleness $\mathscr{R}_0$, so $\mathscr{R} \subseteq \mathscr{R}_0$ (so any open subspace is also formally open). The terms "manifold", "open", etc. refer to $\mathscr{R}$.

If a family U_i of open inclusions $U_i \to M$ and $U_i \to R^n$ witness that M is a manifold, we say that the family is an *atlas* for M, and the individual $U_i \to M$ are called *coordinate charts* (we usually think of $U_i \to R^n$ as a subspace, so its elements are coordinate vectors for their images in M).

The definition of the fundamental neighbour relations $\sim_k$ on R^n, $x \sim_k y$ iff $y - x \in D_k(n)$ may be extended to arbitrary n-dimensional manifolds M (cf. Kock, 1981/2006, §I.17 and §I.19): for x and y in M, we say that $x \sim_k y$ iff there exists a coordinate chart $f : N \to M$, with $N \subseteq R^n$ open, and $x' \sim_k y'$ in N, and with $f(x') = x$, $f(y') = y$.

If $N \subseteq M$ is an open subspace of a manifold M, then clearly N itself is a manifold. One can prove that N is stable under the relation $\sim_k$, meaning that $x \in N$, $x \sim_k y$ in M implies that $y \in N$. The inclusion $i : N \hookrightarrow M$ preserves and reflects $\sim_k$, in the sense that for x and y in N, we have $x \sim_k y$ in N iff $i(x) \sim_k i(y)$ in M.

For many considerations about manifolds, a more coordinate-free notion of coordinate chart suffices, namely open inclusions $U \to M$ such that U admits an open inclusion into an *abstract* n-dimensional vector space $V \cong R^n$. One says that M is *modelled on* V. (So M is an n-dimensional manifold iff it is modelled on R^n iff it is modelled on some n-dimensional vector space V.)

Neighbours, neighbourhoods, monads

For a given manifold M, and for each non-negative integer k, the relation $\sim_k$, as described above, is a reflexive symmetric relation, the kth-order *neighbour* relation $x \sim_k y$. For $k = 1$, it is also denoted just $x \sim y$. Many geometric or combinatorial notions can be described in terms of $\sim$, without any reference to analytic geometry.

The basic object (space) derived from the neighbour relations $\sim_k$ on M is $M_{(k)} \subseteq M \times M$, "the kth neighbourhood of the diagonal",

$$M_{(k)} := \{(x,y) \in M \times M \mid x \sim_k y\}.$$

(It is not itself a manifold.) It comes equipped with two projections to M, $(x,y) \mapsto x$ and $(x,y) \mapsto y$; call these maps proj_1 and proj_2.

Let V be an n-dimensional vector space. In case $M = V$ (or an open subspace thereof), there is a canonical isomorphism

$$M_{(k)} \cong M \times D_k(V), \tag{2.1.1}$$

sending (x,y) to $(x, y-x)$, and under this, $\mathrm{proj}_1 : M_{(k)} \to M$ corresponds to the projection $M \times D_k(V) \to M$.

For given $x \in M$, we define the *k-monad* $\mathfrak{M}_k(x)$ around x to be the set

$$\mathfrak{M}_k(x) := \{y \in M \mid x \sim_k y\} \subseteq M;$$

it is the fibre over x of the bundle $\mathrm{proj}_1 : M_{(k)} \to M$. We also write $\mathfrak{M}(x)$ instead of $\mathfrak{M}_1(x)$. The case $k = 1$ is the most important. The case of arbitrary k will be studied in Section 2.7, and in Chapter 7, and, for $k = 2$, in Chapter 8.

In case $M = V$ (or an open subset thereof), the identification (2.1.1) identifies $\mathfrak{M}_k(x)$ with $\{x\} \times D_k(V)$ (which in turn may be identified with $D_k(V)$ itself). For $0 \in V$, $\mathfrak{M}_k(0) = D_k(V)$.

Note that $x \in \mathfrak{M}_k(x)$ since the relation $\sim_k$ is reflexive, and that

$$y \in \mathfrak{M}_k(x) \text{ iff } x \in \mathfrak{M}_k(y)$$

since the relation $\sim$ is symmetric.

Any map between manifolds preserves the relation $\sim_k$: if $x \sim_k y$ in M, then $f(x) \sim_k f(y)$ in N, where $f : M \to N$; this follows from Corollary 1.4.5.

In the rest of the present chapter (and in most of the book), we deal only with the first-order notions, so $x \sim y$ means $x \sim_1 y$, $\mathfrak{M}(x)$ means $\mathfrak{M}_1(x)$, and "neighbour" means "first-order neighbour". Also, $D(V)$ means $D_1(V)$ (V a finite-dimensional vector space).

Exercise 2.1.1 (1) Any map $\tau : D \to M$ extends to an open subset containing D. (2) Hence for $d \in D$, $\tau(d) \sim \tau(0)$. (3) Prove that if τ_1 and τ_2 are maps $D \to M$, then

$$\tau_1(0) = \tau_2(0) \text{ implies } \tau_1(d) \sim \tau_2(d) \text{ for all } d \in D. \tag{2.1.2}$$

Hint: It suffices to consider the case where M is an open subset of a finite-dimensional vector space V; then $\tau_1(d) = x + d \cdot v_1$ where v_1 is the principal part of τ_1; similarly for τ_2, with the same x.

Exercise 2.1.2 If M is a manifold, then so is $M \times M$. Prove that if $(x,y) \sim (z,z)$ in $M \times M$ for some z, then $x \sim y$ in M. (*Hint:* Use Exercise 1.2.17.)

Infinitesimal simplices and whiskers

Certain configurations derived from the first-order neighbour relation are important enough to deserve special names: a triple (x,y,z) of points in a manifold M is called an *infinitesimal 2-whisker* at x if $x \sim y$ and $x \sim z$:

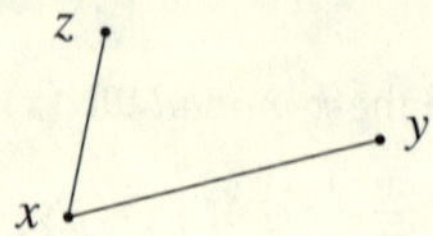

(2.1.3)

where the line segments indicate the neighbour relation. If furthermore $y \sim z$, we shall say that (x,y,z) is an *infinitesimal 2-simplex* at x:

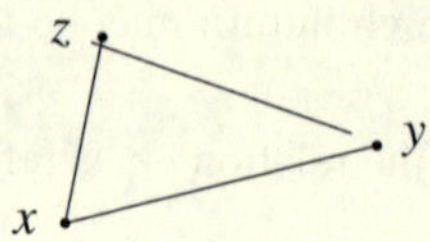

(2.1.4)

More generally, for any natural number k, we have the notions of *infinitesimal k-whisker* and *infinitesimal k-simplex*: an infinitesimal k-whisker in a manifold M is a $k+1$-tuple of points in M, $(x_0, x_1, \ldots, x_k)$, such that $x_0 \sim x_i$ for all $i = 1, \ldots, k$. Note that in a k-whisker, the point x_0 plays a special role. We say that $(x_0, \ldots, x_k)$ is a whisker *at* x_0, or with *base point* x_0.

An *infinitesimal k-simplex* is a $k+1$-tuple of points in M, $(x_0, x_1, \ldots, x_k)$, such that $x_i \sim x_j$ for all $i, j = 0, \ldots, k$. Here, x_0 does not play a special role; but of course, such a k-simplex may also, less symmetrically, be described as an infinitesimal k-whisker $(x_0,\dots,x_k)$ at x_0, with the further property that $x_i \sim x_j$ for all $i, j = 1,\dots,k$.

The infinitesimal k-whiskers in M form a bundle $\pi : Wh_k(M) \to M$ over M, whose fibre over $x_0 \in M$ is the set of infinitesimal k-whiskers at x_0; π is defined by $\pi(x_0,\dots,x_k) = x_0$. The infinitesimal k-simplices form a subbundle $M_{<k>} \subseteq Wh_k(M)$. Its total space $M_{<k>}$ is thus the set of $k+1$-tuples of mutual neighbour points in M, and it is the fundamental object for the theory of (simplicial) combinatorial differential forms.† For $k = 1$, we have $M_{<1>} = M_{(1)}$, the "first neighbourhood of the diagonal", as previously considered.

The symmetric group $\mathfrak{S}_k$ in the letters $1,\dots,k$ acts on the set of infinitesimal k-whiskers, $(x_0,x_1,\dots,x_k) \mapsto (x_0,x_{\sigma(1)},\dots,x_{\sigma(k)})$; it is a fibrewise action in the bundle $Wh_k(M) \to M$, and it restricts to an action on the subbundle $M_{<k>} \to M$. However, the total space $M_{<k>}$ here has a richer symmetry, namely an action by the symmetric group $\mathfrak{S}_{k+1}$ in the letters $0,1,\dots,k$; it is given by $(x_0,x_1,\dots,x_k) \mapsto (x_{\sigma(0)},x_{\sigma(1)},\dots,x_{\sigma(k)})$ for σ a permutation of the $k+1$ letters $0,1,\dots,k$.

For the case where M is a finite-dimensional vector space V, or an open subset thereof, we can exhibit both the "whisker bundle" $Wh_k(M) \to M$ and the "simplex bundle" $M_{<k>} \to M$ more explicitly. There is an isomorphism of bundles over M, $Wh_k(M) \to M \times D(V)^k$, given by sending the infinitesimal k-whisker $(x_0,\dots,x_k)$ into

$$(x_0,(x_1 - x_0,\dots,x_k - x_0)).$$

Equivalently, the fibres of $Wh_k(M) \to M$ may be identified with $D(V)^k$. The subbundle of infinitesimal k-simplices corresponds to the subset $\widetilde{D}(k,V) \subseteq D(V)^k$:

$$Wh_k(M) \cong M \times D(V)^k; \quad \text{and} \quad M_{<k>} \cong M \times \widetilde{D}(k,V). \tag{2.1.5}$$

Neighbours and simplices in general spaces

It is possible to extend the neighbour notion $\sim$ from manifolds to more general spaces. There are two ways to do this, a covariant and a contravariant. Thus for a space A, the covariant determination of $\sim$ is that $a_1 \sim a_2$ if there exists a manifold M and a pair of neighbour points in M, $x_1 \sim x_2$, and a map $f : M \to A$ with $f(x_1) = a_1$ and $f(x_2) = a_2$. In this case, we say that M, x_1, x_2,

† It has many different notations in the literature, even by the present author; in Kock (1981/2006), it is denoted $M_{(1,\dots,1)}$ with k occurrences of 1; in Kock (2000) I used the notation $M_{[k]}$; Breen and Messing (2001) use the notation like $\Delta_M^{(k)}$.

and f *witness* that $a_1 \sim a_2$. If A is itself a manifold, this "new" ("covariant") neighbour relation agrees with the original one.

The contravariant determination of $\sim$ in a general space A says that $a_1 \sim a_2$ if for every map $\phi : A \to M$ to a manifold M, $f(a_1) \sim f(a_2)$. If A is itself a manifold, this "new" ("contravariant") neighbour relation agrees with the original one.

It is trivial to verify that both the covariant and the contravariant $\sim$ is preserved by any map $A \to A'$. Also, it is clear that if $a_1 \sim a_2$ in a for the covariant $\sim$, then also $a_1 \sim a_2$ for the contravariant $\sim$. (In functional-analytic thinking, one would talk about the *strong* and *weak* neighbourhood relation on A.)

We shall not have occasion here to consider the contravariant (weak) determination of $\sim$; the covariant (strong) one will be considered only in Chapter 6. It has to be supplemented by a further determination of the notion of infinitesimal k-simplex; such a simplex is for a general space not just the simple-minded "a $k+1$-tuple of mutual neighbour points", but "a $k+1$-tuple of points which are mutual neighbours by a *uniform* witness"; in other words, a (strong) infinitesimal k-simplex in A is a $k+1$-tuple $(a_0, a_1, \ldots, a_k)$ of points in A such that there exists a manifold M and an infinitesimal $k+1$-simplex $(x_0, x_1, \ldots, x_k)$ in M, and a map $f : M \to A$ with $f(x_i) = a_i$ for $i = 0, 1, \ldots, k$.

It is easy to see that the "witnessing manifolds" M may be taken to be open subsets of finite-dimensional vector spaces (both for the neighbour notion, and for the more general notion of infinitesimal k-simplex).

Also, for the contravariant determination, it is of interest to consider test functions ϕ which are only *locally* defined around a_1 and a_2.

Finally, it may be of interest to consider the category of those spaces where the covariant and contravariant determination of $\sim$ agree; this category will contain all manifolds.

Exercise 2.1.3 Let W be a KL vector space. Let $D(W)$ denote the set of $w \in W$ with $w \sim 0$. Prove that $w \in D(W)$ iff there exists a finite-dimensional vector space V and a linear map $f : V \to W$ with $w = f(d)$ for some $d \in D(V)$.

Affine combinations in manifolds

Let $(y_0, y_1, \ldots, y_n)$ be an infinitesimal n-simplex in a manifold M, and let $(t_0, t_1, \ldots, t_n)$ be an $n+1$-tuple of scalars with sum 1. Then we shall define the affine combination $\sum_{j=0}^{n} t_j \cdot y_j$ as follows. We may pick a bijective map (coordinate chart) from an open subset of M containing y_0 (and hence the other y_is as well) to an open subset of a finite-dimensional vector space V;

we may pick it so that y_0 maps to $0 \in V$. Then in particular, we have a bijection $c : \mathfrak{M}(y_0) \to D(V)$ taking y_0 to 0. Since the y_is are mutual neighbours, $(c(y_1), \ldots, c(y_n)) \in \widetilde{D}(n,V)$. Therefore, by Proposition 1.2.13, the linear combination $\sum_{j=1}^n t_j \cdot c(y_i)$ is in $D(V)$ as well, so we may apply $c^{-1} : D(V) \to \mathfrak{M}(y_0)$ to it, and we define

$$\sum_{j=0}^{n} t_j \cdot y_j := c^{-1}(\sum_{j=1}^{n} t_j \cdot c(y_i)) \in \mathfrak{M}(y_0).$$

We have to argue that this is independent of the choice of the coordinate chart. Another coordinate chart, with values in V_1, say, gives rise to another bijection $c_1 : \mathfrak{M}(y_0) \to D(V_1)$. The composite bijection $c_1 \circ c^{-1} : D(V) \to D(V_1)$ extends by KL to a linear map $\phi : V \to V'$. For $y \in \mathfrak{M}(y_0)$, $c_1(y) = \phi(c(y))$, so c_1 and $\phi \circ c$ agree on $D(V_1)$. Now consider the affine combination in M as defined using c_1; we have

$$c_1^{-1}(\sum_{j=1}^{n} t_j \cdot c_1(y_j)) = c_1^{-1}(\sum_{j=1}^{n} t_j \cdot (\phi \circ c)(y_j))$$

since $y_j \in \mathfrak{M}(y_0)$

$$= c_1^{-1}(\phi(\sum_{j=1}^{n} t_j \cdot c(y_j)))$$

since ϕ is linear

$$= (\phi \circ c)^{-1}(\phi(\sum_{j=1}^{n} t_j \cdot c(y_j)))$$

since $\sum_1^n t_j c(y_j)$ is in $D(V)$ and thus $\phi(\sum_1^n t_j c(y_j))$ in $D(V_1)$

$$= c^{-1}(\sum_{j=1}^{n} t_j \cdot c(y_j))$$

which is the affine combination in M formed using c. This proves that the construction of the point $\sum_{j=0}^n t_j \cdot y_j$ is independent of the choice of coordinate chart. Also, by the very construction, the point constructed is a neighbour of y_0. It also follows from Proposition 1.2.13 that all points obtained by forming affine combinations of the y_is are themselves mutual neighbours. Finally, it is an easy consequence of Proposition 1.5.1 that any map $f : M \to N$ between manifolds preserves the construction.

This is a construction of key importance in most of the present treatise; we summarize its properties in terms of a map: if $(x_0, x_1, \ldots, x_k)$ is an infinitesimal

k-simplex in a manifold M, we get a map

$$[x_0, x_1, \ldots, x_k] : R^k \to M,$$

using affine combinations of the points $x_0, \ldots, x_k$, by the following formula:

$$[x_0, x_1, \ldots, x_k](t_1, \ldots, t_k) := (1 - \sum t_i)x_0 + t_1 x_1 + \ldots + t_k x_k. \tag{2.1.6}$$

(We may also write this combination as $\sum_{i=0}^{k} t_i \cdot x_i$, with $t_0 := 1 - \sum_1^k t_i$.)

Some crucial properties of such affine combinations are then summarized in

Theorem 2.1.4 *Given a $k+1$-tuple $(x_0, x_1, \ldots, x_k)$ of mutual neighbours in a manifold M; then if $t_0, t_1, \ldots, t_k$ is a $k+1$-tuple of scalars with sum 1, the affine combination*

$$\sum_{i=0}^{k} t_i \cdot x_i$$

is well defined in M, and as a function of $(t_1, \ldots, t_k) \in R^k$ defines a map $[x_0, x_1, \ldots, x_k] : R^k \to M$. All points in the image of this map are mutual neighbours. If N is a manifold, or a KL vector space, any map $f : M \to N$ preserves this affine combination,

$$f \circ [x_0, x_1, \ldots, x_k] = [f(x_0), f(x_1), \ldots, f(x_k)].$$

The map $[x_0, x_1, \ldots, x_k] : R^k \to M$ preserves affine combinations.

In the last assertion, the points in R^k of which we are taking affine combinations are not assumed to be mutual neighbours, but their images in M will be.

Definition 2.1.5 *A map $R^k \to M$ coming about in this way from an infinitesimal $k+1$ simplex in M, we call an* infinitesimal k-dimensional parallelepipedum *in M.*

The $k+1$-tuple of points $x_0, \ldots, x_k$ in M may be reconstructed from the map $[x_0, x_1, \ldots, x_k]$, namely as the images of the $k+1$-tuple of points in R^k given by 0 and $e_1, \ldots, e_k$, where e_j is the jth canonical basis vector in R^k.

The construction of affine combinations of neighbour points in a manifold first appeared in Kock (1998).

The infinitesimal (simplicial and cubical) complexes

In R^k, we have a particular 2^k-tuple of points, namely the vertices of the unit cube in R^k. This is the set of points parametrized by the set of subsets H of the set $\{1,\ldots,k\}$; to $H \subseteq \{1,\ldots,k\}$ corresponds the point $\sum_{j\in H} e_j$; to $\emptyset \subseteq \{1,\ldots,k\}$ thus corresponds $0 \in R^k$. For $(x_0,x_1,\ldots,x_k)$ an infinitesimal k-simplex in a manifold M, the map $[x_0,x_1,\ldots,x_k] : R^k \to M$ takes this 2^k-tuple of points in R^k to a 2^k-tuple of points in M, which is a good way to understand the infinitesimal parallelepipedum in geometric terms, see the picture (2.1.7) below.

For $k = 1$, the appropriate expression is "infinitesimal *line segment*". For $k = 2$, the appropriate expression is "infinitesimal *parallelogram*"; here is the picture of parts of the infinitesimal parallelogram spanned by (x_0,x_1,x_2):

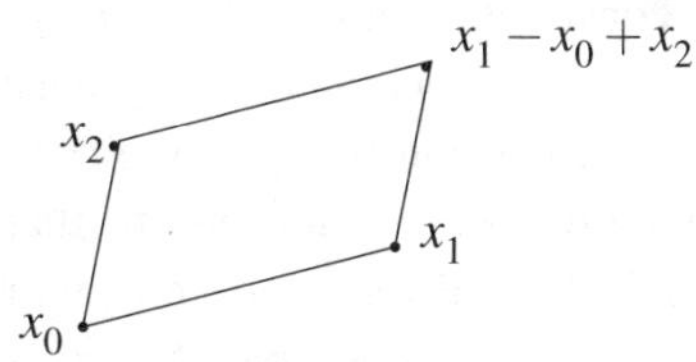

(2.1.7)

Note that all four points here are neighbours, not just those that are connected by lines in the figure.

We use the notation $M_{<k>}$, respectively $M_{[k]}$, for the set of infinitesimal k-simplices, respectively the set of infinitesimal k-dimensional parallelepipeda in M. They organize themselves into a simplicial complex $M_{<\bullet>}$, respectively a cubical complex $M_{[\bullet]}$, cf. Section 2.8 below.

If $\underline{y} = (y_0,y_1,\ldots,y_n)$ is an infinitesimal n-simplex at y_0 in a manifold M, and $\underline{\underline{t}}$ is an $m \times n$ matrix, we can form an infinitesimal m-simplex at y_0, denoted $\underline{\underline{t}} \cdot \underline{y}$ as follows: for each $i = 1,\ldots,m$, let t_{i0} denote $1 - \sum_{j=1}^{n} t_{ij}$. Then $\underline{\underline{t}} \cdot \underline{y}$ is the infinitesimal m-simplex $\underline{x}$ at y_0 whose vertices x_i are $x_0 = y_0$ and $x_i = \sum_{j=0}^{n} t_{ij} \cdot y_j$ (for $i = 1,\ldots,m$).

If $x \in M$, all points in $\mathfrak{M}(x)$ are neighbours of x, by definition (but they need not be mutual neighbours). In particular, if $y \in \mathfrak{M}(x)$, we may form affine combinations of x and y, $(1-t)x + ty$, and such points will then be in $\mathfrak{M}(x)$ as well. This in particular applies to the *central reflection* of y in x, defined as $y \mapsto 2x - y$. It is also called the *mirror image* of y in x. In this way, $\mathfrak{M}(x)$ comes equipped with a canonical involution.

More generally, the multiplicative monoid of R acts on $\mathfrak{M}(x)$, by

$$t * y := (1-t)x + ty. \tag{2.1.8}$$

The action preserves the base point x of the monad.

The formula $(1-t)x+ty$ also provides, for fixed $y \in \mathfrak{M}(x)$, a map $R \to \mathfrak{M}(x)$, which we denote $[x,y] : R \to \mathfrak{M}(x) \subseteq M$. It takes 0 to x and 1 to y.

If an infinitesimal simplex is contained in a monad $\mathfrak{M}(x)$, then any affine combination of the vertices of the simplex is again in the monad. If the vertices are contained in a linear subset of the monad, then so is the affine combination.

Remark 2.1.6 One cannot similarly form affine combinations of the vertices of a whisker in a coordinate-independent way. However, Proposition 1.5.2 gives us a procedure for constructing infinitesimal m-simplices out of infinitesimal n-whiskers. If $\underline{x} = (x_0, x_1, \ldots, x_n)$ is an infinitesimal n-whisker at x_0 in a manifold M, and $\underline{\underline{d}} \in \widetilde{D}(m,n)$, we can construct an infinitesimal m-simplex at x_0, denoted $\underline{\underline{d}} \cdot \underline{x}$ by the following recipe: we may assume that M is embedded as an open subset of a finite-dimensional vector space V, with $x_0 = 0$. Then we make the construction $\underline{\underline{d}} \cdot \underline{v}$ where $\underline{v}$ is the n-tuple of vectors $x_1, \ldots, x_n$. It is in $\widetilde{D}(m,V)$, in particular, each entry is $\sim 0 = x_0 \in M$ and so is in M, by openness. Together with $x_0 = 0$, this m-tuple defines an infinitesimal m-simplex in M. The fact that the construction does not depend on the choice of embedding follows from the second assertion in Proposition 1.5.2, by letting ϕ be the (local) comparison $V_1 \to V_2$ between embeddings of M into V_1 and into V_2.

Remark 2.1.7 Let $f_i : M \to N$ $(i = 1,2)$ be maps between manifolds, and let E be their equalizer, $\{x \in M \mid f_1(x) = f_2(x)\}$. Then E is not necessarily a manifold. However, the neighbour relation $\sim$ on M restricts to a (reflexive symmetric) relation on E; and if $(x_0, x_1, \ldots, x_k)$ is an infinitesimal k-simplex in E (equivalently, an infinitesimal k-simplex in M all of whose vertices are in E), then any affine combination of the x_is, formed in M, is again in E; this follows from the fact that the maps f_1 and f_2 preserve such affine combinations. Similarly if E is the meet $\bigcap E_j$ of such equalizers E_j.

So if the x_is are in E, the map $[x_0, \ldots, x_k] : R^k \to M$ factors through E.

Remark 2.1.8 If M is an *affine* space (say, a vector space), the formula defining $[x_0, x_1, \ldots, x_k] : R^k \to M$ makes sense whether or not the x_is form an infinitesimal simplex, and its existence is a consequence of the fact that R^k is a *free* affine space on the $k+1$ points $0, e_1, \ldots, e_k$.

In particular, if $(x_0, x_1, \ldots, x_k)$ is an infinitesimal k-whisker in an affine space, the map $[x_0, x_1, \ldots, x_k] : R^k \to M$ is defined, and is a parallelepipedum; but it does not qualify as an infinitesimal parallelepipedum in the present usage of the term, unless the infinitesimal whisker happens to be an infinitesimal simplex, $x_i \sim x_j$ for all i, j.

Remark 2.1.9 There are some cases where some affine combinations of a set of points make invariant sense, even though the points are not mutual 1-neighbours. The following example will be of some importance. Consider maps $t : D \to M$ and $\tau : D \times D \to M$, with $t(0) = \tau(0,0)$ $(= m \in M$, say). Consider for any $(d_1, d_2) \in D \times D$ the following points:

$$\tau(d_1, d_2), m, t(d_1 \cdot d_2).$$

They are mutual 2-neighbours, but not in general mutual 1-neighbours. Nevertheless, the affine combination

$$\tau(d_1, d_2) - m + t(d_1 \cdot d_2) \tag{2.1.9}$$

can be proved to be independent of the choice of coordinate system used to define it, and is preserved by any map between manifolds. The proof uses Taylor expansion up to degree 2. Hence for such t and τ, we may define a new $D \times D \to M$, denoted $t \dotplus \tau$ by

$$(t \dotplus \tau)(d_1, d_2) := \tau(d_1, d_2) - m + t(d_1 \cdot d_2).$$

Clearly, τ and $t \dotplus \tau$ agree when $d_1 \cdot d_2 = 0$, i.e. they agree on $D(2) \subseteq D \times D$. Conversely, one can prove that if τ and τ' are maps $D \times D \to M$ which agree on $D(2)$, then there is a unique $t : D \to M$ so that $\tau' = t \dotplus \tau$. This t is then denoted $\tau' \dot{-} \tau$, the *strong difference* of τ' and τ (cf. Kock and Lavendhomme, 1984; Kolar, 1982; Lavendhomme, 1994; Nishimura, 1997; White, 1982). The Kock, Lavendhomme, and Nishimura references are in the context of SDG, and the construction makes sense in the generality of micro-linear spaces, so is more general than the manifold case as discussed presently. The relevant infinitesimal object is denoted $D(2) \vee D$, see Section A.4.

2.2 Framings and 1-forms

Let M be a manifold, and let $x \in M$. A bijection $D(n) \to \mathfrak{M}(x)$ taking 0 to x is called a *frame* at x; if there is given such a frame $k_x : D(n) \to \mathfrak{M}(x)$ at each $x \in M$, we say that we have a *framing* (or *parallelism*) of M. The notion of framing can be expressed more globally: it is an isomorphism

$k : M \times D(n) \to M_{(1)}$ of bundles over M

$$M \times D(n) \xrightarrow[\cong]{k} M_{(1)}$$

with $k(x,0) = (x,x)$, for all x.†

(The set of all frames at points of M forms a bundle over M, the *frame bundle* of M; it is actually a principal fibre bundle, and it will be discussed from this viewpoint in Section 5.6 below. A framing may be construed as a cross-section of this bundle.)

Let k be a framing on the manifold M. Since $D(n)$, as a subset of R^n, consists of *coordinate* vectors, we may think of the inverse of k_x, which we shall call $c_x : \mathfrak{M}(x) \to D(n)$, as providing *coordinates* for the neighbours of x: for $y \in \mathfrak{M}(x)$, $c_x(y) \in D(n) \subseteq R^n$ is "the *coordinate n-tuple* of y in the coordinate system at x given by the framing k". Note that $c_x(x) = 0$ for all x.

We have already in (2.1.1) exhibited an example of a framing, provided M is an open subset of R^n; the framing considered there is the *canonical* framing of $M \subseteq R^n$; it is given by

$$k(x,\underline{d}) = x + \underline{d},$$

for $x \in M$, $\underline{d} \in D(n)$, equivalently $k_x(\underline{d}) = x + \underline{d}$. For this framing, we have $c_x(y) = y - x$ for $y \sim x$.

A comparison with the classical notion of framing on M (trivialization of the tangent bundle of M) follows from the theory in Section 4.3, see in particular Corollary 4.3.6.

One may consider a more "coordinate-free" notion of framing on a manifold M. Let V be a finite-dimensional vector space. Then a *V-framing* is an isomorphism over M, $k : M \times D(V) \to M_{(1)}$ (with $k(x,0) = (x,x)$ for all x). So for each $x \in M$, there is given a bijection $k_x : D(V) \to \mathfrak{M}(x)$ with $k_x(0) = x$. The framings considered previously are thus more precisely to be called *R^n-framings*. If M is an open subset of V, it carries a *canonical* V-framing given by $(x,\underline{d}) \mapsto x + \underline{d}$, $(x \in M, \underline{d} \in D(V))$.

Example 2.2.1 At each point x of the surface M of the earth, except at the poles, we have a coordinate system c_x given as follows. We let $c_x(y) = (\alpha, \beta)$ whenever you reach y by going from x a distance of α metres to the East, and from there β metres to the North. This works, provided y is not too far away

† Such a framing can only exist provided M has dimension n, but we don't need this fact; also, there may be obstructions of global nature to the existence of framings.

from x (so it usually works when x and y are points in the construction site for a house), so at least it works for $y \sim x$.

This "construction-site" framing of this particular and precious M depends on the choice of unit length: metre, say.

Let there be given a V-framing $k : M \times D(V) \to M_{(1)}$, with the corresponding family of "coordinate systems" $c_x : \mathfrak{M}(x) \to D(V) \subseteq V$. We get a function $\omega : M_{(1)} \to V$ described, for $x \sim y$, as follows:

$$\omega(x,y) := c_x(y) = k_x^{-1}(y). \tag{2.2.1}$$

It is called the *framing 1-form* for k. We already have the validity of the first equation (2.2.2) in the following proposition.

Proposition 2.2.2 *The framing 1-form $\omega : M_{(1)} \to V$ satisfies*

$$\omega(x,x) = 0 \tag{2.2.2}$$

for all $x \in M$, and

$$\omega(y,x) = -\omega(x,y) \tag{2.2.3}$$

for all $(x,y) \in M_{(1)}$.

Proof. We shall derive (2.2.3) from (2.2.2) by an argument which is quite general, and which will be used many times over. Since the question is of infinitesimal nature, it suffices to prove the equation in the case where M is an open subset of a finite-dimensional vector space V'. Then since $\mathfrak{M}(x) = x + D(V')$, it follows from KL that the function $\omega(x,-) : \mathfrak{M}(x) \to D(V) \subseteq V$ may be described $\omega(x,y) = \Omega(x; y-x)$, where $\Omega(x;-) : V' \to V$ is a linear map; so collectively, we have a map $\Omega(-;-) : M \times V' \to V$, linear in the variable after the semicolon. In terms of Ω, the assertion (2.2.3) is that $\Omega(x; y-x) = -\Omega(y; x-y)$. Now the desired result follows from the Taylor principle (1.4.2): we may replace the x in front of the semicolon by x; and then the result follows by linearity of Ω in its second argument.

Assume, as in the proof of the proposition, that $M \subseteq V'$. Since frames are bijective maps, it follows that the maps $\Omega(x;-) : V' \to V$, as in the proof of the proposition, are invertible. Let us denote the inverse of $\Omega(x;-)$ by $b(x;-) : V \to V'$; then the data of the framing may be expressed

$$k_x(d) = x + b(x;d).$$

It is convenient here to anticipate a notion and terminology which, in the present context, will be developed more fully in Chapter 3. Let W be a KL vector space, and let M be a manifold.

Definition 2.2.3 *A function* $\omega : M_{(1)} \to W$ *is called a (W-valued)* 1-form *on* M *if* $\omega(x,x) = 0$ *for all* $x \in M$.

Every 1-form satisfies, for all $x \sim y$,

$$\omega(x,y) = -\omega(y,x); \tag{2.2.4}$$

the proof of this is identical to the proof of (2.2.3).

Note that the framing-1-form for a framing $k : M \times D(V) \to M_{(1)}$ is a V-valued 1-form in the sense of this definition.

Definition 2.2.4 *A 1-form* ω *is called a* closed *1-form if for any infinitesimal 2-simplex* (x,y,z) *in* M*, we have*

$$\omega(x,y) + \omega(y,z) = \omega(x,z);$$

equivalently, by (2.2.4), if

$$\omega(x,y) + \omega(y,z) + \omega(z,x) = 0.$$

If $f : M \to W$ is a function, we get a W-valued 1-form df on M by putting

$$df(x,y) := f(y) - f(x).$$

Clearly, df is closed. If ω is a W-valued 1-form on M, we get a W-valued function $d\omega$, defined on the space of infinitesimal 2-simplices x,y,z in M, by putting

$$d\omega(x,y,z) := \omega(x,y) + \omega(y,z) - \omega(x,z).$$

So $d\omega$ is constant 0 iff ω is closed.

Definition 2.2.5 *A 1-form* $\omega : M_{(1)} \to W$ *is called* exact *if* $\omega = df$ *for some* $f : M \to W$*; and then* f *is called a* primitive *of* ω.

Exercise 2.2.6 Let ω be a 1-form on a manifold M, $\omega : M_{(1)} \to W$ (with W a KL vector space). Prove that there exists a unique extension of ω to a map $M_{(2)} \to W$ satisfying (2.2.4) for all $x \sim_2 y$. (*Hint:* It suffices to consider the coordinatized situation $M \subseteq V$, and use Theorem 1.4.2.)

If M is an open subset of a finite-dimensional vector space V, the data of a W-valued 1-form ω on M may be encoded, as in the proof of Proposition 2.2.2, by a function $\Omega : M \times V \to W$,

$$\omega(x,y) = \Omega(x; y-x)$$

linear in the argument after the semicolon. In such a coordinatized situation,

Proposition 2.2.7 *The 1-form ω is closed iff the bilinear $d\Omega(x;-,-) : V \times V \to W$ is symmetric for all $x \in M$.*

Proof. Closedness of ω means the vanishing of $\omega(x,y) + \omega(y,z) - \omega(x,z)$, for all infinitesimal 2-simplices (x,y,z) in M. We calculate this in terms of Ω; we get

$$\begin{aligned} &\Omega(x;y-x) + \Omega(y;z-y) - \Omega(x;z-x) \\ &= \Omega(x;y-x) + \Omega(x;z-y) + d\Omega(x;y-x,z-y) - \Omega(x;z-x), \end{aligned}$$

by Taylor expansion of the middle term. By linearity of $\Omega(x;-)$, three of the terms cancel each other, and we are left with $d\Omega(x;y-x,z-y)$, which is $= d\Omega(x;y-x,z-x)$ by the "Taylor principle". To say that the bilinear $d\Omega(x;-,-)$ vanishes for all $(y-x,z-x) \in \widetilde{D}(2,V)$ is by Proposition 1.3.3 equivalent to saying that $d\Omega(x;-,-)$ is symmetric.

Corollary 2.2.8 *If M is a 1-dimensional manifold, then any 1-form on M is closed.*

Proof. It suffices to consider the case where M is an open subset of a 1-dimensional vector space V. Now any bilinear form $V \times V \to W$ on a 1-dimensional vector space V is symmetric, so the result is immediate from the proposition.

If ω is a W-valued 1-form on M, we may ask for a *partial* primitive for ω on a subset $S \subseteq M$: this means a function $f : S \to W$ such that $\omega(x,y) = f(y) - f(x)$ for all $x,y \in S$ with $x \sim y$ in M.

Proposition 2.2.9 *A 1-form ω on M has a partial primitive on each $\mathfrak{M}(x)$ if and only if ω is closed.*

Proof. If ω is closed, then for each $x \in M$, the function $\omega(x,-)$ is a partial primitive on $S = \mathfrak{M}(x)$, since $\omega(y,z) = \omega(x,z) - \omega(x,y)$ for all y,z in $\mathfrak{M}(x)$ with $y \sim z$; the other implication is trivial.

It is possible to prove that closed 1-forms actually have partial primitives on each $\mathfrak{M}_k(x)$; essentially, one gets such partial primitive by "integration by power series" up to degree k.

The easy proof of the following proposition is left to the reader (use coordinates, and a presentation of ω in terms of Ω, as in the proof of (2.2.3)).

Proposition 2.2.10 *Let ω be a 1-form on a manifold with values in a KL vector space. Then for $x \sim y$, and $t \in R$,*

$$t \cdot \omega(x,y) = \omega(x,(1-t)\cdot x + t \cdot y).$$

2.3 Affine connections

(Affine connections are a special case of connections in a more general sense discussed in Sections 2.5 and 5.2.)

An *affine connection* on a manifold M is a law λ which to any infinitesimal 2-whisker (2.1.3) in M, i.e. to any triple of points (x,y,z) in M with $x \sim y$ and $x \sim z$, associates a fourth point $\lambda(x,y,z)$, subject to the conditions

$$\lambda(x,x,z) = z \tag{2.3.1}$$

and

$$\lambda(x,y,x) = y. \tag{2.3.2}$$

This conception of affine connections first appeared in Kock (1984).

For some of the calculations, it is convenient to change the notation $\lambda(x,y,z)$ into one with x in the middle, as follows:

$$\lambda(x,y,z) = [[y,x,z]].$$

In the following figure, the line segments (whether single or double) indicate the relation of being neighbours (i.e. the relation $\sim$):

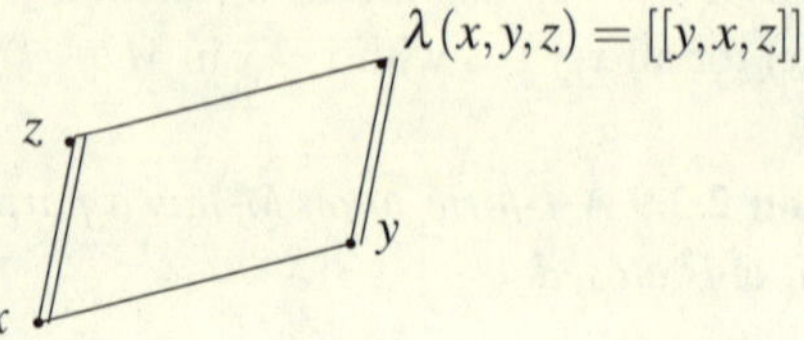

(2.3.3)

The figure† is meant to indicate that the data of λ provides a way of closing an infinitesimal whisker (x,y,z) into a *parallellogram* (one may say that λ provides a notion of *infinitesimal parallelogram*); but note that λ is not required to be symmetric in y and z, which is why in the figure we use different signatures for the line segments connecting x to y and to z, respectively. Let us

† Note the difference between this figure and the figure (2.1.7), in which y and z (there: x_1 and x_2) are assumed to be neighbours, and where the parallelogram *canonically* may be formed.

call an ordered 4-tuple (x,y,z,u) in M a *λ-parallelogram* if $u = \lambda(x,y,z)(= [[y,x,z]])$.

The figure implicitly contains some assertions which need to be proved (proved in Propositions 2.3.2 and 2.3.3 below); namely the assertions

$$y \sim \lambda(x,y,z), \tag{2.3.4}$$

$$z \sim \lambda(x,y,z) \tag{2.3.5}$$

and the "symmetry" assertions

if (x,y,z,u) form a λ-parallelogram

then so do

$$(y,x,u,z), (z,u,x,y), \text{ and } (u,z,y,x). \tag{2.3.6}$$

The symmetry assertions form a set of *three* assertions; they read in equational form

$$\lambda(y,x,\lambda(x,y,z)) = z; \tag{2.3.7}$$

$$\lambda(z,\lambda(x,y,z),x) = y; \tag{2.3.8}$$

$$\lambda(\lambda(x,y,z),z,y) = x. \tag{2.3.9}$$

These assertions look more algebraic in the $[[y,x,z]]$ notation: in so far as these three equations go, $[[y,x,z]]$ behaves like $y \cdot x^{-1} \cdot z$ does in a group; for instance (2.3.7) reads in this notation

$$[[x,y,[[y,x,z]]\,]] = z$$

corresponding to $x \cdot y^{-1} \cdot (y \cdot x^{-1} \cdot z) = z$.

The symmetry assertions can be stated succinctly; the four-group (= the symmetry group of a rectangle, i.e. the group = $\mathbb{Z} \times \mathbb{Z}$) acts on the set of λ-parallelograms. If λ is furthermore symmetric in the sense

$$\lambda(x,y,z) = \lambda(x,z,y),$$

or equivalently $[[y,x,z]] = [[z,x,y]]$, then the eight-group (symmetry group of a square) acts on the set of λ-parallelograms; in this case we call the connection *torsion free* or *symmetric*. To say that λ is torsion free is equivalent to saying that if (x,y,z,u) is a λ-parallelogram, then so is (x,z,y,u). For a symmetric affine connection, some geometric theory of "metric" nature becomes available; we have placed some of this theory in Section 8.2 in the chapter on metric notions, but it can also be read naturally now.

If λ is torsion free, so that we may interchange y and z, we get from (2.3.9) that

$$\lambda(\lambda(x,y,z),y,z) = x; \tag{2.3.10}$$

in fact, the implication goes both ways:

Proposition 2.3.1 *The affine connection λ is torsion free if and only if (2.3.10) holds for all x,y,z (with $x \sim y$ and $x \sim z$).*

Proof. To say the equation holds for all x,y,z is to say that if (x,y,z,u) is a λ-parallelogram, then so is (u,y,z,x). But (u,y,z,x) is a λ-parallelogram iff (x,z,y,u) is so, by the four-group symmetry assertions (2.3.6).

Let us for each $x \in M$ define the map $b_x : \mathfrak{M}(x) \times \mathfrak{M}(x) \to \mathfrak{M}(x)$ by the left-hand side of (2.3.10),

$$b_x(y,z) := \lambda(\lambda(x,y,z),y,z). \tag{2.3.11}$$

By Proposition 2.3.1, the maps b_x measure the extent to which λ fails to be torsion free, i.e. they measure the *torsion* of λ; we may call b_x the *intrinsic torsion of* λ at $x \in M$. At least when the connection comes from a framing, cf. Section 2.4, it is a combinatorial rendering of the *Burgers vector* construction known from materials science (dislocations in crystals); see Kock (1986) and references therein.

The torsion of λ will be construed as a tangent-bundle-valued combinatorial 2-form in Section 4.8, by a re-interpretation of the intrinsic torsion/Burgers vector construction.

For any affine connection λ on M, we have the *conjugate* affine connection $\overline{\lambda}$, given by

$$\overline{\lambda}(x,y,z) := \lambda(x,z,y)$$

or equivalently $\overline{[[y,x,z]]} = [[z,x,y]]$. (So for affine connections, symmetric = torsion free = self-conjugate.)

Christoffel symbols

We proceed to see how affine connections may be described in terms of coordinates. So we consider an affine connection λ on M where M is an open subset of a finite-dimensional vector space V. Then there is a function

$\Gamma : M \times D(V) \times D(V) \to V$ given by

$$\lambda(x,y,z) = y - x + z + \Gamma(x; y-x, z-x). \tag{2.3.12}$$

(Thus Γ is measuring the defect of λ being the canonical affine connection on the affine space V, cf. Example 2.3.9 below.) Since $\lambda(x,x,z) = z$ for all $z \in \mathfrak{M}(x)$, we get that $\Gamma(x; 0, d_2) = 0$ for all $d_2 \in D(V)$. Similarly, $\lambda(x,y,x) = y$ for all $y \in \mathfrak{M}(x)$ implies that $\Gamma(x; d_1, 0) = 0$ for all $d_1 \in D(V)$. By KL for V, $\Gamma(x; -, -)$ therefore extends uniquely to a bilinear map $V \times V \to V$, which we likewise denote $\Gamma(x; -, -)$. The $\Gamma(x; -, -)$s are the *Christoffel symbols* for the connection. The connection is torsion free iff all $\Gamma(x; -, -)$ are *symmetric* bilinear maps $V \times V \to V$.)

Consider an affine connection λ on a manifold M.

Proposition 2.3.2 *For $x \sim y$ and $x \sim z$ in M, we have that*

$$\lambda(x,y,z) \sim y \text{ and } \lambda(x,y,z) \sim z.$$

If furthermore $y \sim z$, $\lambda(x,y,z) \sim x$.

Proof. Since the assertions are coordinate free and local, we may without loss of generality assume that M is an open subset of a finite-dimensional vector space V, so that we may calculate by Christoffel symbols Γ. Then we have

$$\lambda(x,y,z) - y = (z - x) + \Gamma(x; y-x, z-x).$$

This, however, is a linear expression in $(z-x)$, but a linear map $V \to V$ takes $D(V)$ into $D(V)$; since $z - x \in D(V)$, the first assertion follows. The proof of the second assertion is similar. The third assertion follows from Proposition 1.2.18.

We next consider the "four-group" symmetry assertions (2.3.7), (2.3.8), and (2.3.9). Given a manifold M, and given an affine connection λ on it; then

Proposition 2.3.3 *For $x \sim y$ and $x \sim z$ in M, we have that*

$$\lambda(y,x,\lambda(x,y,z)) = z \quad \text{and} \quad \lambda(z,\lambda(x,y,z),x) = y.$$

Proof. To prove the first equation (2.3.7), it again suffices to carry out the proof for the case of M an open subset of a finite-dimensional vector space V; so we express the connection in terms of Γ, as in the proof of the previous proposition. As there, we have $\lambda(x,y,z) = y - x + z + \Gamma(x; y-x, z-x)$.

Therefore

$$
\begin{aligned}
\lambda(y,x,\lambda(x,y,z)) &= x-y+\lambda(x,y,z)+\Gamma(y;x-y,\lambda(x,y,z)-y)\\
&= x-y+(y-x+z+\Gamma(x;y-x,z-x))\\
&\quad +\Gamma(y;x-y,z-x+\Gamma(x;y-x,z-x))\\
&= x-y+(y-x+z+\Gamma(x;y-x,z-x))\\
&\quad +\Gamma(y;x-y,z-x)
\end{aligned}
$$

(because of bilinear occurrence of $x-y \in D(V)$, the nested occurrence of Γ contributes 0)

$$= z+\Gamma(x;y-x,z-x)+\Gamma(y;x-y,z-x);$$

exchanging the y before the semicolon with x by the Taylor principle, the two Γ terms cancel, so that we end up with z, as desired.

The second equation (2.3.8) follows by applying what has already been proved to the conjugate connection. Finally, the third equation (2.3.9) follows purely formally from the first two.

This proves the proposition.

Note that the relationship between λ and Γ in the above may also be written

$$\lambda(x,x+d,x+d') = x+d+d'+\Gamma(x;d,d')$$

for $x \in M \subseteq V$ and d and d' in $D(V)$.

Recall that one can form affine combinations of mutual neighbour points in a manifold. In particular, if x,y,z form an infinitesimal 2-simplex, $y-x+z$ may be formed. We have

Proposition 2.3.4 *Let λ be an affine connection on a manifold M. Then λ is torsion free if and only if for all infinitesimal 2-simplices x,y,z in M, we have $\lambda(x,y,z) = y-x+z$.*

Proof. We may assume that M is embedded as an open subset of a finite-dimensional vector space V. (The affine combination $y-x+z$ does not depend on the embedding, by the general theory in Chapter 1.) We can express λ in terms of the (bilinear) Christoffel symbols $\Gamma(x;-,-) : V \times V \to V$. Then clearly λ is torsion free iff the $\Gamma(x;-,-)$s are symmetric, which in turn, by Proposition 1.3.3, is the case iff $\Gamma(x;-,-)$ vanishes on $\widetilde{D}(2,V)$; but $\Gamma(x;d,d')$ is the discrepancy between $\lambda(x,y,z)$ and $y-x+z$, where $y = x+d$ and $z = x+d'$.

Let us calculate (2.3.11) in terms of Christoffel symbols. We use notation as in the proofs of Propositions 2.3.2 and 2.3.3. First note that $y-\lambda(x,y,z) = (x-z)-\Gamma(x;y-x,z-x)$, which clearly depends linearly on $x-z$. Similarly $z-\lambda(x,y,z) = (x-y)-\Gamma(x;y-x,z-x)$ depends linearly on $x-y$. Now

$$b_x(y,z) = \lambda(\lambda(x,y,z),y,z)$$
$$= y-\lambda(x,y,z)+z+\Gamma(\lambda(x,y,z);y-\lambda(x,y,z),z-\lambda(x,y,z));$$

because the two linear arguments in the Γ-expression depend linearly on $x-z$ and $x-y$, respectively, we may, by the Taylor principle, replace the $\lambda(x,y,z)$ in front of the semicolon by $\lambda(x,x,x)=x$, so that we may continue

$$\begin{aligned}
&= y-\lambda(x,y,z)+z+\Gamma(x;y-\lambda(x,y,z),z-\lambda(x,y,z))\\
&= y-\left[y-x+z+\Gamma(x;y-x,z-x)\right]+z\\
&\quad +\Gamma(x;x-z-\Gamma(x;y-x,z-x),x-y-\Gamma(x;y-x,z-x)).
\end{aligned}$$

Now, we expand the second line using bilinearity of $\Gamma(x;-,-)$, and then nested occurrences of Γ will disappear because they will contain $y-x$ or $z-x$ bilinearly. Also, some ys and zs cancel, for purely additive reasons, and we are left with

$$\begin{aligned}
&= x-\Gamma(x;y-x,z-x)+\Gamma(x;x-z,x-y)\\
&= x+\left[\Gamma(x;z-x,y-x)-\Gamma(x;y-x,z-x)\right].
\end{aligned}$$

So in terms of Christoffel symbols:

$$b_x(y,z)=\lambda(\lambda(x,y,z),y,z)=x+\left[\Gamma(x;z-x,y-x)-\Gamma(x;y-x,z-x)\right]. \tag{2.3.13}$$

If (x,y,z) form an infinitesimal 2-simplex, so that not only $x\sim y$ and $x\sim z$, but also $y\sim z$, then some further constructions can be made: it can easily be proved (e.g. by using Christoffel symbols) that all points obtained from x,y,z by means of λ (including for instance $b_x(y,z)$) are mutual neighbours, and therefore further affine combinations may be made. This in particular applies to the affine combination in the following:

Proposition 2.3.5 *Let λ be an affine connection, and let (x,y,z) be an infinitesimal 2-simplex. Then*

$$b_x(y,z) = 2[y-z+\lambda(y,x,z)]-x.$$

Proof. Note that all the terms in the equation have "objective" significance, i.e. they don't depend on any coordinatization. For the proof of the equation claimed, we may therefore freely use any coordinatization, and do the calculation in terms of Christoffel symbols for λ. We have already calculated $b_x(y,z)$; this is the equation (2.3.13) above. However, since (x,y,z) is an infinitesimal simplex, the bilinear $\Gamma(x;-,-)$ behaves on $(y-x,z-x)$ as if it were alternating, so we have

$$b_x(y,z) = x + 2\Gamma(x;z-x,y-x).$$

On the other hand,

$$y-z+\lambda(y,x,z) = y-z+[x-y+z+\Gamma(y;x-y,z-y)].$$

Because of the linear occurrence of $x-y$ in the Γ expression, we may replace the two other occurrences of y in it by x, and so the equation continues

$$= y-z+[x-y+z+\Gamma(x;x-y,z-x)] = x+\Gamma(x;x-y,z-x)$$
$$= x+\Gamma(x;z-x,y-x),$$

the last equation sign again using that $\Gamma(x;-,-)$ is alternating on $(y-x,z-x)$. The result then follows by an immediate additive calculation.

Problem. Given an affine connection; what is the condition that it extends from a completion procedure for 2-whiskers to a (unambiguous?) completion procedure for k-whiskers ($k \geq 3$)? (Certainly, it suffices that the connection is *integrable* in the sense of Definition 2.4.5.)

Exercise 2.3.6 Let M be a manifold equipped with an affine connection λ. For each infinitesimal 2-whisker (x,y,z) in M (so $x \sim y$ and $x \sim z$), we construct a singular square $[x,y,z]_\lambda : R^2 \to M$ by the recipe

$$(s,t) \mapsto \lambda(x,[x,y](s),[x,z](t)),$$

where $[x,y] : R \to M$ is the infinitesimal line segment given by x,y (thus $[x,y](s) = (1-s)x + sy$; similarly for $[x,z]$). (Note that the singular square $R^2 \to M$ thus described does not qualify as an infinitesimal parallelogram, unless $y \sim z$.)

Prove that the restriction of $[x,y,z]_\lambda$ to each vertical line ($s =$ constant) defines an infinitesimal line segment $R \to M$. Similarly for the restriction of $[x,y,z]_\lambda$ to each horizontal line.

Hint: It suffices to consider the case where M is an open subset of a finite-dimensional vector space V, so that λ may be described by its Christoffel symbols.

Further geometric aspects of torsion-free (= symmetric) affine connections are given in Chapter 8.

Besides the property of being torsion free, there is another geometric property which an affine connection may or may not have, namely the property of being *curvature free* or *flat*. To state it, it is convenient to think of $\lambda(x,y,z) = [[y,x,z]]$ as the result of *transporting* z along the (infinitesimal) line segment from x to y; so x,y is the *active* aspect, the one which *transports*, whereas z is the *passive* one: z is *being transported*. Let us write $\nabla(y,x)$ for the "transport" map $\mathfrak{M}(x) \to \mathfrak{M}(y)$ given by $z \mapsto \lambda(x,y,z)$, to emphasize the active aspect of x,y: $\nabla(y,x)$ *transports* neighbours z of x to neighbours of y; so

$$\nabla(y,x)(z) = \lambda(x,y,z). \tag{2.3.14}$$

With this notation, the equation (2.3.7) says that the transport maps $\nabla(x,y)$ and $\nabla(y,x)$ are inverse of each other.

Recall (cf. (2.1.8)) that the multiplicative monoid $(R,\cdot)$ acts on any monad $\mathfrak{M}(z)$ in a manifold. We have

Proposition 2.3.7 *The transport map* $\nabla(y,x) : \mathfrak{M}(x) \to \mathfrak{M}(y)$ *is equivariant w.r.to the action of* $(R,\cdot)$ *on the monads.*

This is is left as an Exercise using Christoffel symbols:

Exercise 2.3.8 The quantities to be compared are $\lambda(x,y,(1-t)\cdot x+t\cdot z)$ and $(1-t)\cdot y+t\cdot\lambda(x,y,z)$; prove that both these quantities come out as $y+t\cdot z-t\cdot x+t\cdot\Gamma(x;y-x,z-x)$.

If x,y,z form an infinitesimal 2-simplex in M, and $u \sim x$, we may transport u either directly along x,z, or we may transport it in two stages, by first transporting u along x,y, and then transporting along y,z. This will in general give different results. We say that the connection is *flat* or *curvature free* if the result is always the same, i.e. for any infinitesimal 2-simplex (x,y,z) in M,

$$\nabla(z,x) = \nabla(z,y) \circ \nabla(y,x), \tag{2.3.15}$$

as maps $\mathfrak{M}(x) \to \mathfrak{M}(z)$.

We shall in Section 2.4 describe how framings on a manifold give rise to flat affine connections.

Note that the re-interpretation of an affine connection $\lambda(x,y,z)$ as a family of transport maps $\nabla(y,x)$ does not treat y and z on an equal footing; to say that λ is curvature free does not imply that the conjugate connection $\overline{\lambda}$ is also curvature free (it does, of course, in case λ is torsion free, i.e. in case $\lambda = \overline{\lambda}$).

Let us record explicitly in terms of λ (or rather, in terms of $[[-,-,-]]$) what it means for it to be flat, and also, in the same terms, what it means for the conjugate affine connection $\overline{\lambda}$ to be flat. This is entirely a matter of equational rewriting, and we leave the proof as an exercise.

The affine connection λ is flat iff for all infinitesimal 2-simplices (x_0,x_1,x_2) and for all $z \sim x_0$, we have

$$[[x_2,x_1,[[x_1,x_0,z]]\,]] = [[x_2,x_0,z]],$$

and the conjugate affine connection $\overline{\lambda}$ is flat iff for all such (x_0,x_1,x_2,z), we have

$$[[\,[[z,x_0,x_1]],x_1,x_2]] = [[z,x_0,x_2]].$$

(Both these equations are, for the operation $[[y,x,z]] := y \cdot x^{-1} \cdot z$ in a group, aspects of the associative law of the group multiplication.)

Example 2.3.9 In an affine space M (say, a vector space), the ternary operation λ ("parallelogram-formation") given by $\lambda(x,y,z) := y - x + z$ (or $[[y,x,z]] = y - x + z$) is everywhere defined; assume M is a finite-dimensional vector space. Then the restriction of λ to infinitesimal whiskers (x,y,z) is an affine connection. It is torsion free, as well as curvature free. We call it the *canonical* affine connection on the affine space M. It restricts to an affine connection also on any open subset of the affine space in question.

This example explains why the figure above is drawn as a parallelogram; also it explains, I believe, why a structure like λ is called an *affine* connection.

The example can be generalized, as follows. For simplicity, assume that M is an open subset of a finite-dimensional vector space V. Let $\Gamma : M \times V \times V \to M$ be a map, bilinear in the last two arguments. Then the canonical affine connection on M may be deformed by Γ as follows (for $x \sim y$, $x \sim z$):

$$\lambda(x,y,z) := y - x + z + \Gamma(x; y-x, z-x).$$

This affine connection is torsion free iff all $\Gamma(x;-,-)$ are symmetric.

Example 2.3.10 In a (multiplicatively written) group (assumed to be a manifold), the law $\lambda(x,y,z) := yx^{-1}z$, i.e. $[[y,x,z]] = yx^{-1}z$, similarly defines a flat affine connection. The conjugate connection $\overline{\lambda}$ is given by $\overline{\lambda}(x,y,z) = zx^{-1}y$; it is likewise flat. We have that λ is torsion free if the group is commutative. (Conversely, under suitable connectedness assumptions of the group G, known from classical Lie group theory, torsion freeness of λ implies commutativity of the group; in a classical perspective, this is the assertion that a connected Lie group with abelian Lie algebra is commutative.)

Note that the transport laws ∇ for this affine connection λ may be described

$$\nabla(y,x) = \text{ left multiplication by } yx^{-1},$$

and the transport laws $\overline{\nabla}$ for the affine connection $\overline{\lambda}$ may be described

$$\overline{\nabla}(y,x) = \text{ right multiplication by } x^{-1}y.$$

This example generalizes to (Lie) pregroups, in the sense of Kock (1982).

Example 2.3.11 Let M be a smooth surface in the 3-dimensional Euclidean space. Given an infinitesimal 2-whisker (x,y,z) in M, we may in the ambient affine 3-space form $y-x+z$. It will in general not be in M, but we may project it orthogonally back into M – projection in the direction of the surface normal at x will do. Denote the point in M thus obtained by $\lambda(x,y,z)$. This λ construction will be an affine connection on M, evidently torsion free (but usually not flat). (It is known as the *Levi–Civita* or *Riemannian* connection. It will be discussed in Chapter 8 below, for abstract Riemannian manifolds.) Note that if (x,y,z) form an infinitesimal 2-simplex, then $y-x+z$ will already itself be in M (the inclusion of M into 3-space preserves and reflects affine combinations of mutual neighbours), so $\lambda(x,y,z) = y-x+z$.

This recipe in particular provides an affine connection on the (whole) surface of the earth.

Note that this way of getting an affine connection also applies to a smooth curve C in a 2-dimensional Euclidean plane. In this case, the affine connection constructed will not only be torsion free, but also flat; this follows from the theory of differential forms (Chapter 3) below.

2.4 Affine connections from framings

Let V be a finite-dimensional vector space. Given a manifold M and a V-framing k on it, with associated family of infinitesimal coordinatizations $c_x : \mathfrak{M}(x) \to D(V)$, for $x \in M$, then we get an affine connection $\lambda = \lambda_k$ on M by putting

$$\lambda(x,y,z) := k_y(k_x^{-1}(z)) = c_y^{-1}(c_x(z)).$$

Verbally, "$\lambda(x,y,z)$ is that point which in the coordinate system at y has the same coordinates as z does in the coordinate system at x".

(Note that the recipe providing $\lambda_k(x,y,z)$ does not require that $y \sim x$, but it does require that $z \sim x$.)

The transport law $\nabla(y,x) : \mathfrak{M}(x) \to \mathfrak{M}(y)$ for this connection λ is simply $k_y \circ k_x^{-1}$; so for an infinitesimal 2-simplex x, y, z in M,

$$\nabla(z,y) \circ \nabla(y,x) = k_z \circ k_y^{-1} \circ k_y \circ k_x^{-1} = k_z \circ k_x^{-1} = \nabla(z,x),$$

so that λ_k is a *flat* affine connection:

Proposition 2.4.1 *An affine connection λ_k defined by a framing k is flat.*

The affine connection λ_k may have torsion, and the conjugate affine connection may not be flat. We consider affine connections arising in this way in Section 3.7.

A particular case is the "construction-site" framing in Example 2.2.1. It provides a flat affine connection on M = surface of the earth minus the poles. Unlike the framing itself, the connection does not depend on choice of unit measure for length.

For the case of the unit sphere M, we shall describe this framing in terms of spherical coordinates (with θ = distance from "Greenwich Meridian" and ϕ = pole distance, both measured in radians). For $x = (\theta, \phi)$, $k_x : D(2) \to \mathfrak{M}(x) \subseteq M$ is given by

$$k_x(d_1, d_2) = (\theta + \frac{d_1}{\sin\phi}, \phi + d_2),$$

and thus $c_x(\theta + \delta_1, \phi + \delta_2) = (\delta_1 \cdot \sin\phi, \delta_2)$. The framing 1-form $\omega : M_{(1)} \to R^2$ is given by $\omega((\theta, \phi), (\theta + \delta_1, \phi + \delta_2)) = (\delta_1 \cdot \sin\phi, \delta_2)$.

We shall see that this connection has torsion; let us calculate λ in spherical coordinates, in continuation of the calculation above. Let us first note a piece of trigonometry: if $d^2 = 0$, $\sin(\phi + d) = \sin(\phi) + d \cdot \cos(\phi)$, by Taylor expansion, and so

$$\frac{\sin(\phi)}{\sin(\phi + d)} = \frac{\sin(\phi)}{\sin(\phi) + d \cdot \cos(\phi)} = \frac{1}{1 + d \cdot \cot(\phi)} = 1 - d \cdot \cot(\phi). \quad (2.4.1)$$

Let $x = (\theta, \phi)$, $y = (\theta', \phi')$, and $z = (\theta + \delta_1, \phi + \delta_2)$ (with $(\delta_1, \delta_2) \in D(2)$). Then we have $c_x(z) = (\delta_1 \cdot \sin(\phi), \delta_2)$, so

$$\lambda(x,y,z) = k_y(c_x(z)) = (\theta' + \frac{\delta_1 \cdot \sin(\phi)}{\sin(\phi')}, \phi' + \delta_2).$$

If now $y = (\theta', \phi')$ is $\sim x$, we have $(\theta', \phi') = (\theta + d_1, \phi + d_2)$ with $(d_1, d_2) \in D(2)$. Then we have by substituting in the above expression that

$$\lambda(x,y,z) = (\theta + d_1 + \delta_1 \cdot \frac{\sin(\phi)}{\sin(\phi + d_2)}, \phi + d_2 + \delta_2),$$

and by the calculation (2.4.1) above we may continue

$$= (\theta + d_1 + \delta_1 - \delta_1 \cdot d_2 \cdot \cot(\phi), \phi + d_2 + \delta_2).$$

Since $d_2 \cdot \delta_1$ need not be $d_1 \cdot \delta_2$, the expression here is not symmetric in y and z, so the connection is not symmetric (= torsion free).

Let V be a finite-dimensional vector space. For an arbitrary V-framing k on a manifold M, we have the associated affine connection $\lambda = \lambda_k$, as described above

$$\lambda(x,y,z) = k_y(k_x^{-1}(z)),$$

and we have the V-valued framing form $\omega = \omega_k$, as described in Section 2.2

$$\omega(x,y) = k_x^{-1}(y).$$

We shall prove:

Theorem 2.4.2 *Let k be a V-framing on a manifold M, with framing 1-form ω and with associated affine connection λ. Then the following three conditions are equivalent:*

(1) ω is closed

(2) λ is torsion free

(3) for any infinitesimal 2-simplex (x,y,z) in M, $\lambda(x,y,z) = y - x + z$.

Proof. The equivalence of (2) and (3) was proved in Proposition 2.3.4 for arbitrary affine connections. We shall prove the equivalence of (1) and (3). Since the question is local, we may assume that M is an open subset of a finite-dimensional vector space U. Then we may express both λ, k_x, and ω in coordinate terms as follows. For $x \in M$, $v \in D(V)$,

$$k_x(v) = x + b(x;v)$$

where $b(x;-) : V \to U$ is a linear isomorphism, with inverse $\beta(x;-) : U \to V$. Then

$$\omega(x,y) = \beta(x;y-x)$$

(so β is the same as the Ω considered in the proof of Proposition 2.2.2); also, the defining equation of λ, $\lambda(x,y,z) := k_y(k_x^{-1}(z))$ translates into

$$\lambda(x,y,z) := y + b(y;\beta(x;z-x)).$$

Let us, in terms of β, calculate $d\omega(x,y,z)$ for an infinitesimal 2-simplex (x,y,z)

in M. We have

$$\begin{aligned} d\omega(x,y,z) &= \omega(x,y)+\omega(y,z)-\omega(x,z) \\ &= \beta(x;y-x)+\beta(y;z-y)-\beta(x;z-x) \\ &= \beta(x;y-z)+\beta(y;z-y) \end{aligned}$$

by combining the two outer terms and using linearity of $\beta(x;-)$; now do a Taylor expansion on the second term:

$$\begin{aligned} &= \beta(x;y-z)+\beta(x;z-y)+d\beta(x;y-x,z-y) \\ &= d\beta(x;y-x,z-y) \\ &= d\beta(x;y-x,z-x) \end{aligned}$$

using the Taylor principle for the last equality. Summarizing:

$$d\omega(x,y,z) = d\beta(x;y-x,z-x), \tag{2.4.2}$$

for any infinitesimal 2-simplex (x,y,z) in $M \subseteq U$.

Let us also rewrite the expression for $\lambda(x,y,z)$, using Taylor expansions from x in the direction $y-x$; we have

$$\begin{aligned} \lambda(x,y,z) &= y+b(y;\beta(x;z-x)) \\ &= y+b(x;\beta(x;z-x))+db(x;y-x,\beta(x;z-x)) \\ &= y+(z-x)+db(x;y-x,\beta(x;z-x)) \end{aligned}$$

since $b(x;-)$ and $\beta(x;-)$ are inverse of each other. (This, in effect, provides a calculation of the Christoffel symbol for λ.) Using Corollary 1.4.12, with $u=y-x$, $v'=z-x$, we therefore have

$$\lambda(x,y,z) = y-x+z-b(x;d\beta(x;y-x,z-x)), \tag{2.4.3}$$

for any infinitesimal 2-whisker (x,y,z) in M. In particular, we have, for any infinitesimal 2-simplex (x,y,z), that $\lambda(x,y,z)=y-x+z$ iff $b(x;d\beta(x;y-x,z-x))=0$. Since $b(x;-)$ is invertible, this is the case iff $d\beta(x;y-x,z-x)=0$. This in turn is equivalent to closedness of ω, by (2.4.2). This proves the equivalence of (1) and (3) in the theorem.

Exercise 2.4.3 If λ_1 and λ_2 are affine connections on a manifold M, then for any infinitesimal 2-whisker x,y,z in M, we have $\lambda_1(x,y,z) \sim \lambda_2(x,y,z)$. (*Hint:* Express the λs in terms of coordinates and Christoffel symbols, as in (2.3.12).) Deduce that one can form affine combinations of any set of affine connections on M (just take the requisite affine combination of the values of the λ_is).

In particular, for any affine connection λ, one may form the midpoint of λ and $\overline{\lambda}$. It is a torsion-free affine connection.

Deduce that for any affine connection λ and any infinitesimal 2-simplex x, y, z, we have

$$\frac{1}{2}\lambda(x,y,z) + \frac{1}{2}\overline{\lambda}(x,y,z) = y - x + z.$$

(*Hint:* Use Proposition 2.3.4.)

The affine combinations on M in fact form canonically an affine space.

Let M and N be manifolds.

Definition 2.4.4 *A map $f : M \to N$ is called a* local diffeomorphism *or* étale† *if for each $x \in M$, f maps $\mathfrak{M}(x)$ bijectively onto $\mathfrak{M}(f(x))$.*

If M and N are manifolds equipped with affine connections λ_1 and λ_2, respectively, it is clear what it means to say that a map $f : M \to N$ is a *morphism* of manifolds-with-affine-connection, namely that

$$f(\lambda_1(x,y,z)) = \lambda_2(f(x), f(y), f(z))$$

whenever x, y, z form an infinitesimal 2-whisker in M. This is in particular interesting when f is a local diffeomorphism, as in the following definition.

Definition 2.4.5 *An affine connection λ on M is called* integrable *if there is a finite-dimensional vector space V and a local diffeomorphism $f : M \to V$ such that λ corresponds to the canonical affine connection on V via f, more precisely*

$$f(\lambda(x,y,z)) = f(y) - f(x) + f(z)$$

for all $x \sim y$, $x \sim z$ in M.

Clearly, an integrable connection is torsion free.

Proposition 2.4.6 *Let k be a V-framing on M. If the framing 1-form is exact, then the connection $\lambda = \lambda_k$ is integrable.*

Proof. The assumption is that there exists a map $f : M \to V$ such that for $x \sim y$ in M, we have $k_x^{-1}(y) = f(y) - f(x)$. Since k_x^{-1} maps $\mathfrak{M}(x)$ bijectively to $D(V)$, f maps $\mathfrak{M}(x)$ bijectively to $\mathfrak{M}(f(x)) = D(V) + f(x)$, so it is a local diffeomorphism. Now let $x \sim y$, $x \sim z$ in M. Then by definition

† The terminology is only fully justified in case a suitable Inverse Function Theorem is available, "from infinitesimal invertibility to local invertibility".

$k_y^{-1}\lambda(x,y,z) = k_x^{-1}(z)$ which we may rewrite in terms of f,

$$f(\lambda(x,y,z)) - f(y) = f(z) - f(x),$$

which is equivalent to the desired relationship between f and λ.

Note that the proof gives a little more than asserted in the statement: namely that a primitive of the framing 1-form will serve as the map witnessing integrability of the connection.

2.5 Bundle connections

(A formulation of a general connection notion in terms of groupoids is given in Section 5.2 below. This is the context in which *curvature* of connections is best described.)

We use the term *bundle over M* for any map $\pi : E \to M$; E is the *total space* and M the *base* of the bundle; the *fibre* over $x \in M$ is the subspace $E_x := \pi^{-1}(x) \subseteq E$. A *map* of bundles over M is a map between the total spaces making the obvious triangle commute.

Examples of bundles over M are $\mathrm{proj}_1 : M_{(1)} \to M$, and product bundles like $\mathrm{proj} : M \times F \to M$. The map (2.1.1) above is a map of bundles over M.

For a bundle $\pi : E \to M$ over a manifold, we have a notion of *connection* (due to Joyal, unpublished), which we describe now. We shall use the term *bundle connection*, to distinguish it from the related notion of *connection in a groupoid*, to be considered in Section 5.2 below. Affine connections can be construed as bundle connections, and also as groupoid connections.

So given a bundle $\pi : E \to M$, with M a manifold, a *bundle connection* in it is a law ∇ which to any $(x,y) \in M_{(1)}$ and $e \in E_x$ associates an element $\nabla(y,x)(e) \in E_y$, subject to the requirement

$$\nabla(x,x)(e) = e \tag{2.5.1}$$

for any $x \in M$ and any $e \in E_x$.

Remark 2.5.1 The reader may prefer to think of ∇ as providing an *action* (left action) of the graph $M_{(1)} \rightrightarrows M$ on $E \to M$. We may say that $\nabla(y,x)$ *acts* on e to give $\nabla(y,x)(e)$, or that $\nabla(y,x)$ *transports* $e \in E_x$ to $\nabla(y,x)(e) \in E_y$. This leads to another way of describing connections in $E \to M$ (satisfying (2.5.2)), which is to consider the groupoid $\mathfrak{S}(E \to M)$ whose objects are the fibres of $E \to M$, and whose arrows are the bijections between such fibres; then $\nabla(y,x)$ may be seen as an arrow in this groupoid. This is the viewpoint in Section 5.2

below. Sometimes, we shall want a symbol, like $\dashv$ for the action itself, and write $\nabla(y,x) \dashv e$ for $\nabla(y,x)(e)$ (or even $(y,x) \dashv e$ or ${}^{yx}e$).

The following property can then be verified in many situations, using coordinates (see e.g. Example 2.5.2 below):

$$\nabla(x,y)(\nabla(y,x)(e)) = e \tag{2.5.2}$$

for any $(x,y) \in M_{(1)}$ and any $e \in E_x$.

If (x,y,z) is an infinitesimal simplex in M, and ∇ is a bundle connection in a bundle $E \to M$, we have for any $e \in E_x$ two elements in E_z which we may want to compare, namely the direct transport of e from x to z, or the transport via y; we say that the bundle connection is *flat* or *curvature free* if these two elements are always the same, i.e. if for all infinitesimal simplices (x,y,z) and all $e \in E_x$,

$$\nabla(z,x)(e) = \nabla(z,y)(\nabla(y,x)(e)), \tag{2.5.3}$$

or equivalently, if

$$\nabla(z,y) \circ \nabla(y,x) = \nabla(z,x)$$

as maps $E_x \to E_z$. This is in analogy (and in fact generalizes) the notion of flatness of affine connections.

There is a property which can be verified in many situations, using coordinates (see Example 2.5.2 below); namely that under the same assumptions (that x,y,z form an infinitesimal 2-simplex, and $e \in E_x$)

$$\nabla(z,y)(\nabla(y,x)(e)) \sim e. \tag{2.5.4}$$

We shall, however, not make (2.5.2) or (2.5.4) part of the definition of bundle connection.

Note that $\nabla(z,x)(e) \sim e$, so flatness of ∇ implies (2.5.4).

Example 2.5.2 Let U and V be finite-dimensional vector spaces, and consider a bundle connection in the bundle $U \times V \to U$. Since $\nabla(x,x) \dashv (x,v) = (x,v)$, it follows that the connection is of the form

$$\nabla(y,x) \dashv (x,v) = (y, v + L(x,v;y-x)),$$

with $L : U \times V \times V \to V$, linear in the last variable. We now calculate $\nabla(z,y) \dashv \nabla(y,x) \dashv (x,v)$. It is of the form (z,w) for some $w \in V$, and this w, we calculate in terms of the function L. We have

$$\begin{aligned} w &= v + L(x,v;y-x) + L(y, v + L(x,v;y-x); z-y) \\ &= v + L(x,v;y-x) + L(y,v;z-y) + dL(y,v;L(x,v;y-x), z-y) \end{aligned}$$

by taking the directional derivative of L in its second variable in the direction $L(x,v;y-x)$; we now use the Taylor principle and replace the first occurrence of x by y, so that we have

$$\begin{aligned} &= v + L(y,v;y-x) + L(y,v;z-y) + dL(y,v;L(x,v;y-x), z-y) \\ &= v + L(y,v;z-x) + dL(y,v;L(x,v;y-x), z-y) \end{aligned}$$

using linearity of L in the argument after the semicolon. Finally, the last term has a linear occurrence $y-x$, so that we may replace the last letter y by x, and after this replacement, we have

$$= v + L(y,v;z-x) + dL(y,v;L(x,v;y-x), z-x).$$

Each of the last two terms now depends linearly on $z-x$; summarizing, $w = v + \tilde{L}(z-x)$, with $\tilde{L}$ linear. Then also $(z,w) - (x,v) \in U \times V$ depends linearly on $z-x$ and therefore $(z,w) \sim (x,v)$. This proves that (2.5.4) holds.

This example may be modified to give that (2.5.4) holds for bundles $E \to M$ which locally are product bundles $U' \times V'$ with U' and V' manifolds.

An affine connection λ on M may be seen as a bundle connection ∇ in the bundle $M_{(1)} \to M$ (= the bundle whose fibre over x is $\mathfrak{M}(x)$); namely via the formula

$$\nabla(y,x)(z) := \lambda(x,y,z). \tag{2.5.5}$$

The equation (2.5.1) holds by virtue of (2.3.1). Then (2.5.2) holds, by (2.3.7) (put $e = z$). (Essentially the same formula gives that an affine connection may be interpreted as a bundle connection in the tangent bundle $TM \to M$, see Proposition 4.3.7 below.)

The notion of *torsion* which one has for affine connections, makes no sense for general bundle connections.

The fibres E_x of a bundle $E \to M$ may be equipped with some structure; they may be vector spaces (so $E \to M$ is a vector bundle); or they may be pointed spaces (so the chosen points in the E_xs define a cross-section of $E \to M$ – vice versa, a cross-section of $E \to M$ makes the E_xs into pointed spaces).

In these cases, it makes sense to say that a bundle connection ∇ in $E \to M$ *preserves* the structure in question: this means that each individual $\nabla(y,x)$:

$E_x \to E_y$ preserves the structure. Thus, if $E \to M$ is a vector bundle, and ∇ preserves the vector space structure, one says that ∇ is a *linear* connection. (Note that the connection studied in Example 2.5.2 is a linear connection iff the function $L : U \times V \times V \to V$ is linear in the second variable.) If $E \to M$ is a bundle of groups, or algebras, and ∇ preserves the group structure (respectively the algebra structure), one says that ∇ is a *group connection*, respectively an *algebra* connection, etc.

The bundle $M_{(1)} \to M$ has the monads $\mathfrak{M}(x)$ for its fibres; they are pointed sets, $\mathfrak{M}(x)$ having as base point x. Equivalently, there is given a cross-section of $M_{(1)} \to M$, namely the diagonal map $x \mapsto (x,x)$. When an affine connection λ in M is seen as a bundle connection in $M_{(1)} \to M$, it preserves the pointed structure: $\nabla(y,x)(x) = y$. This is equation (2.3.2).

Example 2.5.3 An ordinary differential equation $y' = F(x,y)$, as in the Calculus Books, may be seen as a bundle connection. Let $M \subseteq R$ be an open set (an “open interval”, say) and assume F is defined for $(x,y) \in M \times R$. Then $y' = F(x,y)$ defines a bundle connection in the bundle $M \times R \to M$, as follows: for x and x_1 in M, with $x \sim x_1$, put

$$\nabla(x_1,x)(x,y) := (x_1, y + F(x,y) \cdot (x_1 - x)). \tag{2.5.6}$$

If $x_1 = x$, the right-hand side clearly is (x,y), so (2.5.1) holds.

What has been proved here can also be formulated: the map $\nabla(x_1,x)$ provides an invertible map from the fibre over x to the fibre over x_1, with $\nabla(x,x_1)$ as inverse. This leads to the abstract groupoid-theoretic viewpoint on connections, as will be discussed in Section 5.2.

In the present case, the fibres may be identified with R (= the y-axis). With this identification, $\nabla(x_1,x)$ may be seen as an invertible map (diffeomorphism) $R \to R$, with formula $y \mapsto y + F(x,y) \cdot (x_1 - x)$.

Conversely, let $M \subseteq R$ be open, and consider a bundle connection ∇ in the bundle $M \times R \to M$. For $(x,y) \in M \times R$ and $d \in D$, we have $\nabla(x+d,x)(x,y)$ in the fibre over $x+d$, thus it is of the form $(x+d, \eta(d))$ for some function $\eta : D \to R$. By (2.5.1), $\eta(0) = y$, so by KL, η is of the form $\eta(d) = y + d \cdot F$ for some constant $F \in R$. Now let (x,y) vary; the constant $F \in R$ then becomes a function of (x,y), so we have $F : M \times R \to R$. The given bundle connection is now the one given as above by the differential equation $y' = F(x,y)$.

So there is a bijective correspondence between bundle connections in $M \times R \to M$ and ordinary (first-order) differential equations $y' = F(x,y)$ (with $F : M \times R \to R$).

2.6 Geometric distributions

We give examples from differential geometry below, but since some of the theory is pure combinatorics (graph theory, in fact), we begin with a formal combinatorial treatment, which also makes sense in other contexts than SDG.

So we consider a set M and a reflexive symmetric relation $\sim$ on M. For $x \in M$, we put $\mathfrak{M}(x) = \{y \in M \mid y \sim z\}$ (in graph theory, this is sometimes called the *star* of x).

A *pre-distribution* (relative to $\sim$) is a reflexive symmetric relation $\approx$ which refines $\sim$ in the sense that $x \approx y$ implies $x \sim y$. We put $\mathfrak{M}_{\approx}(x) = \{y \in M \mid y \approx x\}$. We have $\mathfrak{M}_{\approx}(x) \subseteq \mathfrak{M}(x)$. A pre-distribution (relative to a given $\sim$) may be described in terms of this family of subsets.

This synthetic rendering of geometric distributions appeared in Kock (2003).

Definition 2.6.1 *A pre-distribution $\approx$ on M (relative to $\sim$) is called* involutive *if for all x, y, z in M,*

$$[x \approx y\,,\, x \approx z \text{ and } y \sim z] \quad \textit{imply} \quad y \approx z. \tag{2.6.1}$$

A relevant picture is the following; single lines indicate the neighbour relation $\sim$, double lines indicate the assumed "strong" neighbour relation $\approx$:

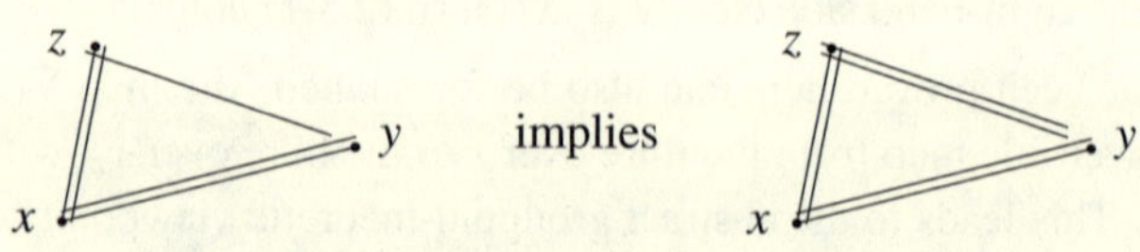

This idea is reminiscent of (one of) Ehresmann's (1954) formulations of the notion of foliation on M, given in terms of two topologies on M, one refining the other. With suitable understanding of the words: to say that $\approx$ is involutive is to say that it is *relatively* transitive (relative to $\sim$).

Note that if r is an equivalence relation on M, then the relation

$$x \sim_r y \text{ iff } x \sim y \text{ and } x\, r\, y \tag{2.6.2}$$

is an involutive pre-distribution (relative to $\sim$).

An *integral set* for a pre-distribution $\approx$ on M (relative to $\sim$) is a subset $F \subseteq M$ such that for any $x \in F$,

$$\mathfrak{M}(x) \cap F \subseteq \mathfrak{M}_{\approx}(x),$$

or, equivalently, such that

$$\big[x \in F, y \in F \text{ and } x \sim y\big] \quad \text{imply} \quad x \approx y. \tag{2.6.3}$$

Clearly a subset of an integral set for $\approx$ is again integral. Also, M itself is an integral set iff $\approx$ equals $\sim$.

The sets $\mathfrak{M}_{\approx}(x)$ need not be integral subsets; rather

Proposition 2.6.2 *For any pre-distribution $\approx$, the following two conditions are equivalent:*
(1) $\approx$ is involutive
(2) all sets of the form $\mathfrak{M}_{\approx}(u)$ ($u \in M$) are integral subsets.

Proof. Assume (1), and let $x \in \mathfrak{M}_{\approx}(u)$, $y \in \mathfrak{M}_{\approx}(u)$, and $x \sim y$. Then $u \approx x$, $u \approx y$, and $x \sim y$, but these three conditions imply $x \approx y$, by involutivity of $\approx$. So (2.6.3) holds for $F = \mathfrak{M}_{\approx}(u)$. Assume conversely (2), and assume $x \approx y$, $x \approx z$, and $y \sim z$. With $F := \mathfrak{M}_{\approx}(x)$, we have $y \in F$, $z \in F$, and $y \sim z$. Since F is integral, we conclude $y \approx z$.

We now return to the context of SDG, so that M is a manifold, and the relation $\sim$ is the (first-order) neighbour relation $M_{(1)} \subseteq M \times M$.

The celebrated Frobenius Theorem which we are about to state is an integration result, and it depends on topological notions for its formulation, notably on a suitable notion of a *connected* set; recall from the Appendix (Section A.6) that this may be taken to mean *path connected*, or (cf. Remark A.6.3) it may be taken in a sense that depends on the notion of *open* set.

Consider a pre-distribution $\approx$ on a manifold M.

Definition 2.6.3 *A* leaf *Q through $x \in M$ is an integral subset which is connected, and which is maximal in the following sense: any other integral connected subset F containing x is contained in Q.*

By maximality, it is clear that a leaf through x is unique if it exists. In this case, we may denote it $Q(x)$. And $x \in Q(x)$ (apply maximality and use $F = \{x\}$).

Proposition 2.6.4 *Assume that for every $x \in M$, there is a leaf $Q(x)$ through x. Then if $y \in Q(x)$, we have that $Q(y) = Q(x)$, and the family of leaves form a partition of M.*

Proof. The set $Q(x)$ is a connected integral subset, and it contains y by assumption. By the maximality of the leaf $Q(y)$, we therefore have $Q(x) \subseteq Q(y)$.

Since $x \in Q(x)$, we therefore also have $x \in Q(y)$. So $Q(y)$ is a connected integral subset containing x, and therefore, by maximality of $Q(x)$, $Q(y) \subseteq Q(x)$. So $Q(x) = Q(y)$. From this immediately follows that the Qs form a partition (parametrized by the space M).

Proposition 2.6.5 *Assume that $\approx$ is involutive, and assume that for every $x \in M$, there exists a leaf $Q(x)$ through x. Assume that each $\mathfrak{M}_{\approx}(x)$ is connected. Then $\mathfrak{M}_{\approx}(x) = \mathfrak{M}(x) \cap Q(x)$.*

Proof. Since $\approx$ is involutive, we get by Proposition 2.6.2 that the set $\mathfrak{M}_{\approx}(x)$ is an integral subset; it contains x, and it is connected by assumption. By maximality of $Q(x)$, we therefore have $\mathfrak{M}_{\approx}(x) \subseteq Q(x)$. Also $\mathfrak{M}_{\approx}(x) \subseteq \mathfrak{M}(x)$, so the inclusion $\subseteq$ follows. The converse holds, since $Q(x)$ is an integral subset for $\approx$.

We shall describe the notion of *distribution* in contrast to *pre-distribution* (in classical terms, a pre-distribution is rather like "a distribution with singularities"). This depends on the notion of *linear subsets* of monads, generalizing the notion (Chapter 1) of linear subsets of $D(V)$ when V is a finite-dimensional vector space.

Proposition 1.5.3 justifies the phrase "hence any" in the following

Definition 2.6.6 *Let M be a manifold, and let $x \in M$. A subset $S \subseteq \mathfrak{M}(x)$ is called a* linear subset *if for some (hence any) bijection $f : \mathfrak{M}(x) \to D(V)$ (with $f(x) = 0$, V a finite dimensional vector space), S maps bijectively to a linear subset of $D(V)$.*

The linear subset is said to have dimension m if the linear subspace of V occurring in the definition, has dimension m.

In Chapter 4, we shall see that V and the bijection $\mathfrak{M}(x) \to D(V)$ may be canonically chosen, namely by taking V to be the *tangent space* $T_x(M)$.

The definition may be alternatively formulated: an inclusion map $j : S \to \mathfrak{M}(x)$ makes S a *linear subset* if there is a finite dimensional vector space V and a finite dimensional linear subspace U of V such that the following total diagram is a pull back (with f a bijection with $f(x) = 0$)

$$\begin{array}{ccccc}
S & \longrightarrow & D(U) & \longrightarrow & U \\
{\scriptstyle j}\downarrow & & \downarrow & & \downarrow{\scriptstyle \subseteq} \\
\mathfrak{M}(x) & \xrightarrow{\ f\ } & D(V) & \longrightarrow & V
\end{array}$$

(the right hand square here is always a pull back, by Proposition 1.2.4, so the total of the above diagram is a pull back if and only if the left hand square is a pull back).

A pre-distribution is called a *distribution* if the subsets $\mathfrak{M}_{\approx}(x) \subseteq \mathfrak{M}(x)$ are *linear* subsets, in the sense thus described. The examples we present below are all distributions in this sense. We assume that R is connected; from Corollary A.6.4 in the Appendix it then follows that each $\mathfrak{M}_{\approx}(x)$ is connected.

The justification for the terminology (i.e. the comparison with the classical notion of distribution, and with the property of being involutive) is not trivial; see Theorem 3.6.2. See also Section 4.10. In fact, to formulate the classical notion of involutiveness, some theoretical superstructure is needed: either the notion of vector fields and their Lie brackets, or the differential algebra of differential forms (exterior derivative and wedge product).

Example 2.6.7 Let ω be a 1-form (R-valued, for simplicity) on a manifold M. It defines a pre-distribution,

$$x \approx y \text{ iff } x \sim y \text{ and } \omega(x,y) = 0.$$

The relation $\approx$ is reflexive because $\omega(x,x) = 0$, and it is symmetric because of $\omega(x,y) = -\omega(y,x)$ by (2.2.4). It is a distribution if ω is suitably regular. In any case, this pre-distribution is involutive if ω is closed (the converse is not true, see Example 2.6.8 below). For, if ω is closed and x, y, z form an infinitesimal 2-simplex, then

$$\omega(x,y) + \omega(y,z) = \omega(x,z),$$

so if two of these three entries are 0, then so is the third; in other words, if two of the assertions $x \approx y$, $y \approx z$, and $x \approx z$ hold, then so does the third.

Example 2.6.8 If a pre-distribution $\approx$ comes about from a 1-form ω, as in Example 2.6.7, and if $f : M \to R$ is a function with invertible values, then $\omega(x,y) = 0$ iff $(f \cdot \omega)(x,y) = 0$ (here, $f \cdot \omega$ denotes the 1-form given by $(f \cdot \omega)(x,y) := f(x) \cdot \omega(x,y)$). So the pre-distribution defined by $f \cdot \omega$ is likewise $\approx$. Now, closedness of ω does not imply closedness of $f \cdot \omega$. Therefore, a non-closed 1-form (like $f \cdot \omega$) may happen to define a pre-distribution which is involutive.

A 1-form which defines an involutive distribution is called an *integrable* 1-form. The terminology derives from the Frobenius Theorem quoted below.

If θ is an integrable 1-form, it makes sense to ask for a function $g : M \to R$ with invertible values such that $g \cdot \theta$ is closed. Such g is then called an *integrating factor* for the 1-form θ.

Example 2.6.9 Let M be a manifold, and let $G_k(M)$ be the space of pairs (x,S), where $x \in M$ and $S \subseteq \mathfrak{M}(x)$ is a k-dimensional linear subset. This is known classically (through a different construction, namely via the tangent bundle of M, see Chapter 4 below) to be a manifold (a "fibrewise Grassmannian" of the tangent bundle). It carries a canonical distribution: for $(x,S) \sim (x',S')$, we say that $(x,S) \approx (x',S')$ if $x' \in S$ (which can be proved to be equivalent to $x \in S'$). Distributions of this type are studied extensively by Lie and by Klein; they use the term "vereinigte Lage", see Lie (1977, p. 38) or Klein (1926, pp. 239, 269, 275, 285). Such distributions are not involutive.

The following is an integration result of classical differential topology; here, we take it as an axiom (whose validity in the various topos models may be investigated, or may be reduced to more basic integration assumptions, like existence of primitives; see McLarty (1983), and Faran (1998) for some investigations in this direction). Recall that leaves are unique if they exist.

Theorem 2.6.10 (Frobenius Theorem) *If $\approx$ is an involutive k-dimensional distribution on M, then for any $x \in M$, there exists a leaf $Q(x)$ through x.*

It follows that the $Q(x)$s form a partition of M; and

$$\mathfrak{M}(x) \cap Q(x) = \mathfrak{M}_{\approx}(x). \tag{2.6.4}$$

Each $\mathfrak{M}_{\approx}(x)$ is connected; so these assertions are consequences of the purely combinatorial results in Propositions 2.6.4 and 2.6.5.

The second assertion here may also be formulated like this:

Let r denote the equivalence relation on M corresponding to the partition into leaves; then the pre-distribution $\sim_r$ equals $\approx$, where $x \sim_r y$ is defined by "$x \sim y$ and $x\, r\, y$", and in particular, the pre-distribution $\sim_r$ is an (involutive) distribution.

Example 2.6.11 Consider (like in Example 2.5.3) an ordinary first-order differential equation

$$y' = F(x,y),$$

as in the Calculus Books; as known from these books, the equation gives rise to a "direction field": through each point $e = (x,y) \in R \times R$, one draws a "little" line segment $S(x,y)$ with slope $F(x,y)$.

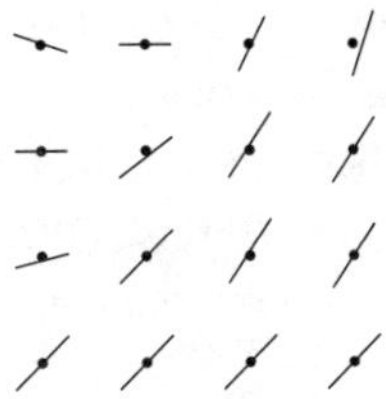

The family of subsets $S(x,y)$ drawn may be construed as the sets $\mathfrak{M}_{\approx}(x,y)$ for a distribution $\approx$ in the plane R^2 given (for $(x,y) \sim (x_1,y_1)$) by

$$(x_1,y_1) \approx (x,y) \text{ iff } y_1 - y = F(x,y)\cdot(x_1 - x).$$

Equivalently, for $(d_1,d_2) \in D(2)$,

$$(x+d_1,y+d_2) \approx (x,y) \text{ iff } d_2 = F(x,y)\cdot d_1.$$

This distribution is involutive. (In fact, every 1-dimensional distribution is involutive, cf. Proposition 2.6.16 below, but presently, we give proofs entirely in terms of elementary calculus.) For, assume $(x+d_1,y+d_2) \approx (x,y)$ and $(x+\delta_1,y+\delta_2) \approx (x,y)$ with $(x+d_1,y+d_2) \sim (x+\delta_1,y+\delta_2)$, so $((d_1,d_2),(\delta_1,\delta_2)) \in \widetilde{D}(2,2)$. By assumption we have

$$d_2 = F(x,y)\cdot d_1 \tag{2.6.5}$$

and

$$\delta_2 = F(x,y)\cdot\delta_1.$$

Subtracting these two equations, we get

$$\delta_2 - d_2 = F(x,y)\cdot(\delta_1 - d_1);$$

the desired $\approx$ relation is

$$\delta_2 - d_2 = F(x+d_1,y+d_2)\cdot(\delta_1 - d_1).$$

Subtracting these two equations gives that the desired equation is equivalent to

$$0 = \left[F(x+d_1,y+d_2) - F(x,y)\right]\cdot(\delta_1 - d_1),$$

which in turn by Taylor expansion says

$$0 = (\frac{\partial F}{\partial x}\cdot d_1 + \frac{\partial F}{\partial y}\cdot d_2)\cdot(\delta_1 - d_1) \tag{2.6.6}$$

(all the partial derivatives evaluated at (x,y)). Now some of the "arithmetic" of $\widetilde{D}(2,2)$ enters; if $((d_1,d_2),(\delta_1,\delta_2)) \in \widetilde{D}(2,2)$, the products d_1^2, $d_1\cdot\delta_1$, and $d_1\cdot d_2$ are all 0. Using this, the expression (2.6.6) reduces to $\partial F/\partial y \cdot d_2 \cdot \delta_1$. Substituting (2.6.5) and using $d_1\cdot\delta_1 = 0$ gives 0, as desired.

This example may of course be generalized to the case where $F(x,y)$ is defined only for x ranging in an open subset $M \subseteq R$. Locally, the leaves asserted by the Frobenius Theorem are then graphs of local solutions $y(x)$ for the differential equation $y' = F(x,y)$. We recognize here the "graphical method" for solving first-order differential equations, known from the Calculus Books.

Consider the similar situation for functions in two variables. Then we will encounter distributions which are not involutive.

Example 2.6.12 Let $M \subseteq R^2$ be an open subset of the plane. The data (= the right-hand side) of a first-order partial differential equation on M of the form

$$\frac{\partial z}{\partial x} = F(x,y,z)$$

$$\frac{\partial z}{\partial y} = G(x,y,z)$$

may be construed as a distribution $\approx$ on $M \times R \subseteq R^3$, in analogy with Example 2.6.11, as follows: for $(x,y,z) \sim (x_1,y_1,z_1)$, we put

$$(x,y,z) \approx (x_1,y_1,z_1) \quad \text{iff} \quad z_1 - z = F(x,y,z)\cdot(x_1 - x) + G(x,y,z)\cdot(y_1 - y).$$

Here is a picture (from Burke, 1985) of some of the $S(x,y,z)$s (for suitable F,G):

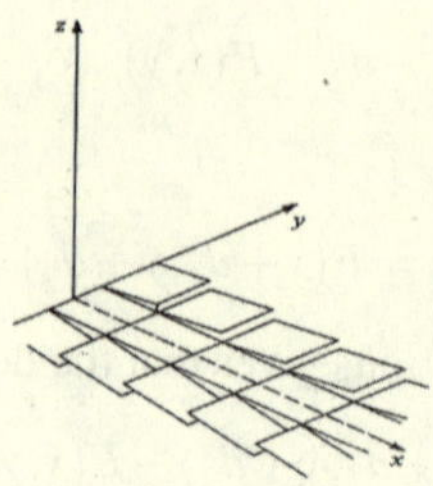

(2.6.7)

The familiar condition for solvability of the partial differential equation considered in Example 2.6.12 (cf. e.g. Spivak, 1979, Vol. 1, chapter 6, p. 253) is that

$$\frac{\partial F}{\partial y} + \frac{\partial F}{\partial z}\cdot G = \frac{\partial G}{\partial x} + \frac{\partial G}{\partial z}\cdot F. \tag{2.6.8}$$

(In particular, if F and G do not depend on z, the condition is that $\frac{\partial F}{\partial y} = \frac{\partial G}{\partial x}$.)

Proposition 2.6.13 *The distribution $\approx$ is (combinatorially) involutive iff (2.6.8) holds.*

Proof. Assume first that the equation (2.6.8) holds. Let $((d_1,d_2,d_3),(\delta_1,\delta_2,\delta_3)) \in \widetilde{D}(2,3)$. The proof begins much like the one in Example 2.6.11. In analogy with (2.6.6), we have to prove that if

$$d_3 = F(x,y,z)\cdot d_1 + G(x,y,z)\cdot d_2, \tag{2.6.9}$$

and similarly for the δ_i (with $((d_1,d_2,d_3),(\delta_1,\delta_2,\delta_3)) \in \widetilde{D}(2,3)$), then

$$\begin{aligned}(\frac{\partial F}{\partial x}\cdot d_1 + \frac{\partial F}{\partial y}\cdot d_2 + \frac{\partial F}{\partial z}\cdot d_3)(\delta_1 - d_1)\\ +(\frac{\partial G}{\partial x}\cdot d_1 + \frac{\partial G}{\partial y}\cdot d_2 + \frac{\partial G}{\partial z}\cdot d_3)(\delta_2 - d_2) = 0\end{aligned}$$

where all the partial derivatives are to be evaluated at (x,y,z). Multiplying out, and using the arithmetic of $\widetilde{D}(2,3)$ (e.g. $d_i\cdot d_j = 0$, $d_i\cdot\delta_i = 0$), the left-hand side ends up being

$$\frac{\partial F}{\partial y}\cdot d_2\cdot\delta_1 + \frac{\partial F}{\partial z}\cdot d_3\cdot\delta_1 + \frac{\partial G}{\partial x}\cdot d_1\cdot\delta_2 + \frac{\partial G}{\partial z}\cdot d_3\cdot\delta_2.$$

Now we substitute the expression (2.6.9) assumed for d_3; using the arithmetic, some more terms cancel, and we end up with

$$\frac{\partial F}{\partial y}\cdot d_2\cdot\delta_1 + \frac{\partial F}{\partial z}\cdot G\cdot d_2\cdot\delta_1 + \frac{\partial G}{\partial x}\cdot d_1\cdot\delta_2 + \frac{\partial G}{\partial z}\cdot F\cdot d_1\cdot\delta_2$$

(where also F and G likewise are to be evaluated at (x,y,z)). Finally use the fact that $d_2\cdot\delta_1 = -d_1\cdot\delta_2$ (again part of the arithmetic of $\widetilde{D}(2,3)$), this may be rewritten

$$d_1\cdot\delta_2\cdot[-\frac{\partial F}{\partial y} - \frac{\partial F}{\partial z}\cdot G + \frac{\partial G}{\partial x} + \frac{\partial G}{\partial z}\cdot F].$$

The square bracket here is 0 by the assumption (2.6.8).

The converse of course depends on a sufficient supply of infinitesimal 2-simplices; this "sufficient supply" is here secured by the cancellation principles deriving from the KL axiom for $\widetilde{D}(2,2)$:

Assume the distribution $\approx$ defined by F and G is involutive in the combinatorial sense. To see that the expression in (2.6.8) is 0, it suffices by the quoted case of KL, to prove that for all $((d_1,d_2),(\delta_1,\delta_2)) \in \widetilde{D}(2,2)$, and all (x,y,z)

$$d_1\cdot\delta_2\cdot[-\frac{\partial F}{\partial y} - \frac{\partial F}{\partial z}\cdot G + \frac{\partial G}{\partial x} + \frac{\partial G}{\partial z}\cdot F] = 0. \tag{2.6.10}$$

(Here, again, the functions F, $\frac{\partial F}{\partial y}$, etc. are to be evaluated at the given (x,y,z).) So take $d_3 := F(x,y,z)\cdot d_1 + G(x,y,z)\cdot d_2$ and $\delta_3 := F(x,y,z)\cdot\delta_1 + G(x,y,z)\cdot\delta_2$. From $((d_1,d_2),(\delta_1,\delta_2)) \in \widetilde{D}(2,2)$ and Proposition 1.2.7 it follows that

$$((d_1,d_2,d_3),(\delta_1,\delta_2,\delta_3)) \in \widetilde{D}(2,3),$$

and by construction $(x,y,z) \approx (x+d_1, y+d_2, z+d_3)$ and similarly with δ_is instead of d_is. It follows from the combinatorial involution condition that

$$(x+d_1,y+d_2,z+d_3) \approx (x+\delta_1,y+\delta_2,z+\delta_3),$$

and now the calculation in the first half of the proof gives the validity of (2.6.10). This proves the proposition.

Using the proposition, one may provide "analytic" examples of non-involutive distributions on R^3. For instance, let us in the above recipe put $F(x,y,z) = y$, $G(x,y,z) = 0$. This is the analytic aspect of the picture (2.6.7). (Since F and G here are independent of z, only (some of) the $S(x,y,z)$ for $z = 0$ are drawn; (some of) the rest may be obtained by vertical translation from these.)

Using this kind of "elementary calculus", it is easy to construct examples of manifolds M with distributions $\approx$ which are as far from being involutive as possible, in the sense that for any two points in M, there exists a connected integral set C in M containing both points; typically C is a curve. I believe such distributions are sometimes called *totally an-holonomic*.

Bundle connections as geometric distributions

In the previous subsection, we noted that the same "analytic" material $y' = F(x,y)$ manifested itself both as a bundle connection, and as a distribution. The Calculus Books take the distribution (= direction field) as the primary conceptual formulation (rather than the bundle-connection formulation), since it allows pictures to be drawn, in analogy with the representation of a function $R \to R$ in terms of its *graph* $\subseteq R \times R$.

More generally, bundle connections may always be represented geometrically in terms of their "graphs":

Let E and M be manifolds. If $\pi : E \to M$ is a bundle, a *(pre-)distribution transverse to the fibres* is a (pre-)distribution $\approx$ on E such that for each $e \in E$, π maps $\mathfrak{M}_{\approx}(e)$ bijectively to $\mathfrak{M}(\pi(e))$. The distributions given by differential equations, as in Examples 2.6.11 and 2.6.12, are transverse to the fibres.

We give here a recipe which from a bundle connection produces a distribution which is transverse to the fibres in this sense.

Given a bundle connection ∇ on a bundle $\pi : E \to M$ with E and M manifolds, for each $e \in E_x$, we get a subset $S(e)$, namely the set of elements $e' \in E$ which can be reached from e by transport by ∇, more precisely, by transport by $\nabla(y,x)$ for some $y \sim x$.

Therefore, a bundle connection may be encoded by a law S which to each $e \in E$ associates a subset $S(e) \ni e$, mapping by π bijectively to $\mathfrak{M}(\pi(e))$. Assume the condition that $\nabla(x,y) \circ \nabla(y,x) = \text{id}$ (cf. (2.5.2)) for ∇. Then this family S of subsets $S(e)$ may also be expressed as the sets $\mathfrak{M}_{\approx}(e)$ for a certain pre-distribution $\approx$ on E, with $e \approx e_1$ iff $e_1 \in S(e)$ (iff $\nabla(x_1,x)(e) = e_1$ where $x = \pi(e)$ and $x_1 = \pi(e_1)$); the symmetry of the relation $\approx$ follows from (2.5.2).

We have already considered a special case: the bundle connection associated to the differential equation $y' = F(x,y)$ (cf. Example 2.5.3) gives, by this recipe, rise to the geometric distribution which we considered in Example 2.6.11.

This is the most geometric way of presenting a bundle connection in a bundle $E \to M$: *drawings* can actually be made to represent it (in this sense, it is analogous to representing a function by its *graph*). The direct way from the family S of such subsets to the connection ∇ is that $\nabla(y,x)(e)$ ($e \in E_x$) is the unique element in $S(e)$ which by π maps to y. Conversely, given the bundle connection ∇, $S(e)$ is the set $S(e) = \{\nabla(y,x)(e) \mid y \in \mathfrak{M}(x)\}$ (for $e \in E_x$).

The relationship between the bundle connection ∇ in $\pi : E \to M$ and the distribution (transverse to the fibres of π) may be expressed in terms of $\approx$ as follows. For $e \in E_x$ and $x \sim y$, we have for all $e' \in E_y$

$$\nabla(y,x)(e) = e' \text{ iff } e \approx e'.$$

We may summarize:

Proposition 2.6.14 *Given a bundle $E \to M$ with E and M manifolds, then there is a bijective correspondence between bundle connections in $E \to M$, and pre-distributions on E transverse to the fibres.*

The pre-distribution will under mild assumptions actually be a distribution.

Recall the very weak condition (2.5.4) for a bundle connection. Let the bundle connection ∇ and the distribution $\approx$ correspond to each other. Then we have

Proposition 2.6.15 *If the connection ∇ is flat then the distribution $\approx$ is involutive; conversely if $\approx$ is involutive, then ∇ is flat, provided ∇ satisfies (2.5.4).*

Proof. Assume ∇ flat, and let e, e', e'' be an infinitesimal 2-simplex in E above the infinitesimal 2-simplex x, y, z. Assume $e \approx e'$ and $e' \approx e''$. We have to prove $e \approx e''$, i.e. $\nabla(z,x)e = e''$. The assumptions give $\nabla(y,x)(e) = e'$ and $\nabla(z,y)(e') = e''$. By flatness of ∇,

$$\nabla(z,x)e = \nabla(z,y)\nabla(y,x)e = \nabla(z,y)e' = e'',$$

so $e \approx e''$.

Conversely, assume $\approx$ is involutive. Let x, y, z be an infinitesimal 2-simplex in M. To prove $\nabla(z,y)\nabla(y,x)e = \nabla(z,x)e$ for arbitrary e in E_x, we consider $e' := \nabla(y,x)e$ and $e'' = \nabla(z,y)e'$. Then $e \approx e'$ and $e' \approx e''$, and also, by virtue of (2.5.4), $e \sim e''$; so e, e', e'' form an infinitesimal 2-simplex in E, and to this infinitesimal 2-simplex we can now apply the assumption of involutiveness: we conclude $e \approx e''$. So $\nabla(z,x)e = e''$; but e'' is by construction $\nabla(z,y)\nabla(y,x)e$.

Generally, bundle connections on bundles $\text{proj}_1 : M \times R \to M$ correspond to first-order *partial* differential equations on M, whereas first-order *ordinary* differential equations more appropriately are seen as vector fields, and their study belongs to kinematics, rather than to (static) geometry. (From this perspective, $y' = F(x,y)$ is really a *partial* differential equation!)

1-dimensional distributions

We shall here prove

Proposition 2.6.16 *Any 1-dimensional distribution on a manifold M is involutive.*

We have not been completely explicit about the notion of *dimension*; it is best dealt with in terms of the tangent spaces $T_x(M)$ to M, see Chapter 4. Here we shall be content with giving the definition in case M is an open subset of a finite-dimensional vector space V. Then for any distribution $\approx$ on M, one may locally find functions $f : M \times V \to W$ (with W a finite-dimensional vector space), linear in the second argument, such that (for $x \sim y$)

$$x \approx y \text{ iff } f(x; y-x) = 0.$$

So $\mathfrak{M}_{\approx}(x) = \mathfrak{M}(x) \cap (x + N(f(x;-)))$, where $N(f(x;-))$ is the kernel of the linear map $f(x;-) : V \to W$. To say that the distribution is 1-dimensional is to say that all these kernels are 1-dimensional linear subspaces of V.

Assume now that x, y, z form an infinitesimal 2-simplex, and that $x \approx y$ and $x \approx z$. We have to prove $y \approx z$, i.e. to prove $f(y; z-y) = 0$. By Taylor expansion of f in the direction of $y-x$, we have

$$f(y;z-y) = f(x;z-y) + df(x;y-x,z-y)$$
$$= f(x;(z-x)-(y-x)) + df(x;y-x,z-x)$$

(using the Taylor principle to replace the last occurrence of y by x)

$$= df(x;y-x,z-x)$$

because the first term vanishes, using linearity of $f(x;-)$, and using $x \approx y$, $x \approx z$. Now both the arguments $y-x$ and $z-x$ are in $N(f(x;-))$, by assumption, and also are in $D(V)$, hence (Proposition 1.2.4), they are in $D(N(f(x;-))$; also they are mutual neighbours in the 1-dimensional $N = N(f(x;-))$, by the assumption $y \sim z$. But any bilinear function on a 1-dimensional vector space is symmetric, and so vanishes on a pair of mutual neighbours in $D(N)$, cf. Proposition 1.3.3. This proves the proposition.

If a vector field X on M is suitably regular (classically, one needs just to say "nowhere vanishing"), then it defines a 1-dimensional distribution on M, with $x \approx y$ iff $y = X(x,d)$ for some d. Such distribution is then involutive, with the "field lines" of the vector field as the integral manifolds. Similarly, given two vector fields X and Y which are suitably pointwise "linearly independent", they define a 2-dimensional distribution (not necessarily involutive), but the description of this in combinatorial terms is not completely straightforward. Involutivity can in this case be expressed in terms of the Lie bracket of the two vector fields, cf. Section 4.10. Similarly for several vector fields.

2.7 Jets and jet bundles

One way of summarizing an important aspect of the synthetic method that we use here, is to say: *the jet notion is assumed representable* (namely represented by the monads $\mathfrak{M}_k(x)$). Therefore, the monad notions (or the equivalent neighbour notions $\sim_k$) on manifolds make the theory of jets rather simple and combinatorial. The basic definition is:

Definition 2.7.1 *Let M be a manifold, and N an arbitrary space. For $x \in M$, a* k-jet at x with values in N *is a map $f : \mathfrak{M}_k(x) \to N$; x is called the* source *of the jet, $f(x)$ the* target.

The space of k-jets with source in M and target in N is denoted $J^k(M,N)$. It comes with a map $J^k(M,N) \to M \times N$ ("anchor"), associating to a jet the pair consisting of its source and target; the fact that these individual jets can be comprehended into one space $J^k(M,N)$ is argued in Remark 7.3.1 below.

It is clear that $J^k(M,N)$ depends functorially on N, and that the anchor $J^k(M,N) \to M \times N$ is natural in N.

It can be proved, by using coordinate charts, that if not only M, but also N is a manifold, then $J^k(M,N)$ is a manifold as well; see e.g. (Saunders, 2008). For instance, $J^k(R,R)$ is a manifold of dimension $k+2$, since a k-jet from $a \in R$ to $b \in R$ is given by a Taylor polynomial $b + c_1 \cdot (x-a) + \ldots + c_k/k! \cdot (x-a)^k$, and is then described by the $k+2$-tuple $(a,b,c_1,\ldots,c_k)$ of numbers $\in R$.

Particularly important are 1-jets from $0 \in R$ to $x \in M$; such are called *tangent vectors* at x; since $\mathfrak{M}(0)$ in R is just D, we have

Definition 2.7.2 *A* tangent vector *at $x \in M$ is a map $\tau : D \to M$ with $\tau(0) = x$.*

It makes sense even without requiring M to be a manifold.

In the context of schemes (algebraic geometry), D is the spectrum of the ring of dual numbers, and this conception of tangent vectors is, in the scheme context, classical in algebraic geometry, cf. Mumford (1965/1988, III.4, p. 338), who calls D a "sort of disembodied tangent vector". In axiomatic terms, this notion of tangent vector was considered by Lawvere in 1967; he made the crucial observation that tangent vectors of M may be comprehended into a tangent *bundle* M^D, using cartesian closedness of $\mathscr{E}$; this was seminal for the development of the present form of synthetic differential geometry. The tangent bundle $M^D = T(M)$ will be studied in Chapter 4. Similarly,

Definition 2.7.3 *A* cotangent vector *(or just, a* cotangent*) at $x \in M$ is a 1-jet from $x \in M$ to $0 \in R$.*

Thus, a cotangent at $x \in M$ is a map $\mathfrak{M}(x) \to R$ with $x \mapsto 0$. We also call such things *combinatorial*, or *simplicial* cotangents, to distinguish them from *classical* cotangents, which are linear functionals $T_xM \to R$, as described in Chapter 4 below. The existence of a cotangent *bundle* (combinatorial or classical) requires not only cartesian closedness, but *local* cartesian closedness of $\mathscr{E}$. (In the context of algebraic geometry, however, the cotangent bundle is usually constructed prior to the tangent bundle.)

The combinatorics of jets in the present context depends on the following facts about the neighbour relations, and the corresponding facts about the resulting monads. We recall these facts:

- For each k, we have a reflexive symmetric relation $\sim_k$; these relations are preserved by any map $f : M \to N$ between manifolds.

• For fixed M, we have $x \sim_0 y$ iff $x = y$, and

$$x \sim_k y \text{ implies } x \sim_l y \text{ for } k \leq l \tag{2.7.1}$$

$$x \sim_k y \text{ and } y \sim_l z \text{ implies } x \sim_{k+l} z. \tag{2.7.2}$$

(So if we define "dist$(x,y) \leq k$ iff $x \sim_k y$", this "dist"-function is almost like an $\mathbb{N}$-valued metric. And all maps are contractions.)

If M is an m-dimensional manifold and $x \in M$, there exists (not canonically) a bijection $\mathfrak{M}_k(x) \to D_k(m)$ taking x to 0. For, take an open subset U of M containing x and equipped with a local diffeomorphism g to an open subset of R^m, taking x to 0. The restriction of g to $\mathfrak{M}_k(x)$ is a bijection of the required kind.

Also, we claim that any map $j : \mathfrak{M}_k(x) \to N$ (where N is another manifold) extends to a map $U \to N$, for some open subset U of M containing x. It suffices to consider the case where N is an open subset of R^n. Using the local diffeomorphism g as above, this may be translated into the problem of extending a map $\gamma : D_k(m) \to R^n$ to an open subset of R^m; but γ extends to the whole of R^m, by the KL axiom; by this axiom, it extends in fact to a polynomial of degree $\leq k$.

Thus a map $j : \mathfrak{M}_k(x) \to N$ extends to a map $J : U \to N$ with U open in M; in particular such U is a manifold, and since J preserves $\sim_k$, it follows that J, and hence j, maps $\mathfrak{M}_k(x)$ into $\mathfrak{M}_k(j(x))$.

This proves

Proposition 2.7.4 *Let M and N be manifolds, and let $x \in M$, $y \in N$. A* k-jet *from x to y is the same thing as a map $j : \mathfrak{M}_k(x) \to \mathfrak{M}_k(y)$ preserving base points.*

If $f : M \to N$ is a map between manifolds, and $x \in M$, we get a k-jet from x to $f(x)$, namely the *restriction* of f to $\mathfrak{M}_k(x)$. This jet may pedantically be denoted $j_k^x(f)$.

In a similar spirit: there is, for each non-negative integer $l \geq k$, a restriction map

$$J^l(M,N) \to J^k(M,N) \quad l \geq k,$$

natural in N. These restriction maps come about by using (2.7.1), which implies $\mathfrak{M}_k(x) \subseteq \mathfrak{M}_l(x)$, so that a map $\mathfrak{M}_l(x) \to N$ may be restricted to $\mathfrak{M}_k(x)$.

Note that $J^0(M,N) \cong M \times N$. Thus, the anchor may be seen as a restriction map $J^k(M,N) \to J^0(M,N)$.

From Proposition 2.7.4 in particular follows that a 1-jet from $x \in M$ to $y \in N$ is a map $j : \mathfrak{M}_1(x) \to \mathfrak{M}_1(y)$. Recall from (2.1.8) that 1-monads like $\mathfrak{M}_1(x) = \mathfrak{M}(x)$ carry a canonical action by the multiplicative monoid $(R, \cdot)$.

Proposition 2.7.5 *Let j be a 1-jet $x \to y$ where $x \in M$ and $y \in N$, where M and N are manifolds. Then j preserves the action by $(R, \cdot)$.*

Proof. Any 1-jet $\mathfrak{M}(x) \to \mathfrak{M}(y) \subseteq N$ may be extended to a map $f : U \to N$ defined in an open neighbourhood U of x. Now f preserves affine combinations of mutual neighbour points. In particular, for $x' \in \mathfrak{M}(x)$, it preserves the affine combination $(1-t) \cdot x + t \cdot x'$, and that combination defines the action by $t \in R$ on $x' \in \mathfrak{M}(x)$.

Exercise 2.7.6 Consider a k-jet f from $x \in M$ to $y \in N$ (M and N manifolds). Let $l < k$ be an integer. Call f *l-contracting* if $f : \mathfrak{M}_k(x) \to N$ is constant on $\mathfrak{M}_l(x) \subseteq \mathfrak{M}_k(x)$ (this constant value is then y). Prove that a $k-1$-contracting k-jet factors through $\mathfrak{M}_1(y)$. (*Hint:* It suffices to consider the case $M = R^m$, $N = R^n$, and argue in terms of degrees of polynomials.) Such $k-1$-contracting k-jets are of importance in connection with the theory of *symbols* of differential operators, cf. Chapter 7 below (where we use the term "*annular*" for such contracting jets).

Degree calculus for vector-space-valued jets

We consider n-jets on a manifold M with values in a KL vector space W, $f : \mathfrak{M}_n(x) \to W$, where $\mathfrak{M}_n(x) \subseteq M$ is the n-monad around $x \in M$. If $k \leq n$, we say that f *vanishes to order* $k+1$ at x if the restriction of f to $\mathfrak{M}_k(x) \subseteq \mathfrak{M}_n(x)$ has constant value $0 \in W$.

The following proposition is a slight generalization of Proposition 1.5.5; it is deduced from it by choosing a coordinate chart with $x = 0$, so that $\mathfrak{M}_n(x)$ gets identified with $D_n(V)$.

Proposition 2.7.7 *Let W_1, W_2, and W_3 be KL vector spaces; let $* : W_1 \times W_2 \to W_3$ be a bilinear map. Let M be a manifold. Let k and l be non-negative integers, and let $n \geq k+l+1$. If a function $f : \mathfrak{M}_n(x) \to W_1$ vanishes on $\mathfrak{M}_k(x)$ and $g : \mathfrak{M}_n(x) \to W_2$ vanishes on $\mathfrak{M}_l(x)$, then the map $f * g : \mathfrak{M}_n(x) \to W_3$ vanishes on $\mathfrak{M}_{k+l+1}(x)$.*

Jet bundles

A *section* of a bundle $\pi : E \to M$ is a map $s : M \to E$ with $\pi \circ s$ the identity map of M, i.e. $\pi(s(x)) = x$ for all $x \in M$. A *partial* section defined on a subset $S \subseteq M$ is a map $s : S \to E$ with $\pi(s(x)) = x$ for all $x \in S$.

If $\pi : E \to M$ is a bundle with M a manifold, a *section k-jet* at $x \in M$ is a partial section defined on $\mathfrak{M}_k(x)$, i.e. a k-jet $f : \mathfrak{M}_k(x) \to E$ such that

$\pi(f(y)) = y$ for all $y \in \mathfrak{M}_k(x)$. Equivalently, it is a partial section defined on a k-monad. Such section k-jets form a bundle $J^k(E) \to M$ (or more precisely, $J^k(\pi) \to M$) over M whose fibre over x is the section k-jets at x. (To see that there is indeed such a bundle, one needs that if α and β denote the two projections $M_{(k)} \to M$, the bundle $J^k(E) \to M$ may be described in categorical terms as $\beta_* \alpha^*(E) \in \mathscr{E}/M$; see Section 7.3, Remark 7.3.1.)

Note that an element in $J^k(E)$ over $x \in M$ may be described as a law which to each $y \sim_k x$ associates an element in E_y.

For $l \geq k$, we have restriction maps $J^l(E) \to J^k(E)$, which are maps of bundles over M. Since $J^0(E) = E$, we have in particular $J^k(E) \to E$. So $J(E)$ is not only a bundle over M, but a bundle over E.

It can be verified, by working in coordinates (see e.g. Saunders, 1989), that if $E \to M$ is locally a product $M \times F \to M$ (with F a manifold), then $J^k(E)$ is again a manifold, and $J^k(E) \to E$ as well as $J^k(E) \to M$ are locally products.

Exercise 2.7.8 Consider the product bundle $M \times N \to M$ (where M is a manifold). Prove that the set of section k-jets of this bundle may be identified with the set of k-jets from points in M to points in N. Thus, the notion of k-jet may be subordinated to the notion of *section* k-jet (where we have taken the opposite course).

Corresponding to the functorality in N of $J^k(M,N)$, there is a functorality of $J^k(\pi)$ with respect to maps between bundles over M, with M fixed. Recall that if $\pi_1 : E_1 \to M$ and $\pi_2 : E_2 \to M$ are bundles over M, then a map between these bundles is a map $g : E_1 \to E_2$ with $\pi_2 \circ g = \pi_1$. It is now clear that if $j : \mathfrak{M}_k(x) \to E_1$ is a section jet of E_1, then $g \circ j$ is a section jet of E_2, with the same source x. This recipe defines a map $J^k(g) : J^k(E_1) \to J^k(E_2)$ of bundles over M. The functorial structure is evidently compatible with the restriction maps $J^k(E_i) \to J^h(E_i)$ for $h \leq k$.

If a bundle $E \to M$ has some fibrewise algebraic structure, say is a vector bundle, then the bundle $J^k(E) \to M$ inherits the same kind of structure. Thus, for instance, a fibrewise addition $+$ in $E \to M$ gives rise to a fibrewise addition in $J^k(E) \to M$: given two elements j_1 and j_2 in the same fibre, so they are section k-jets at a common point, $x \in M$, say. Then we may form $j_1 + j_2$ by the evident recipe $(j_1 + j_2)(y) = j_1(y) + j_2(y)$ for $y \in \mathfrak{M}_k(x)$, and this is again a section jet at x.

If $g : E_1 \to E_2$ is a map of such algebraically structured bundles over M, and g preserves the fibrewise algebraic structure, then so does $J^k(g) : J^k(E_1) \to J^k(E_2)$.

We are in particular interested in vector bundles $E \to M$, and this case will be discussed in Chapter 7 below.

Prolongation

If $\pi : E \to M$ is a bundle and f a section, we get canonically a section $\overline{f}$ of the bundle $J^k E \to M$, called the *prolongation of f*. It is defined by

$$\overline{f}(x) = f \mid \mathfrak{M}_k(x);$$

surely $f \mid \mathfrak{M}_k(x)$ (the restriction of f to $\mathfrak{M}_k(x) \subseteq M$) is a k-jet section of $E \to M$ at x, and thus an element of the x-fibre of $J^k E \to M$.

In general, there is no need to have a special notation $\overline{f}$; we may just write f for the prolongation just defined.

Non-holonomous jets

The notion of non-holonomous jet is a generalization of the notion of jet; for contrast, jets, as considered above, are sometimes called *holonomous* jets. The present synthetic rendering of this (Ehresmann) notion is from Kock (1985).

Just as the notion of k-jet at $x \in M$ in the present context depends on the notion of k-monad $\mathfrak{M}_k(x)$ around x, the notion of non-holonomous jet depends on the notion of *non-holonomous monad* around x. This will be a set equipped with a map to M, but this map will not be an inclusion.

For simplicity, we give the definition of non-holonomous jet in as general a form as possible – which is much more general than the cases we shall consider later. Let M be a manifold. Let $k_1, \ldots, k_r$ be a sequence of non-negative integers. Then we consider the set $M_{(k_1,\ldots,k_r)} \subseteq M^{r+1}$ consisting of $r+1$-tuples

$$X = (x_0, x_1, \ldots, x_r) \text{ with } x_0 \sim_{k_1} x_1 \sim_{k_2} \cdots \sim_{k_r} x_r.$$

Note that if $r = 1$, this is the k_1th neighbourhood $M_{(k_1)}$ considered previously. Assigning to such $r+1$-tuple its first entry x_0 makes $M_{(k_1,\ldots,k_r)}$ into a bundle over M. The fibre over $x = x_0$ of this bundle is called the (non-holonomous) $(k_1,\ldots,k_r)$-monad around x and is denoted $\mathfrak{M}_{k_1,\ldots,k_r}(x)$. The map which for an $r+1$-tuple X, as above, picks out the last entry x_r will be denoted ε (for "extremity"). Note that for $X = (x_0, \ldots, x_r)$, $x_0 \sim_{k_1+\ldots+k_r} \varepsilon(X)$, by virtue of (2.7.2). Thus there is a map from $M_{k_1,\ldots,k_r}$ to $M_{k_1+\ldots+k_r}$ which preserves the "centre" x_0 as well as the "extremity" x_r. The restriction of this map to fibres over x defines a map $\mathfrak{M}_{k_1,\ldots,k_r}(x) \to \mathfrak{M}_{k_1+\ldots+k_r}(x)$ from the non-holonomous monad to the "corresponding" holonomous one. It preserves the extremity map

ε – which for the case of the holonomous monad is just the inclusion of this monad as a subset of M.

By restricting along the maps $\mathfrak{M}_{k_1,\dots,k_r}(x) \to \mathfrak{M}_{k_1+\dots+k_r}(x)$, one gets (as x ranges over M) a map $J^{k_1+\dots+k_r}(M,N) \to J^{(k_1,\dots,k_r)}(M,N)$. For many N (in particular, for N a manifold) this map is injective, and this allows us to say that a non-holonomous jet j at x_0 is *holonomous* if $j(x_0,x_1,\dots,x_r)$ only depends on (x_0 and) x_r.

There is also a natural notion of non-holonomous section jet in a bundle $\pi : E \to M$; a $(k_1,\dots,k_r)$-section jet at $x \in M$ in such a bundle is a map $j : \mathfrak{M}_{k_1,\dots,k_r}(x) \to E$ with the property that $\pi \circ j = \varepsilon$. Note that if $r = 1$, ε is the inclusion $\mathfrak{M}_{k_1} \subseteq M$, and then $\pi \circ j = \varepsilon$ just says that j is a section of π over this subset, so we recover the notion of section jet.

We denote the set of such non-holonomous section jets by $J^{k_1,\dots,k_r}(E)$. It is in a natural way a bundle over M: assign to the j above its source x_0.

The classical way to introduce non-holonomous jets (cf. e.g. Libermann, 1971) is by iteration; we shall describe this, and its relationship to our approach. Consider a bundle $\pi : E \to M$ where M is a manifold. Then we have described the bundle $J^k(E)$ over M, whose fibre over x is the section k-jets of E at x. Then we may form the bundle of section l-jets of $J^k(E)$, i.e. we form $J^l(J^k(E))$ – again a bundle over M.

The following proposition is purely a matter of logic ("lambda conversion", "exponential adjointness"); there are of course similar propositions for $r > 2$, but we stick to the case $r = 2$, i.e. to a *pair* of non-negative integers k and l:

Proposition 2.7.9 *There is an isomorphism of bundles over M*

$$J^k(J^l(E)) \cong J^{k,l}(E).$$

Proof. Given $j \in J^k(J^l(E))_x$, so $j : \mathfrak{M}_k(x) \to J^l(E)$ is a map; applying it to a $y \sim_k x$ yields an element $j(y) \in J^l(E)_y$, since j is a section. So $j(y) : \mathfrak{M}_l(y) \to E$ is a map; applying it to a $z \sim_l y$ yields an element $j(y)(z) \in E_z$, since $j(y)$ is a section. We then define $\hat{j} : \mathfrak{M}_{k,l}(x) \to E$ by putting $\hat{j}(x,y,z) := j(y)(z)$, for $(x,y,z) \in \mathfrak{M}_{k,l}(x)$. Then $\pi(\hat{j}(x,y,z)) = \pi(j(y)(z)) = z = \varepsilon(x,y,z)$, so $\hat{j}$ is indeed a section (k,l)-jet. Conversely, given a section (k,l)-jet $\hat{j}$ at x. We define $\mathfrak{M}_k(x) \to J^l(E)$ by $y \mapsto \hat{j}(x,y,-)$; for fixed y, this expression describes a map $\mathfrak{M}_l(y) \to E$, which is indeed a section jet since for $z \in \mathfrak{M}_l(y)$, $\hat{j}(x,y,-)(z) \in E_z$, by $\pi \circ \hat{j} = \varepsilon$. The two processes are evidently inverse of each other.

Bundle connections in terms of sections of jet bundles

A bundle connection ∇ in $E \to M$ gives rise to a section s of the bundle $J^1(E) \to E$:

$$s(e) := [y \mapsto \nabla(y,x)(e)]$$

for $e \in E_x$ and $y \in \mathfrak{M}(x)$. Vice versa, a section s of $J^1(E) \to E$ gives rise to a ∇ satisfying at least (2.5.1):

$$\nabla(y,x)(e) := s(e)(y)$$

for $e \in E_x$ and $y \sim x$.

A bundle connection on $E \to M$ can therefore alternatively be formulated as a section of the bundle $J^1(E) \to E$ with certain properties; cf. e.g. Palais *et al.* (1965, p. 84). This formulation has the advantage of applying to a much wider class of objects than manifolds, with a suitable reformulation of the jet notion. This approach is the one taken by Nishimura (1998), in a synthetic context, and by Libermann (1971) in the context of classical differential geometry.

Let $E \to M$ be a bundle, and let ∇ be a bundle connection in it. We have already seen that it gives rise to (in fact, can be identified with) a cross-section of $J^1(E) \to E$. But ∇ defines also a cross-section ∇^2 of $J^{1,1}(E) \to E$, a cross-section ∇^3 of $J^{1,1,1}(E) \to E$, etc., by "iteration". Thus, given $e \in E_x$ and given $(x,y,z) \in \mathfrak{M}_{1,1}(x)$, we define $\nabla^2(x,y,z)(e)$ in E_z by†

$$\nabla^2(x,y,z)(e) := \nabla(z,y)(\nabla(y,x)(e)).$$

One says that the $\nabla^2 : E \to J^{1,1}E$ is *holonomous* if it factors through the inclusion $J^2E \to J^{1,1}E$.

If $E \to M$ is locally a product $M \times F \to M$, then under mild assumptions on F, or on ∇, one can prove that ∇^2 is holonomous (i.e. factors through $J^2E \to J^{1,1}E$) iff ∇ is flat.

2.8 Infinitesimal simplicial and cubical complex of a manifold

Let M be a manifold. There is by the very construction a bijective correspondence between the set of infinitesimal k-simplices and the set of infinitesimal k-dimensional parallelepipeda in M (cf. Theorem 2.1.4). So why do we need *two* notions? The answer is that, when k ranges, we get two different kinds of complexes, a simplicial and a cubical, by using the infinitesimal simplices and infinitesimal parallelepipeda, respectively. In the picture (2.1.7) above, we can

† The left–right conventions could be made more elegant here.

clearly see *four* faces of the parallelogram, just as we can see the *three* faces of the simplex in (2.1.4).

More precisely: the sets $M_{<k>}$ of infinitesimal k-simplices form, as k ranges, a simplicial complex $M_{<\bullet>}$, whereas the family of sets $M_{[k]}$ of infinitesimal parallelepipeda form a cubical complex $M_{[\bullet]}$; both of these complexes are equipped with some further "symmetry" structure. We shall be explicit about these combinatorial structures: for $M_{<\bullet>}$ it is very easy, the ith face operator $\partial_i : M_{<k>} \to M_{<k-1>}$ $(i = 0,\ldots,k)$ just omits the ith entry x_i in an infinitesimal k-simplex $(x_0,x_1,\ldots,x_k)$, and the jth degeneracy operator $\sigma_j : M_{<k-1>} \to M_{<k>}$ $(j = 0,\ldots,k-1)$ repeats the jth entry. It is actually a *symmetric* simplicial set (in the sense of Grandis, 2001, say), due to the action of $\mathfrak{S}_{k+1}$ on $M_{<k>}$.

The notion of cubical complex is made explicit in Hilton and Wylie (1960) or Grandis and Mauri (2003), say. We describe the combinatorial structure, including the symmetry structure, and also a structure we call "subdivision" on $M_{[\bullet]}$, by describing these structures on the complex $S_{[\bullet]}(M)$ of *all* singular cubes in M. A *singular k-cube* in M is here understood as an arbitrary map $\gamma : R^k \to M$. Thus, an infinitesimal k-dimensional parallelepipedum is in particular a singular k-cube. The set of singular k-cubes in M is denoted $S_{[k]}(M)$; by construction, $M_{[k]} \subseteq S_{[k]}(M)$, and the cubical complex $M_{[\bullet]}$ is a subcomplex. (The reason we prefer to say "infinitesimal parallelepipedum" rather than "infinitesimal singular cube" is to emphasize that infinitesimal parallelepipeda are *affine* maps (preserve affine combinations), whereas a singular cube is not assumed to preserve any such kind of structure, as indicated by the derogative word "singular".)

Let us describe explicitly the faces of an infinitesimal k-dimensional parallelepipedum $[x_0,x_1,\ldots,x_k]$ as infinitesimal $k-1$-dimensional parallelepipeda. We have

$$\partial_i^0([x_0,x_1,\ldots,x_k]) = [x_0,x_1,\ldots,\widehat{x_i},\ldots,x_k] \tag{2.8.1}$$

and

$$\partial_i^1([x_0,x_1,\ldots,x_k]) = [x_i,x_1-x_0+x_i,\ldots,\widehat{i},\ldots,x_k-x_0+x_i]. \tag{2.8.2}$$

Note that the entries like $x_k - x_0 + x_i$ are affine combinations (of mutual neighbour points), so they make sense in a manifold M; $x_k - x_0 + x_i$ is the fourth vertex (opposite x_0) of a parallelogram whose three other vertices are x_0, x_i, and x_k (cf. the picture 2.1.7).

Now for the "subdivision" structure: first, for any $a,b \in R$, we have an affine map $R \to R$, denoted $[a,b] : R \to R$; it is the unique affine map sending 0 to

a and 1 to b, and is given by $t \mapsto (1-t)\cdot a + t\cdot b$. Precomposition with it, $\gamma \mapsto \gamma \circ [a,b]$ defines an operator $S_{[1]}(M) \to S_{[1]}(M)$ which we denote $\mid [a,b]$, and we let it operate on the right, thus

$$\gamma \mid [a,b] := \gamma \circ [a,b].$$

Heuristically, it represents the restriction of γ to the interval $[a,b]$. Note that $\gamma = \gamma \mid [0,1]$, since $[0,1] : R \to R$ is the identity map.

More generally, for $a_i, b_i \in R$ for $i = 1,\ldots,k$, we have the map

$$[a_1,b_1] \times \ldots \times [a_k,b_k] : R^k \to R^k.$$

It induces by precomposition an operator $S_{[k]}(M) \to S_{[k]}(M)$, which we similarly denote $\gamma \mapsto \gamma \mid [a_1,b_1] \times \ldots \times [a_k,b_k]$.

Given an index $i = 1,\ldots,k$, and $a,b \in R$. We use the abbreviated notation $[a,b]_i$ for the map $[0,1] \times \ldots \times [a,b] \times \ldots \times [0,1]$ with the $[a,b]$ appearing in the ith position; the correponding operator is denoted $\gamma \mapsto \gamma \mid_i [a,b]$. Given a,b and $c \in R$, and an index $i = 1,\ldots,k$, then we say that the ordered pair

$$\gamma \mid_i [a,b], \quad \gamma \mid_i [b,c]$$

form a subdivision of $\gamma \mid_i [a,c]$ *in the ith direction.*

There are also compatibilities between the subdivision relation and the face maps; they are described in Section A.8; they apply to the whole of $S_{[\bullet]}(M)$.

The face operators, both in the simplicial and in the cubical case, are used to describe the exterior derivative of combinatorial differential forms, see Section 3.2 below.

We shall not make explicit use of degeneracy operators, but rather have ad hoc notions of degenerate simplices and degenerate singular cubes.

For the simplicial complex $M_{<\bullet>}$, we use the term "degenerate simplex" not just for one which is the value of one of the degeneracy operators, but for any simplex $(x_0,\ldots,x_k)$ with two vertices equal. We say that an infinitesimal singular parallelepipedum $[x_0,x_1,\ldots,x_k]$ is degenerate if for some $i \neq 0$, $x_i = x_0$.

The simplicial complex $M_{<\bullet>}$ was described in Kock (1981/2006, §I.18; 2000), and, in a more general situation (schemes) in Breen and Messing (2001); for affine schemes, it was known by Joyal in 1979 (unpublished). The cubical complex $M_{[\bullet]}$ was described in Kock (2007, 2008).

3
Combinatorial differential forms

We present three related combinatorial manifestations of the notion of differential form. The main one of these manifestations is the simplicial one, introduced in Kock (1981/2006) (following an unpublished idea of Bkouche and Joyal); in Breen and Messing (2001), differential forms in this sense are called *combinatorial* differential forms; we use "combinatorial differential form" in a wider sense, namely as a name that also applies to combinatorial forms in their cubical or whisker manifestation, which we introduce here. For 1-forms, the three notions agree, and so we may unambiguously use the term "combinatorial 1-form". Differential forms, in the classical terms as certain functions on the tangent bundle, are dealt with in Chapter 4.

The infinitesimal notions in the present chapter are "first order", thus $x \sim y$ means $x \sim_1 y$, $\mathfrak{M}(x)$ means $\mathfrak{M}_1(x)$, etc.

3.1 Simplicial, whisker, and cubical forms

We consider differential forms on a manifold M, with values in a KL vector space W (the most important case is $W = R$).

We recall what the axiom says in this case. Note that since $\widetilde{D}(k,n) \subseteq R^{k \cdot n}$, we may view the elements $\underline{\underline{d}} \in \widetilde{D}(k,n)$ as $k \times n$ matrices (k rows, n columns). Using this, the axiom says

any map $\widetilde{D}(k,n) \to W$ *is of the form* $\underline{\underline{d}} \mapsto \Sigma_{\underline{\underline{a}}} \det(\underline{\underline{a}}) \cdot v_{\underline{\underline{a}}}$ *for unique* $v_{\underline{\underline{a}}} \in W$

where $\underline{\underline{a}}$ ranges over all $l \times l$ submatrices of $\underline{\underline{d}}$ ($0 \leq l \leq k$, $0 \leq l \leq n$, $v_{\underline{\underline{a}}} \in W$).

The vector space R itself satisfies this, by the KL axiom assumed for $\widetilde{D}(k,n)$. Hence also any vector space of the form R^X satisfies the axiom.

From the description of maps $\widetilde{D}(k,n) \to W$ provided by the axiom follows in particular that, for $k \leq n$, a map $\widetilde{D}(k,n) \to W$, which vanishes on any matrix

$\underline{\underline{d}} \in \widetilde{D}(k,n)$ with a zero row, is a linear combination (with coefficients from W) of the $k \times k$ subdeterminants of $\underline{\underline{d}}$, hence it is k-linear alternating; in fact, it extends uniquely to a k-linear alternating map $(R^n)^k \to W$.

Definition 3.1.1 A simplicial differential k-form *on M with values in W is a law ω, which to any infinitesimal k-simplex $\underline{x}$ in M associates an element $\omega(\underline{x}) \in W$, subject to the "normalization" requirement that $\omega(\underline{x}) = 0$ if any two of the vertices of $\underline{x}$ are equal.*

Thus, ω is a map

$$\omega : M_{<k>} \to W$$

vanishing on degenerate k-simplices, where we take "degenerate" to mean "two of the vertices are equal".

The normalization requirement stated in the definition can be weakened: only *certain* of the degenerate simplices need to be mentioned, see Proposition 3.1.6 below.

It can be proved that such ω is automatically $\mathfrak{S}_{k+1}$-alternating, meaning

$$\omega(\underline{x}) = \operatorname{sign}(\sigma)\omega(\underline{x}') \tag{3.1.1}$$

where $\underline{x}'$ comes about by permuting the vertices of $\underline{x} = (x_0, \ldots, x_k)$ by a permutation $\sigma \in \mathfrak{S}_{k+1}$, cf. Theorem 3.1.5 below.

Definition 3.1.2 A whisker differential k-form *on M, with values in W, is a law ω, which to any infinitesimal k-whisker $\underline{x} = (x_0, x_1, \ldots, x_k)$ in M associates an element $\omega(\underline{x}) \in W$, subject to the requirements that*
(1) it is $\mathfrak{S}_k$-alternating, *meaning*

$$\omega(\underline{x}) = \operatorname{sign}(\sigma)\omega(\underline{x}') \tag{3.1.2}$$

where $\underline{x}'$ comes about by permuting the vertices $x_1, \ldots, x_k$ by a permutation $\sigma \in \mathfrak{S}_k$; and
(2) $\omega(\underline{x}) = 0$ if $x_i = x_0$ for some $i \geq 1$.

Thus, ω is a map $Wh_k(M) \to W$ with certain properties.

From the alternating property required of whisker differential forms follows immediately that $\omega(x_0, x_1, \ldots, x_k) = 0$ if $x_i = x_j$ for some $i, j \geq 1$, $i \neq j$; combined with (2), it follows that $\omega(x_0, x_1, \ldots, x_k) = 0$ if any two of the vertices are equal. Hence a whisker differential k-form restricts to a simplicial differential k-form along the inclusion $M_{<k>} \subseteq Wh_k(M)$.

Recall from Section 2.1 that any infinitesimal k-simplex $(x_0,x_1,\ldots,x_k)$ in a manifold M gives rise to a map $[x_0,x_1,\ldots,x_k] : R^k \to M$, using affine combinations in M; this map is what we called the *infinitesimal parallelepipedum spanned by* the infinitesimal simplex.

Definition 3.1.3 *A* cubical differential k-form *on M, with values in W, is a law ω, which to any infinitesimal k-dimensional parallelepipedum P in M associates an element $\omega(P) \in W$, subject to the normalization requirement that $\omega(P) = 0$ if the spanning k-simplex $(x_0,x_1,\ldots,x_k)$ has $x_0 = x_i$ for some $i \geq 1$.*

Thus, ω is a map $M_{[k]} \to W$ with a certain "normalization" property.

It can be proved that such ω is automatically $\mathfrak{S}_{k+1}$-alternating, meaning

$$\omega(\underline{x}) = \operatorname{sign}(\sigma)\omega(\underline{x}') \tag{3.1.3}$$

where $\underline{x}'$ comes about by permuting the vertices of the spanning simplex $\underline{x} = (x_0,\ldots,x_{k+1})$ by a permutation $\sigma \in \mathfrak{S}_{k+1}$, cf. Theorem 3.1.5 below.

Since any map $f : N \to M$ between manifolds preserves the neighbour relation, it is clear that a simplicial or whisker k-form ω on M gives rise to a (simplicial, resp. whisker) k-form $f^*(\omega)$ on N, by the standard formula for contravariant functorality of cochains on a simplicial complex; explicitly

$$f^*(\omega)(x_0,x_1,\ldots,x_k) := \omega(f(x_0), f(x_1),\ldots, f(x_k))$$

for $(x_0,x_1,\ldots,x_k)$ an infinitesimal k-simplex (resp. an infinitesimal k-whisker) in N. Since f by Theorem 2.1.4 preserves those affine combinations that define infinitesimal parallelepipeda out of infinitesimal simplices, it follows that also for a cubical k-form ω on M, we get a cubical k-form $f^*(\omega)$ on N.

If ω is a (simplicial, cubical, or whisker) differential k-form, we also say that it is a (simplicial, cubical, or whisker) form *of degree k*.

Proposition 3.1.4 *All the values of a simplicial, whisker, and cubical differential k-form with values in W are ~ 0, provided $k \geq 1$.*

Proof. We do the simplicial case only, leaving the analogous proofs of the other cases to the reader. We may assume that $M \subseteq V$, as in the proof of Theorem 3.1.5, so that a k-form may be encoded in terms of $\Omega : M \times \widetilde{D}(k,V) \to W$, and, as there, Ω extends to a map $\Omega : M \times V^k \to W$, k-linear and alternating in the last k arguments. The map Ω (whose domain is a manifold!), together with $x_0 \sim x_1$ witnesses that $\omega(x_0,x_1,\ldots,x_k) \sim \omega(x_0,x_0,\ldots,x_k) = 0$, for the covariant determination of $\sim$ in W.

Recall that there is a bijective correspondence $M_{<k>} \cong M_{[k]}$ between, on the one side, infinitesimal k-simplices $(x_0, x_1, \ldots, x_k)$ in M, and on the other side infinitesimal k-dimensional parallelepipeda $[x_0, x_1, \ldots, x_k]$ in M.

Since the "inputs" of cubical k-forms thus are in bijective correspondence with the inputs of simplicial k-forms, it is not surprising that the notions are coextensive: this is a matter of seeing that the normalization requirements are equivalent, which is not hard to do, see Proposition 3.1.9 below. So why have two different names and notions? The point is that the *coboundaries* (simplicial and cubical) are different; both these coboundaries express the exterior derivative of differential forms in geometric language (modulo a factor $k+1$, for the simplicial case). This is discussed in Section 3.2.

If ω is a simplicial k-form, we may use the same notation ω for the corresponding cubical form; thus, if $X = (x_0, x_1, \ldots, x_k)$ is an infinitesimal k-simplex and $[X] = [x_0, x_1, \ldots, x_k]$ the corresponding infinitesimal parallelepipedum, $\omega(X) = \omega([X])$, where the "ω" on the left denotes the simplicial form, and on the right, it denotes the corresponding cubical form.

Theorem 3.1.5 *Any simplicial differential k-form (with values in a KL vector space W) is $\mathfrak{S}_{k+1}$-alternating. Also, any cubical k-form (with values in W) is $\mathfrak{S}_{k+1}$-alternating.*

Here, "$\mathfrak{S}_{k+1}$-alternating" for a simplicial form ω means that for any $\sigma \in \mathfrak{S}_{k+1}$, and any infinitesimal k-simplex $\underline{x}$,

$$\omega(\sigma \underline{x}) = \operatorname{sign}(\sigma)\, \omega(\underline{x}).$$

For cubical forms, it means the same sign change whenever the vertices of the generating simplex of an infinitesimal parallelepipedum are permuted according to σ. Note that this is a symmetry of one degree higher than the usual $\mathfrak{S}_k$-alternating property of differential k-forms, or of whisker k-forms. This will in particular be significant for differential 1-forms, where $\mathfrak{S}_1$-alternating says nothing, but $\mathfrak{S}_2$-alternating says $\omega(x,y) = -\omega(y,x)$.

Proof. There is no harm in assuming that M is an open subset of a finite-dimensional vector space V. Then both a simplicial and a cubical k-form ω may, by (2.1.5), be presented in coordinates by a function $\Omega : M \times \widetilde{D}(k,V) \to W$,

$$\omega(x_0, x_1, \ldots, x_k) = \Omega(x_0; x_1 - x_0, \ldots, x_k - x_0). \tag{3.1.4}$$

The assumption that the output value of ω vanishes if one of the input x_is ($i \geq 1$) equals x_0 implies that $\Omega(x_0; -, \ldots, -) : \widetilde{D}(k,V) \to W$ vanishes if one of the inputs is zero. By KL, therefore, $\Omega(x_0;-,\dots,-)$ extends uniquely to a k-linear alternating function $V^k \to W$. Let us denote its value on $(u_1,\dots,u_k) \in V^k$ by the same symbol Ω, $\Omega(x_0;u_1,\dots,u_k)$, with Ω being k-linear and alternating in the arguments after the semicolon.

From this follows immediately that $\omega(x_0,x_1,\dots,x_k)$ is $\mathfrak{S}_k$-alternating, i.e. the value changes sign if x_i and x_j are swapped, $i,j \geq 1$. The fact that ω is $\mathfrak{S}_{k+1}$-alternating, i.e. that the value of ω also changes sign when some x_i is swapped with x_0, requires a separate argument. We have to prove that

$$\omega(x_0,\dots,x_i,\dots) = -\omega(x_i,\dots,x_0,\dots).$$

We may assume $i = 1$, and for simplicity of notation, let us omit the remaining arguments. Let us calculate $\omega(x_1,x_0)$ in terms of the function Ω. We have $\omega(x_1,x_0) = \Omega(x_1;x_0 - x_1) = \Omega(x_0;x_0 - x_1)$ by the Taylor principle; by linearity of $\Omega(x_0;-)$ this equals $-\Omega(x_0;x_1 - x_0) = -\omega(x_0,x_1)$. This proves the theorem.

A similar analysis in coordinate terms of a whisker differential k-form likewise reveals that it is given by a function $\Omega : M \times V^k \to W$ which for fixed $x \in M$ is k-linear and alternating in the remaining arguments $\in V$. (Here, the alternating property of Ω comes from the requirement (3.1.2) in the definition.) In other words, in coordinate terms the data for a simplicial k-form and a whisker k-form are the same. So there is a bijective correspondence between simplicial and whisker k-forms on M. The correspondence does not depend on the coordinatization; for, the passage from a whisker k-form $\overline{\omega}$ to the corresponding simplicial k-form ω just amounts to restriction along the inclusion $i : M_{<k>} \subseteq Wh_k(M)$,

$$\omega = \overline{\omega} \circ i.$$

We observe that in the proof of the theorem, we only used that the value of ω vanishes if $x_i = x_0$ for some $i \geq 1$ to argue that the function $\Omega(x_0;\dots)$ was k-linear alternating. And from the alternating property of $\Omega(x_0;\dots)$ immediately follows that the value of ω vanishes if $x_i = x_j$, $i,j \geq 1, i \neq j$; this proves that for a function $\omega : M_{<k>} \to W$ to be a differential form, it suffices that $\omega(x_0,\dots,x_k) = 0$ whenever $x_i = x_0$ for some $i \geq 0$. From the point of view of the proof, there is nothing special about the index 0; we may therefore conclude the following:

Proposition 3.1.6 *For a function $\omega : M_{<k>} \to W$ to be a simplicial differential form, it suffices that there is some index $j \in \{0,\dots,k\}$ such that $\omega(x_0,\dots,x_k) = 0$ whenever $x_i = x_j$ for some $i \neq j$.*

The proof of the above theorem also gives

Theorem 3.1.7 *Any simplicial differential k-form on a manifold M (with values in a KL vector space W) extends uniquely to a whisker differential k-form.*

Remark 3.1.8 It is possible to give a more explicit characterization of the whisker form $\hat{\omega}$ extending a given simplicial k-form ω; namely it is characterized by validity, for all $\underline{\underline{d}} \in \widetilde{D}(k,k)$, of

$$\det(\underline{\underline{d}}) \cdot \hat{\omega}(x_0, x_1, \dots, x_k) = \omega(\underline{\underline{d}} \cdot (x_0, x_1, \dots, x_k)) \tag{3.1.5}$$

(recall the notation $\underline{\underline{d}} \cdot (x_0; x_1, \dots, x_k)$ from Remark 2.1.6).

Since for $\underline{\underline{d}} \in \widetilde{D}(k,k)$, the determinant is $k!$ times the product of the diagonal entries, the $\hat{\omega}$ can also be characterized by the validity (for all $\underline{\underline{d}} \in \widetilde{D}(k,k)$) of

$$k!\, d_{11} \cdot \ldots \cdot d_{kk} \cdot \hat{\omega}(x_0; x_1, \dots, x_k) = \omega(\underline{\underline{d}} \cdot (x_0, x_1, \dots, x_k)). \tag{3.1.6}$$

We now turn to the comparison between simplicial and cubical forms.

Proposition 3.1.9 *A map (simplicial cochain) $\omega : M_{<k>} \to W$ is a simplicial differential form if and only if the corresponding map (cubical cochain) $\bar{\omega}$: $M_{[k]} \to W$ is a cubical differential form.*

Proof. This is a question of comparing the normalization conditions which qualify a cochain (simplicial, resp. cubical) as a (simplicial, resp. cubical) differential form. The one for simplicial forms is evidently stronger, but is equivalent to a weaker one, by Proposition 3.1.6, and this weaker condition (with $j = 0$) is equivalent to the normalization required for cubical forms.

We can summarize the discussion of the three manifestations of differential forms in

Theorem 3.1.10 *There are natural bijections between the sets of simplicial, whisker, and cubical k-forms on M (with values in a KL vector space W).*

The naturality comes from the fact that the correspondences are induced by the inclusion $M_{<k>} \subseteq Wh_k$, and by the bijection between infinitesimal simplices and infinitesimal parallelepipeda in M, and both these correspondences are natural in M.

There is a natural way to multiply a combinatorial W-valued k-form ω on M with a function $g : M \to R$,

$$(g \cdot \omega)(x_0, x_1, \dots, x_k) := g(x_0) \cdot \omega(x_0, x_1, \dots, x_k). \tag{3.1.7}$$

This applies for simplicial, whisker, or cubical forms. (From the Taylor principle, it is easy to see that the preferred role of x_0 here is spurious.) It is immediate to see that if $f : N \to M$ is a map, then

$$f^*(g \cdot \omega) = (g \circ f) \cdot f^*(\omega). \tag{3.1.8}$$

We shall now be interested in R-valued k-forms on the manifold $M = R^k$. On R^k, we have a canonical simplicial k-form Vol, the *volume form*, given by

$$\mathrm{Vol}(x_0, x_1, \ldots, x_k) = \det(x_1 - x_0, \ldots, x_k - x_0);$$

heuristically, it represents the (signed) volume of the parallelepipedum $[X]$ spanned by the simplex $X = (x_0, x_1, \ldots, x_k)$. The corresponding cubical form is likewise denoted Vol.

Proposition 3.1.11 *Any simplicial, resp. cubical, R-valued k-form on an open subset N of R^k is of the form $g \cdot \mathrm{Vol}$ for a unique function $g : N \to R$.*

Proof. As in (3.1.4), ω is given by a function $\Omega : N \times (R^k)^k \to R$, multilinear and alternating in the last k arguments. But a k-linear alternating $(R^k)^k \to R$ is given by a constant times the determinant function. Thus, for an infinitesimal k-simplex $(x_0, x_1, \ldots, x_k)$ in N

$$\omega(x_0, x_1, \ldots, x_k) = g(x_0) \cdot \det(x_1 - x_0, \ldots, x_k - x_0),$$

where $g(x_0)$ is the constant g corresponding to the k-linear alternating $\Omega(x_0; \ldots) : (R^k)^k \to R$.

The volume form on R (i.e. the case $k = 1$) is usually denoted dx; so $dx\,(x, y) = y - x$. Every 1-form on an open subset N of R is, by the proposition, of the form $f(x)\,dx$ for a unique function $f : N \to R$.

Because of the prominent role of the volume form expressed by the proposition, it is natural to look for expressions for k-forms on open subsets $N \subseteq R^k$ given as $\alpha^*(\mathrm{Vol})$ where $\alpha : N \to R^k$. Consider such $\alpha : N \to R^k$. Then we have a function $d\alpha : N \times R^k \to R^k$, linear in the second argument, such that for $x \sim y$ in N,

$$\alpha(y) = \alpha(x) + d\alpha(x; y - x)$$

($d\alpha(x; -)$ is the differential of α at x, and is given by the $k \times k$ Jacobi matrix of α at x). We get a map $J\alpha : N \to R$, by putting

$$J\alpha(x) = \det(d\alpha(x; -))$$

(the Jacobi determinant of α at x).

Proposition 3.1.12 *Let $N \subseteq R^k$ be an open subset, and let $\alpha : N \to R^k$ be a map. Then*

$$\alpha^*(\mathrm{Vol}) = J\alpha \cdot \mathrm{Vol}.$$

Proof. Let $(x_0, x_1, \ldots, x_k)$ be an infinitesimal k-simplex in N. Then

$$\begin{aligned}\alpha^*(\mathrm{Vol})(x_0, x_1, \ldots, x_k) &= \mathrm{Vol}(\alpha(x_0), \alpha(x_1), \ldots, \alpha(x_k))\\ &= \det(\alpha(x_1) - \alpha(x_0)), \ldots, \alpha(x_k) - \alpha(x_0))\\ &= \det(d\alpha(x_0; x_1 - x_0), \ldots, d\alpha(x_0; x_k - x_0))\end{aligned}$$

by the defining property of the differential $d\alpha(x_0; -)$

$$= \det(d\alpha(x_0; -)) \cdot \det(x_1 - x_0, \ldots, x_k - x_0)$$

by the product rule for determinants

$$\begin{aligned}&= J\alpha(x_0) \cdot \mathrm{Vol}(x_0, x_1, \ldots, x_k)\\ &= (J\alpha \cdot \mathrm{Vol})(x_0, x_1, \ldots, x_k).\end{aligned}$$

This proves the proposition.

If $w \in W$ in a vector space W, we get a W-valued k-form on R^k denoted $w \cdot \mathrm{Vol}$,

$$(w \cdot \mathrm{Vol})(x_0, x_1, \ldots, x_k) := \mathrm{Vol}(x_0, x_1, \ldots, x_k) \cdot w$$

(a scalar multiplied on the vector w); such forms are called *constant* (W-valued) differential forms on R^k. With this understanding, the following holds not only for R-valued k-forms on R^k, but for W-valued k-forms on R^k as well, provided W is a KL vector space:

Theorem 3.1.13 *Let M be a manifold, and let ω be a simplicial k-form on M. Then for any infinitesimal k-simplex $X = (x_0, x_1, \ldots, x_k)$ in M, and the corresponding map $[X] : R^k \to M$,*

$$[X]^*(\omega) = \omega(X) \cdot \mathrm{Vol},$$

in particular, $[X]^*(\omega)$ is a constant differential form. Likewise, if ω is a cubical k-form on M, $[X]^*(\omega) = \omega([X]) \cdot \mathrm{Vol}$.

Proof. We need to prove that for any infinitesimal k-simplex Y in R^k,

$$([X]^*(\omega))(Y) = \omega(X) \cdot \mathrm{Vol}(Y). \tag{3.1.9}$$

We may assume that M is an open subset of some finite-dimensional vector space V, and that $x_0 = 0$. Then the map $[X]$ is a linear map $R^k \to V$. The infinitesimal k-simplex Y in R^k is of the form $y + \underline{\underline{d}}$ for some $y \in R^k$ and $\underline{\underline{d}} \in \widetilde{D}(k,k)$. The given k-form ω is, as in the proof of Theorem 3.1.5, given by a function $\Omega : M \times V^k \to R$, k-linear and alternating in the last k variables,

$$\omega(z_0, z_1, \ldots, z_k) = \Omega(z_0; z_1 - z_0, \ldots, z_k - z_0).$$

The left-hand side of (3.1.9) may now be expressed as follows (here, $\underline{d}_j$ denotes the jth row of the $k \times k$ matrix $\underline{\underline{d}}$)

$$([X]^*(\omega))(Y) = \omega([X](y), [X](y + \underline{d}_1), \ldots, [X](y + \underline{d}_k))$$

$$= \omega([X](y), [X](y) + [X](\underline{d}_1), \ldots, [X](y) + [X](\underline{d}_k))$$

because $[X]$ is linear,

$$= \Omega([X](y); \underline{d}_1 \cdot X, \ldots, \underline{d}_k \cdot X),$$

where $\underline{d}_j \cdot X$ is the linear combination of the k vectors x_j in X using as coefficients the entries from $\underline{d}_j$. From the (generalized) product rule for determinants (Proposition 1.2.19), applied to the k-linear alternating function $\Omega([X](y); \ldots)$, we see that we may rewrite this as

$$\det(\underline{\underline{d}}) \cdot \Omega([X](y); x_1, \ldots, x_k). \tag{3.1.10}$$

On the other hand, the right-hand side of (3.1.9) is $\det(\underline{\underline{d}}) \cdot \omega[X]$, which is

$$\det(\underline{\underline{d}}) \cdot \Omega(0; x_1, \ldots, x_k).$$

This is almost the expression in (3.1.10), except that the arguments in front of the semicolon in Ω are not the same. To see the desired equality is now in essence k applications of the Taylor principle; more explicitly, we make k Taylor expansions of Ω in its non-linear variable in front of the semicolon. Note that $[X](y)$ is a linear combination of the x_is, each of which is in $D(V)$. Taking directional derivative in the direction of x_1, we replace x_1 by 0, at the cost of introducing a remainder $d\Omega(0; x_1, \ldots)$, but since x_1 already appears among the linear arguments of Ω, and hence of $d\Omega$, the remainder term vanishes. Next take directional derivative in the direction of x_2 and do the same argument, etc. This proves the equation (3.1.9) and thus the theorem. (The second assertion is equivalent to the first, it only differs in notation.)

We have in Section 2.2 seen examples of 1-forms: a V-framing on a manifold gives rise to a V-valued 1-form, the framing 1-form. If ω is a 1-form, the law $d\omega$ given by $(x, y, z) \mapsto \omega(x, y) + \omega(y, z) - \omega(x, z)$ is an example of a

2-form (and to say that it vanishes is equivalent to saying that ω is closed). The process $\omega \mapsto d\omega$, for simplicial as well as cubical forms, is the subject of Section 3.2 below.

Simplicial forms with values in a pointed space

Note that Definition 3.1.1 does not utilize the algebraic structure of W, except that a base point 0 has to be specified in order to state the "normalization condition". So it makes sense to talk about simplicial differential forms with values in any pointed space $(W,*)$; thus W could be a group and $*$ the neutral element; this generalization will be important later on. We assume that the pointed space is a space locally diffeomorphic to a KL vector space; there is no harm in assuming that the base point is 0, so that the analysis of a k-form ω in terms of a function Ω, as in the proof of Theorem 3.1.5, is applicable. Under these circumstances, we have

Proposition 3.1.14 *Let ω be a simplicial k-form on M with values in the pointed space $(W,0)$, $k \geq 2$. Then if $a \sim b$ in M, the value of ω on any infinitesimal k-simplex in the image of $[a,b] : R \to M$ is 0.*

Proof. We shall do this for the case $k = 2$ only. We may assume that $M = V$, a finite-dimensional vector space and that $a = 0 \in V$, and thus $b \in D(V)$. Then $[a,b](s) = s \cdot b$ for all $s \in R$. Thus the task is to prove that $\omega(s \cdot b, t \cdot b, u \cdot b) = 0 \in W$ for all $s,t,u \in R$. Using the coordinate description of ω in terms of Ω as in the proof of Theorem 3.1.5, we should prove

$$\Omega(s \cdot b; (t-s) \cdot b, (u-s) \cdot b) = 0.$$

But this is clear, since Ω is bilinear in the two arguments after the semicolon, and since $b \in D(V)$.

Note: It is true that any 2-form on R is 0; in particular, this holds for the 2-form $[a,b]^*(\omega)$. But note that $\omega(s \cdot b, t \cdot b, u \cdot b)$ is not an application instance of $[a,b]^*(\omega)$ unless s,t,u are mutual neighbours.

3.2 Coboundary/exterior derivative

For any manifold M, we have described both the simplicial complex $M_{<\bullet>}$ of infinitesimal simplices, and the cubical complex $M_{[\bullet]}$ of infinitesimal parallelepipeda. Both these complexes depend in a functorial way on M. For any abelian group W, we may consider the associated cochain complexes of W-valued cochains, using the standard coboundary formulas from algebraic topology. These cochain complexes depend contravariantly on M.

We shall prove that simplicial differential forms with values in a KL vector space W constitute a subcomplex of the complex of simplicial cochains (i.e. that it is stable under the simplicial coboundary operator); and similarly differential forms, in their cubical guise, constitute a subcomplex of the complex of cubical cochains. And in fact, in both cases, we recover essentially the classical exterior derivative of differential forms. More precisely, for the cubical case, we recover the exterior derivative on the nose; for the simplicial case, we recover it, for k-forms, modulo a factor $(k+1)$.

Consider first the coboundary formation for cochains on the simplicial complex $M_{<\bullet>}$. A W-valued k-cochain ω on it is any map $M_{<k>} \to W$; the (simplicial) coboundary $d\omega$ of it is a $k+1$-cochain, whose value on a $k+1$-simplex $\underline{x}$ is given by the classical alternating sum with $k+2$ terms (one term for each face of the simplex $\underline{x}$) (here, $\underline{x}$ denotes $(x_0, x_1, \ldots, x_{k+1})$):

$$d\omega(x_0,\ldots,x_{k+1}) = \omega(x_1,\ldots,x_{k+1}) + \sum_{i=1}^{k+1}(-1)^i\omega(x_0,x_1,\ldots,\widehat{x_i},\ldots,x_{k+1}). \tag{3.2.1}$$

Note that for $k=0$ and $k=1$, this is consistent with the usage in Section 2.2. From standard simplicial theory, we get that $d \circ d = 0$.

Recall that a simplicial cochain ω is called a (simplicial) *differential k-form* if it satisfies the "normalization" condition that it vanishes on simplices where two vertices are equal; it suffices that the weaker condition of Proposition 3.1.6 holds, i.e. $\omega(x_0,x_1,\ldots,x_k) = 0$ if $x_j = x_0$ for some $j > 0$.

Proposition 3.2.1 *If the simplicial cochain $\omega : M_{<k>} \to W$ satisfies the normalization condition, then so does $d\omega : M_{<k+1>} \to W$.*

Equivalently, the simplicial differential k-forms form a subcomplex of the cochain complex of the simplicial complex $M_{<\bullet>}$. In particular, $d \circ d = 0$.

Proof. Assume ω is a simplicial k-form. By Proposition 3.1.6 it suffices to see that $d\omega$ takes value 0 on any $k+1$-dimensional infinitesimal simplex $(x_0,x_1,\ldots,x_{k+1})$, which is degenerate by virtue of $x_i = x_0$ for some $i > 0$. In the formula (3.2.1), all terms in $\sum_1^{k+1}$, except the ith, vanish by virtue of the normalization condition for ω. So it suffices to see that

$$\omega(x_1,\ldots,x_i,\ldots,x_{k+1}) + (-1)^i\omega(x_0,\ldots,\widehat{x_i},\ldots,x_{k+1}) = 0.$$

In the first term here, we make $i-1$ transpositions to bring x_i to the first position, and using that ω is alternating, we thus get for the sum

$$(-1)^{i-1}\omega(x_i,x_1,\ldots,\widehat{x_i},\ldots,x_{k+1}) + (-1)^i\omega(x_0,\ldots,\widehat{x_i},\ldots,x_{k+1});$$

now using the fact that $x_0 = x_i$, we see that the two terms are equal except for sign, and thus cancel.

Remark 3.2.2 Let us note that the terms appearing in the formula (3.2.1) for the coboundary of a simplicial W-valued differential form are actually mutual neighbours in W (for the covariant determination of the $\sim$-relation in W, as described in Section 2.1). Let us indicate the proof of this for the case of 1-forms. Let ω be a W-valued 1-form on a manifold M. Let us prove for instance that $\omega(x,y) \sim \omega(y,z)$ in W. Since the question is local on M, we may assume that M is an open subset of a finite-dimensional vector space V, and so, as in (3.1.4), ω may be described $\omega(x,y) = \Omega(x; y-x)$ for some $\Omega : M \times V \to W$, linear in the second variable. Then

$$\omega(x,y) - \omega(y,z) = \Omega(x; y-x) - \Omega(y; z-y) = \Omega(y; y-x) - \Omega(y; z-y),$$

using the Taylor principle to replace x by y in the first term. By linearity of $\Omega(y; -)$, this equals

$$\Omega(y; (y-x) - (z-y)) = \Omega(y; 2(y-x) - (z-x)).$$

Now since (x,y,z) form an infinitesimal 2-simplex, $(y-x, z-x) \in \widetilde{D}(2,V)$, and so any linear combination of these vectors is in $D(V)$. Therefore $\Omega(y; 2(y-x) - (z-x))$, with $2(y-x) - (z-x) \in D(V)$, gives a witness that $\omega(x,y) \sim \omega(y,z)$.

For an infinitesimal 2-simplex (x,y,z) in M, it is not in general true that $(x,y) \sim (y,z)$ in the manifold $M \times M$.

We consider next the coboundary formula for W-valued cochains on the cubical complex $M_{[\bullet]}$. Such a k-cochain is a map $\omega : M_{[k]} \to W$; the (cubical) coboundary of it is a cubical $k+1$-cochain, whose value on a $k+1$-cube P is given by the classical alternating sum with $2(k+1)$ terms (one term for each face of P) (here, $P = [x_0, x_1, \ldots, x_{k+1}]$ is an infinitesimal $k+1$-dimensional parallelepipedum; recall the description of the faces, (2.8.1) and (2.8.2), in singular cubes, in particular of infinitesimal parallelepipeda):

$$\sum_{i=1}^{k+1} (-1)^i \Big\{ \omega[x_0, x_1, \ldots, \widehat{i}, \ldots, x_{k+1}] - \omega[x_i, x_i - x_0 + x_1, \ldots, \widehat{i}, \ldots, x_i - x_0 + x_{k+1}] \Big\}. \tag{3.2.2}$$

Note that for $k = 0$ and $k = 1$, this is consistent with the usage in Section 2.2.

Recall that a cubical cochain ω is called a cubical differential k-form if it satisfies the "normalization" condition that it vanishes on any infinitesimal parallelepipedum whose spanning simplex is a degenerate whisker, $x_j = x_0$ for some $j > 0$.

Proposition 3.2.3 *If the cubical cochain $\omega : M_{[k]} \to W$ satisfies the normalization condition, then so does $d\omega : M_{[k+1]} \to W$.*

Equivalently, the cubical differential k-forms form a subcomplex of the cochain complex of the cubical complex $M_{[\bullet]}$.

Proof. Assume ω is a cubical k-form. Consider a $k+1$-dimensional infinitesimal parallelepipedum $[x_0 x_1, \dots, x_{k+1}]$, which is degenerate by virtue of $x_j = x_0$ with $j > 0$. In the formula (3.2.2), the curly bracket expressions with $i \neq j$ vanish because each of the two terms in the brackets vanish individually, by the normalization condition assumed for ω. In the jth curly bracket, the two terms cancel each other since $x_j = x_0$ implies $x_j - x_0 + x_1 = x_1, \dots, x_j - x_0 + x_{k+1} = x_{k+1}$.

Recall the bijective correspondence between simplicial and cubical differential forms. We want to compare their coboundaries, as described in (3.2.1) and (3.2.2), respectively. Let us denote the simplicial coboundary by d and the cubical coboundary by $\tilde{d}$, for the sake of comparison.

Theorem 3.2.4 *Let ω be a simplicial differential k-form on a manifold M, and let $\tilde{\omega}$ be the corresponding cubical differential form. Then*

$$\tilde{d}\tilde{\omega} = (k+1)\widetilde{(d\omega)}.$$

Proof. Since the correspondence $\omega \leftrightarrow \tilde{\omega}$ does not depend on coordinates, there is no harm in proving the assertion in a coordinatized situation, so assume that M is an open subset of a finite-dimensional vector space V. As in the proof of Theorem 3.1.5, ω is then given by a function $\Omega : M \times V^k \to W$, k-linear and alternating in the k last arguments:

$$\omega(x_0, x_1, \dots, x_k) = \Omega(x_0; x_1 - x_0, \dots, x_k - x_0),$$

and the corresponding $\tilde{\omega}$ is given by

$$\tilde{\omega}([x_0, x_1, \dots, x_k]) = \Omega(x_0; x_1 - x_0, \dots, x_k - x_0)$$

(same expression!).

We begin by calculating the cubical coboundary $\tilde{d}\tilde{\omega}[x_0, x_1, \dots, x_{k+1}]$ in terms of Ω; this actually leads to a classical coordinate formula. For, consider the ith term ($i = 1, \dots, k+1$) in (3.2.2), expressed in terms of Ω and its directional derivatives; it is

$$\pm\{\Omega(x_0; x_1 - x_0, \dots, \widehat{i}, \dots, x_{k+1} - x_0) \\ - \Omega(x_i; x_1 - x_0, \dots, \widehat{i}, \dots, x_{k+1} - x_0)\}. \tag{3.2.3}$$

Let as usual $d\Omega(x_0;u,\ldots)$ denote the directional derivative at x_0 of $\Omega(x;\ldots)$, as a function of x, in the direction of u. Then the difference (3.2.3) is, by Taylor expansion,

$$= \pm d\Omega(x_0;x_i - x_0, x_1 - x_0, \ldots, \widehat{x_i - x_0}, \ldots, x_k - x_0),$$

so that we have

$$\tilde{d}\tilde{\omega} = \sum_{i=1}^{k+1} (-1)^{i+1} d\Omega(x_0;x_i - x_0, x_1 - x_0, \ldots, \widehat{x_i - x_0}, \ldots, x_{k+1} - x_0). \quad (3.2.4)$$

Note that $d\Omega(x_0;\ldots)$ is $k+1$-linear in the $k+1$ arguments after the semicolon, and alternating w.r.t. the last k of these (only the whole sum is alternating in all $k+1$ arguments, and is a classical formula, cf. e.g. Madsen and Tornehave, 1997, p. 15). However, in the present case, the x_is form an infinitesimal $k+1$-simplex, so that $(x_1 - x_0, \ldots, x_{k+1} - x_0) \in \widetilde{D}(k+1,V)$, and this implies that any $k+1$-linear map $V^{k+1} \to W$ behaves on this $k+1$-tuple *as if* it were alternating (see the end of Section 1.3). In particular, $d\Omega(x_0;\ldots)$ does so, and this in turn implies that the $k+1$ terms in the sum (3.2.4) *are equal*, so that (3.2.4) may be rewritten

$$\tilde{d}\tilde{\omega} = (k+1) d\Omega(x_0;x_1 - x_0, x_2 - x_0, \ldots, x_{k+1} - x_0). \quad (3.2.5)$$

We next proceed to calculate the simplicial coboundary $d\omega(x_0,\ldots,x_{k+1})$ in terms of Ω and its directional derivatives; it is

$$\begin{aligned} d\omega(x_0,x_1,\ldots,x_{k+1}) &= \sum_{i=0}^{k+1} (-1)^i \omega(x_0,\ldots,\widehat{x_i},\ldots,x_{k+1}) \\ &= \Omega(x_1;x_2 - x_1, \ldots, x_{k+1} - x_1) \\ &\quad + \sum_{i=1}^{k+1} (-1)^i \Omega(x_0;x_1 - x_0, \ldots, \widehat{x_i - x_0}, \ldots, x_{k+1} - x_0). \end{aligned} \quad (3.2.6)$$

Here, the first term is special. We rewrite it by Taylor expansion from x_0 in the direction $x_1 - x_0$. This term then becomes

$$\Omega(x_0;x_2 - x_1, \ldots, x_{k+1} - x_1) + d\Omega(x_0;x_1 - x_0, x_2 - x_1, \ldots, x_{k+1} - x_1). \quad (3.2.7)$$

We write, for $i = 2,\ldots,k+1$, $x_i - x_1$ as $(x_i - x_0) - (x_1 - x_0)$ and expand the first term in (3.2.7) using k-linearity of $\Omega(x_0;-,\ldots,-)$. This should give 2^k terms, but fortunately, many of them contain $x_1 - x_0$ more than once; since $x_1 - x_0 \in D(V)$, such terms vanish. So the first term in (3.2.7) equals

$$\Omega(x_0;x_2 - x_0, \ldots, x_{k+1} - x_0) - \sum_{i=2}^{k+1} \Omega(x_0;x_2 - x_0, \ldots, x_1 - x_0, \ldots, x_{k+1} - x_0)$$

where in the ith term, $x_1 - x_0$ replaces $x_i - x_0$. The $k+1$ terms we obtained here cancel the $k+1$ terms in the last line of (3.2.6) because $\Omega(x_0; -, \ldots, -)$ is alternating. This means that the expression (3.2.6) reduces to $d\Omega(x_0; x_1 - x_0, x_2 - x_1, \ldots, x_{k+1} - x_1)$. The k last occurrences of x_1 may be replaced by x_0s, by the Taylor principle, and then we are left with the expression in (3.2.5) except for the factor $k+1$; and this proves the theorem.

Let us call a simplicial or cubical k-form ω *closed* if $d(\omega) = 0$, where d refers to the complex of simplicial, resp. cubical, forms. From the theorem follows that a simplicial k-form is closed iff its corresponding cubical form is closed. Also, we see that for (simplicial) 1-forms ω, the closedness notion introduced already in Definition 2.2.4 is consistent with the present more general usage for simplicial k-forms.

3.3 Integration of forms

Recall that if $S \subseteq R$ is an open subset, and $f : S \to R$ is a function, there is a unique function $f' : S \to R$, the *derivative* of f, such that for all $d \in D$ and $x \in S$

$$f(x+d) = f(x) + d \cdot f'(x).$$

An *anti-derivative* for a function $f : S \to R$ is a function $F : S \to R$ such that $F' = f$.

Some of the following makes sense also for R-valued functions which are defined only on open, connected, simply connected subsets S of R (whatever that is supposed to mean in the present context), but for simplicity, we consider functions defined on all of R, $f : R \to R$ ("entire functions").

With some modifications, the codomain R could be replaced by a KL vector space W.

Integration theory in SDG is not very well developed; in most places, like in Kock (1981/2006), the theory depends on anti-derivatives. The present text is no exception, and the theory here is even more primitive than in Kock (1981/2006), since we do not consider any kind of order relation $\leq$ on R, hence we do not consider *intervals* as domains of functions to be integrated; we consider only *entire* functions $R \to R$, for simplicity. So we put

Integration Axiom *Every function $f : R \to R$ has an anti-derivative, and any two such differ by a unique constant.*

From uniqueness of anti-derivatives follows of course that if f and g are functions $R \to R$ such that $f(0) = g(0)$ and $f' = g'$, then $f = g$; but more generally, if f and g are functions $R^k \to R$ with $f(0) = g(0)$ and $\partial f/\partial x_i = \partial g/\partial x_i$

for all $i = 1, \dots, k$, then $f = g$. This follows by an easy iteration argument from the 1-dimensional case.

One immediate application of the Integration Axiom is a "Hadamard Remainder" formula. There are traditionally many ways to formulate information about the remainders $R(x)$ in a Taylor expansion of a function. In its simplest form, the Hadamard Lemma says that for any smooth $f : R \to R$,

$$f(t) - f(0) = t \cdot h(t)$$

for some smooth (and actually unique) $h : R \to R$. (Note that the left-hand side here is a zero-order Taylor remainder.) The construction of h is by an integral,

$$h(t) = \int_0^1 f'(t \cdot u)\, du;$$

likewise for the higher Taylor remainders $R_n(t)$ in one variable, there is a standard integral formula for them, and the Hadamard observation is that this integral depends smoothly on the parameter t; see e.g. Lavendhomme (1996, Propositions 1.3.4 and 1.3.13) for this in synthetic/axiomatic context.

The integration axiom for R allows us to define integrals of functions in terms of anti-derivatives (as calculus students often do, after having paid the required lip service to Riemann sums). Namely, let $f : R \to R$ be a function. If a and b are elements in R, we define $\int_a^b f$ to be $F(b) - F(a)$ for some anti-derivative F of f; it is independent of choice of anti-derivative, since anti-derivatives are unique modulo a constant. (The definition makes sense whether or not $a \leq b$; in fact, we have not assumed any preorder $\leq$ on R at all; and we have $\int_a^b f = -\int_b^a f$.) The integrals $\int_a^b f$ defined by this recipe, we call *scalar* integrals, for contrast with the integrals of differential forms (like curve integrals) to be considered later.

This is as in calculus, and as there, one proves the subdivision rule

$$\int_a^c f = \int_a^b f + \int_b^c f \tag{3.3.1}$$

for any a, b, c in R. Similarly for the substitution rule (substitution *along* g),

$$\int_a^b (f \circ g) \cdot g' = \int_{g(a)}^{g(b)} f \tag{3.3.2}$$

for any function $g : R \to R$. It is easy to see that, for fixed a, the map $b \mapsto \int_a^b f$ is an antiderivative of f ("Fundamental Theorem of Calculus" Part 1).

For functions $f : R^2 \to R$, one can iterate the integration procedure, and define, in terms of anti-derivatives of functions in one variable, the iterated integral $\int_a^b \int_c^d f$ for any a, b, c, d in R, and the usual rules hold (Fubini Theorem, etc.). More generally, if $f : R^k \to R$ is a function, and $a_1, \ldots, a_k, b_1, \ldots, b_k \in R$, we define the iterated integral $\int_{a_1}^{b_1} \int_{a_2}^{b_2} \ldots \int_{a_k}^{b_k} f$ in the expected way, by iteration of one-dimensional integrals.

Unlike for one-variable scalar integrals, the notation $\int_{a_1}^{b_1} \int_{a_2}^{b_2} \ldots \int_{a_k}^{b_k} f$ for this iterated integral is known not to be practical, since it does not tell us which of the k variables of f is to range over which "interval"; one may by convention say that x_k ranges over the interval $[a_k, b_k]$ in the innermost integral, ..., x_1 ranges over $[a_1, b_1]$ in the outermost one. Rather than depending on such implicit conventions, the following notation is known to work well (and we adopt it):

$$\int_{a_1}^{b_1} \int_{a_2}^{b_2} \cdots \int_{a_k}^{b_k} f(x_1, \ldots, x_k)\, dx_k \ldots dx_1;$$

but the reader should be warned that $dx_k \ldots dx_1$ does not denote any differential k-form, in particular, it should not be confused with $dx_k \wedge \ldots \wedge dx_1$.

One reason why iterated integrals in general are not sufficient for calculus, is the fact that they are only defined over rectangular boxes (and a few other simple kinds of regions), and that therefore a theory of substitution in such integrals cannot be well formulated, say substitution along an arbitrary map $R^k \to R^k$; such a map may destroy rectangular shape.

However, the success of the substitution rule for one-variable integrals does leave some trace on the theory of iterated integrals, namely substitution along maps $\alpha : R^k \to R^k$ of the form $\alpha = \alpha_1 \times \ldots \times \alpha_k : R^k \to R^k$, where each α_i is a map $R \to R$. Then the one-variable rule generalizes into

$$\begin{aligned} &\int_{a_1}^{b_1} \int_{a_2}^{b_2} \cdots \int_{a_k}^{b_k} f(\alpha(\underline{x})) \cdot \alpha_1'(x_1) \cdot \ldots \cdot \alpha_k'(x_k)\, dx_k \ldots dx_1 \\ &\qquad = \int_{\alpha_1(a_1)}^{\alpha_1(b_1)} \cdots \int_{\alpha_k(a_k)}^{\alpha_k(b_k)} f(t_1, \ldots, t_k)\, dt_k \ldots dt_1 \end{aligned} \tag{3.3.3}$$

(here $\underline{x}$ is short for $(x_1, \ldots, x_k)$).

Curve integrals

We shall deal with k-surface integrals of k-forms later, but for simplicity of exposition, we begin with the case $k = 1$, curve integrals of 1-forms.

The scalar integrals $\int_a^b f$ allow us to define curve integrals of 1-forms. We consider first 1-forms on R itself.

Let θ be an R-valued 1-form on R. It is of the form $\theta(s,t) = (t - s) \cdot g(s)$ for a unique function $g : R \to R$. The 1-form θ thus given is traditionally denoted

$g(x)\,dx$, (or $g(u)\,du$ or $g(v)\,dv$, ...). Thus, in combinatorial terms, for $s \sim t$ in R,

$$(g(x)\,dx)(s,t) = g(s) \cdot (t-s),$$

or equivalently, for all $d \in D$,

$$(g(x)\,dx)(s,s+d) = d \cdot g(s).$$

In particular, if g is constant 1, we have the 1-form dx given by $(dx)(s,t) = t-s$. It is the volume form for R.† The notation dx for this form is consistent with the exterior-derivative use of d; the symbol x is a traditional notation for the identity map $R \to R$, and the exterior derivative of the identity map is exactly the 1-form dx. More generally, we have the equality of 1-forms $df = f'(x)dx$, for $f : R \to R$ a 0-form (= a function).

Even more generally, for $\alpha : R \to R$ a function,

$$\alpha^*(g(x)\,dx) = g(\alpha(x)) \cdot \alpha'(x)\,dx. \tag{3.3.4}$$

Let ω be an R-valued 1-form on a manifold M, and let $\gamma : R \to M$ be a map (a "path, or curve, in M") (or even, as in Section 2.1: a "singular 1-cube in M"). Then we get a 1-form $\gamma^*(\omega)$ on R. We may write

$$\gamma^*(\omega) = g(x)\,dx,$$

for some uniquely determined function $g : R \to R$, and we put

$$\int_\gamma \omega := \int_0^1 g$$

this is the *(curve) integral of ω along the path γ.*

There are some special curve integrals which can immediately be calculated, namely curve integrals along infinitesimal 1-dimensional parallelepipeda: recall (2.1.6) that if $x \sim y$ in the manifold M, we have a canonical "infinitesimal" curve $[x,y] : R \to M$, defined using affine combinations of the neighbour points x and y in M.

Proposition 3.3.1 *Let ω be a 1-form on M, and let $x \sim y$. Then*

$$\int_{[x,y]} \omega = \omega(x,y).$$

† The 1-form dx is also the Maurer–Cartan form for the additive group $(R,+)$; for R^n with $n \geq 2$, there is no comparison between the volume form (an R-valued n-form) and the Maurer–Cartan form (an R^n-valued n-form).

Proof. Consider fixed $x \sim y$. By Theorem 3.1.13, $[x,y]^*(\omega) = \omega(x,y) \cdot \mathrm{Vol}$ (a constant form) so $\int_{[x,y]} \omega = \int_0^1 \omega(x,y)$; but the scalar integral from 0 to 1 of a constant equals that constant.

Let ω be a 1-form on M. If $\alpha : R \to R$ is a function, and $\gamma : R \to M$ a curve in M, we want to compare the curve integrals of ω along γ and along $\gamma \circ \alpha$. Assume as before that $\gamma^*(\omega) = g(x)\,dx$. Then $(\gamma \circ \alpha)^*(\omega) = \alpha'(x) \cdot g(\alpha(x))\,dx$, and so by (3.3.4) and (3.3.2) we have

$$\int_{\gamma\circ\alpha} \omega = \int_0^1 (g \circ \alpha) \cdot \alpha' = \int_{\alpha(0)}^{\alpha(1)} g. \tag{3.3.5}$$

We shall in particular consider the case of an *affine* map α.

Let a and b be elements of R. Since R is a free affine space on two generators 0 and 1, there is a unique affine map $R \to R$ with $0 \mapsto a$ and $1 \mapsto b$. This map is denoted $[a,b]$. (It is given by $t \mapsto (1-t) \cdot a + t \cdot b$; if $a \sim b$, this is consistent with the use for infinitesimal 1-dimensional parallelepipeda.) The map $[0,1] : R \to R$ is the identity map.

Applying (3.3.5) to the map $\alpha = [a,b]$, we therefore have

$$\int_{\gamma\circ[a,b]} \omega = \int_a^b g, \tag{3.3.6}$$

where as before $\gamma^*(\omega) = g(x)\,dx$.

From this, and the subdivision rule for scalar integrals (3.3.1), we therefore get the subdivision rule for curve integrals of ω:

$$\int_{\gamma\circ[a,c]} \omega = \int_{\gamma\circ[a,b]} \omega + \int_{\gamma\circ[b,c]} \omega \tag{3.3.7}$$

for any a, b, and c in R. In particular, since $\int_\eta \omega = 0$ for any constant path η, we have

$$\int_{\gamma\circ[a,b]} \omega = -\int_{\gamma\circ[b,a]} \omega.$$

Here is a consequence of (3.3.6): consider a 1-form $\omega = f(x)\,dx$ on R, where $f : R \to R$ is a function. Then for $[a,b] : R \to R$

$$\int_{[a,b]} f(x)\,dx = \int_a^b f;$$

the left-hand side is a curve integral in the sense of differential forms, whereas the right-hand side is a scalar integral, defined in terms of anti-derivatives of f. The equation says that the two expressions in this equation are two aspects of the same quantity, and the traditional notation for this quantity straddles between these expressions; it is the notation

$$\int_a^b f(x)\,dx,$$

which on the one hand succeeds in mentioning the 1-form $f(x)\,dx$, but also the "lower and upper" "endpoints" a and b of the "interval" $[a,b]$ of the integration. (The word "endpoint" is used here only to assist the intuition; we have not assumed any partial or preorder $\leq$ on R with respect to which we have the interval $[a,b]$ as a *subset* of R.)

For fixed ω, we consider the process $\gamma \mapsto \int_\gamma \omega$ as a map ("functional") $\Omega : P(M) \to R$, where $P(M) = M^R$ is the set of paths $R \to M$ in M. (Note that $P(M)$ appears elsewhere in the present text under the name $S_{[1]}(M)$, the set of singular 1-cubes on M.)

By (3.3.7) (subdivision law), Ω has the property

$$\Omega(\gamma \circ [a,c]) = \Omega(\gamma \circ [a,b]) + \Omega(\gamma \circ [b,c]). \tag{3.3.8}$$

Following essentially Meloni and Rogora (1988), we call a functional $\Omega : P(M) \to R$ with this property a 1-dimensional *observable* on M. (The notion of k-dimensional observable for general k is given in Definition A.8.2.)

Theorem 3.3.2 *The process which to a combinatorial 1-form on M associates the observable $\gamma \mapsto \int_\gamma \omega$ provides a bijection between the set of 1-forms on M, and the set of 1-dimensional observables on M.*

Proof. Given a 1-dimensional observable Ω on M, we define a 1-form ω on M by putting

$$\omega(x,y) := \Omega([x,y]). \tag{3.3.9}$$

Since Ω is 0 on constant paths, it follows that $\omega(x,x) = 0$, so that ω is indeed a (combinatorial) 1-form. We prove first, with ω thus defined, that $\Omega(\gamma) = \int_\gamma \omega$ for any path $\gamma : R \to M$. We do this by proving more generally that

$$\Omega(\gamma \circ [0,t]) = \int_{\gamma \circ [0,t]} \omega. \tag{3.3.10}$$

Both sides take value 0 when $t = 0$. We prove that furthermore both sides, as functions f of $t \in R$, satisfy the equation

$$d \cdot f'(t) = \Omega([\gamma(t), \gamma(t+d)]) \tag{3.3.11}$$

for all $d \in D$, so that the result follows from uniqueness of primitives and by cancellation of universally quantified ds. We get, with $f(t) := \Omega(\gamma \circ [0,t])$ that

$$f(t+d) - f(t) = \Omega(\gamma \circ [0,t+d]) - \Omega(\gamma \circ [0,t]) = \Omega(\gamma \circ [t,t+d]),$$

the last equation by the subdivision law for Ω. Now any map preserves affine combination of mutual neighbour points (Theorem 2.1.4); therefore $\gamma \circ [t,t+d] = [\gamma(t), \gamma(t+d)]$, so

$$\Omega(\gamma \circ [t,t+d]) = \Omega([\gamma(t), \gamma(t+d)]),$$

proving that the left-hand side of (3.3.10) satisfies the differential equation (3.3.11). For the right-hand side, we similarly get, by the subdivision rule for curve integrals, that

$$\int_{\gamma\circ[t+d,0]} \omega - \int_{\gamma\circ[0,t]} \omega = \int_{\gamma\circ[t,t+d]} \omega = \int_{[\gamma(t),\gamma(t+d)]} \omega = \omega(\gamma(t), \gamma(t+d)),$$

using Proposition 3.3.1. But this is by definition of ω the same as $\Omega([\gamma(t), \gamma(t+d)])$. So also the right-hand side of (3.3.10) satisfies the differential equation (3.3.11). This proves that $\Omega = \int \omega$.

Conversely, let us start with a differential 1-form ω, and let Ω denote the observable $\int \omega$. The process (3.3.9) provides here the 1-form $\tilde{\omega}$ given by $\tilde{\omega}(x,y) = \Omega([x,y]) = \int_{[x,y]} \omega$. But by Proposition 3.3.1, this is the same as $\omega(x,y)$, proving that $\tilde{\omega} = \omega$; the theorem is proved.

Higher surface integrals

We do not only want to construct curve integrals out of 1-forms, but to construct surface integrals (over k-dimensional surfaces) out of k-forms. We here understand "k-dimensional surface" in a manifold to be parametrized by R^k, in other words, a k-dimensional surface is here the same thing as a singular k-cube $R^k \to M$. In particular, the k-surfaces in M, as k ranges, form the cubical complex $S_{[\bullet]}(M)$, as described in the Appendix, Section A.8.

If ω is a (cubical-combinatorial†) k-form on a manifold M, and $\gamma : R^k \to M$ a singular k-cube in M, we shall define the *integral* of ω *along* γ, to be denoted $\int_\gamma \omega$. It should be functorial w.r.t. maps $f : M \to N$ between manifolds, i.e. if

† The definitions themselves also make sense for simplicial forms, but the "coboundary" formulae, Stokes' Theorem, etc. to be developed, work more seamlessly for the cubical formulation. For $k = 1$, simplicial = cubical, and we recover the notions leading to Theorem 3.3.2.

θ is a k-form on N, and γ a singular k-cube on M, the integral should satisfy

$$\int_{\gamma} f^*(\theta) = \int_{f\circ\gamma} \theta. \tag{3.3.12}$$

Therefore, the essence resides in defining integrals $\int_{\mathrm{id}} \omega$, where id is the "generic k-cube", i.e. the identity map $R^k \to R^k$, and where ω is a k-form on R^k. Recall that the identity map $R^k \to R^k$ may be described as the map $[0,1]\times\ldots\times[0,1]$, so $\int_{\mathrm{id}} \omega$ is also denoted $\int_{[0,1]\times\ldots\times[0,1]} \omega$.

To define these integrals, recall from Proposition 3.1.11 that an R-valued k-form ω on R^k may be written $\widetilde{\omega}\cdot\mathrm{Vol}$ for a unique function $\widetilde{\omega} : R^k \to R$. Therefore, we can define

$$\int_{[0,1]\times\ldots\times[0,1]} \omega := \int_0^1 \ldots \int_0^1 \widetilde{\omega}(t_1,\ldots,t_k)\, dt_k \ldots dt_1. \tag{3.3.13}$$

The right-hand side is an ordinary iterated scalar integral, as described above; its description depends only on the possibility of forming anti-derivatives of functions $R \to R$.

The fact that (3.3.12) holds is then an immediate consequence of the fact that given a k-form θ on N and

$$R^k \xrightarrow{\ \gamma\ } M \xrightarrow{\ f\ } N,$$

then $\gamma^*(f^*(\theta)) = (f\circ\gamma)^*(\theta)$.

From Theorem 3.1.13, we deduce

Proposition 3.3.3 *Let ω be a (cubical-combinatorial) k-form on a manifold M. Then for any infinitesimal k-simplex $(x_0,x_1,\ldots,x_k)$ in M, we have*

$$\int_{[x_0,x_1,\ldots,x_k]} \omega = \omega[x_0,x_1,\ldots,x_k].$$

Proof. It is immediate that for any constant λ (like $\omega([x_0,x_1,\ldots,x_k])$), we have the second equality sign in $\int_{\mathrm{id}} \lambda\cdot\mathrm{Vol} = \int_0^1 \ldots \int_0^1 \lambda\, dt_k \ldots dt_1 = \lambda$. So the proposition follows from the theorem quoted.

Consider a k-form $\omega = g\cdot\mathrm{Vol}$ on R^k, where $g : R^k \to R$ is a function. By definition, we have $\int_{\mathrm{id}} \omega = \int_0^1 \ldots \int_0^1 g(t_1,\ldots,t_k)\, dt_k \ldots dt_1$. But we have more generally:

Proposition 3.3.4 *Consider the map $\alpha : R^k \to R^k$ given, as in (3.3.3), by $\alpha = [a_1, b_1] \times \ldots \times [a_k, b_k]$, and let $\omega = g \cdot \mathrm{Vol}$. Then*

$$\int_\alpha \omega = \int_{a_1}^{b_1} \cdots \int_{a_k}^{b_k} g(t_1, \ldots, t_k)\, dt_k \ldots dt_1. \tag{3.3.14}$$

Proof. We have by (3.3.12) and by the formula assumed for ω that

$$\begin{aligned}\int_\alpha \omega &= \int_{\mathrm{id}} \alpha^*(\omega) = \int_{\mathrm{id}} \alpha^*(g \cdot \mathrm{Vol}) \\ &= \int (g \circ \alpha) \cdot \alpha^*(\mathrm{Vol}),\end{aligned}$$

now the determinant of the linear part of the affine map α is here clearly $\prod(b_i - a_i)$, so Proposition 3.1.12 allows us to continue

$$= \int_{\mathrm{id}} (g \circ \alpha) \cdot \Pi(b_i - a_i) \cdot \mathrm{Vol} = \int_0^1 \cdots \int_0^1 (g \circ \alpha) \Pi(b_i - a_i)$$

by the basic definition (3.3.13),

$$\begin{aligned}&= \int_0^1 \cdots \int_0^1 g(\alpha_1(x_1), \ldots, \alpha_k(x_k)) \cdot \Pi(b_i - a_i)\, dx_k \ldots dx_1 \\ &= \int_{a_1}^{b_1} \cdots \int_{a_k}^{b_k} g(t_1, \ldots, t_k)\, dt_k \ldots dt_1\end{aligned}$$

by (3.3.3).

We can now prove:

Proposition 3.3.5 *Let θ be a k-form on a manifold M. Then the functional $\gamma \mapsto \int_\gamma \theta$ satisfies the subdivision property.*

Proof. Let $\gamma : R^k \to M$ be a map (a singular k-cube). With the notation of Section A.8, we need to prove that for $i = 1, \ldots, k$ and $a, b, c \in R$, we have

$$\int_{\gamma|_i[a,c]} \theta = \int_{\gamma|_i[a,b]} \theta + \int_{\gamma|_i[b,c]} \theta.$$

Let $\gamma^*(\theta) = g \cdot \mathrm{Vol}$ (where g is a function $R^k \to R$). Then

$$\int_{\gamma|_i[a,b]} \theta = \int_{[a,b]_i} \gamma^*(\theta) = \int_0^1 \cdots \int_a^b \cdots \int_0^1 g(t_1, \ldots, t_k)\, dt_k \ldots dt_1,$$

by Proposition 3.3.4, and similarly for $\int_{\gamma|_i[a,c]}$ and $\int_{\gamma|_i[b,c]}$. The result now follows from the subdivision rule $\int_a^c = \int_a^b + \int_b^c$ for one-variable integrals (which immediately gives similar rules for subdivision of k-variable integrals).

The notion of k-dimensional *observable* on a manifold M is defined in the Appendix, Definition A.8.2: a function $S_{[k]}(M) \to R$ which is alternating and satisfies the subdivision law, as explained there.

Some of the results proved lead to

Theorem 3.3.6 *(1) For θ a k-form on a manifold M, the functional $\gamma \mapsto \int_\gamma \theta$ is an observable on M.*

(2) Let γ be an infinitesimal k-dimensional parallelepipedum; then

$$\int_\gamma \theta = \theta(\gamma).$$

Proof. The functional described has the subdivision property, by Proposition 3.3.5, and the fact that it is alternating is an immediate consequence of Proposition 3.1.12 applied to the affine map α which interchanges the ith and jth coordinate. Finally, the last assertion is contained in Proposition 3.3.3.

3.4 Uniqueness of observables

We shall in this section prove that a k-dimensional observable is completely determined by its value on infinitesimal parallelepipeda. Since the k-dimensional observables on a manifold M clearly form a linear subspace of the space of all functions $M^{R^k} \to R$, it suffices to prove

Theorem 3.4.1 *Let Ψ be a k-dimensional observable on M which takes value 0 on all infinitesimal k-dimensional parallelepipeda. Then Ψ is constant 0.*

Proof. Let $\gamma : R^k \to M$ be a singular cube. Consider the observable $\Phi := \gamma^*(\Psi)$ on R^k. Because any map, in particular γ, preserves infinitesimal parallelepipeda (Theorem 2.1.4), it follows that Φ vanishes on all infinitesimal k-parallelepipeda in R^k. We shall prove that Φ vanishes on all *rectangles* in R^k; by a *rectangle* in R^k, we understand a singular cube $\alpha : R^k \to R^k$ which is not only affine, but whose linear part is given by a *diagonal* matrix, $\alpha = \|a \mid A\|$, with A a diagonal matrix (see Section A.8 for the matrix calculus for affine maps); the diagonal entries of this matrix are called the *sides* of the rectangle.† If we can prove that Φ vanishes on all rectangles, then Φ will certainly vanish on the identity map $\|0 \mid I,\|$ where I is the identity matrix. So $\Phi(\mathrm{id}) = \gamma^*(\Psi)(\mathrm{id}) = \Psi(\gamma)$, which therefore is 0.

The theorem will therefore be proved by proving the following two lemmas.

† Note that the map α occurring in Proposition 3.3.4 is such a rectangle; there is also a converse statement.

Lemma 3.4.2 *Assume that a k-dimensional observable* Φ *on* R^k *vanishes on all infinitesimal k-dimensional parallelepipeda in* R^k. *Then it vanishes on all rectangles with infinitesimal sides.*

(Note that an infinitesimal parallelepipedum need not be a rectangle; in fact, for $k \geq 2$, those infinitesimal k-dimensional parallelepipeda in R^k which are also rectangles have the property that *any* observable vanishes on them. See Remark 3.4.6 at the end of the section.)

Lemma 3.4.3 *Assume that a k-dimensional observable* Φ *on* R^k *vanishes on all rectangles with infinitesimal sides. Then it vanishes on all rectangles.*

Proof of Lemma 3.4.2. We prove in fact more, namely that Φ vanishes on all singular cubes of the form $[x_0, x_1, \ldots, x_k]$ where $x_i \sim x_0$ for all i. (Note that such a cube is not necessarily an infinitesimal parallelepipedum, since we have not required $x_i \sim x_j$; nonetheless $[x_0, x_1, \ldots, x_k]$ makes sense, since it here takes values in an *affine* space, namely R^k; cf. Remark 2.1.8.) There is no harm in assuming that $x_0 = 0$. The k-tuple $(x_1, \ldots, x_k)$ is therefore an element of $D(k)^k$. Let $\phi : D(k)^k \to R$ be the function $(x_1, \ldots, x_k) \mapsto \Phi([0, x_1, \ldots, x_k])$. It is an easy consequence of the subdivision property for Φ that ϕ vanishes if one of the x_is is 0. It therefore follows from the KL axioms that ϕ extends to a k-linear map $(R^k)^k \to R$, and this map (likewise denoted ϕ) is alternating because of the alternating property assumed for Φ. By the assumption that Φ vanishes on infinitesimal k-parallelepipeda, it follows that ϕ vanishes on $\widetilde{D}(k,k) \subseteq D(k)^k$. But a k-linear alternating map $(R^k)^k \to R$ is completely determined by its restriction to $\widetilde{D}(k,k)$. So ϕ is the zero map.

Proof of Lemma 3.4.3. This is by a downward induction, starting from k: by assumption, Φ vanishes on rectangles with all k sides infinitesimal. Assume we have already proved that Φ vanishes on all rectangles with the first i sides infinitesimal; we prove that it then also vanishes on rectangles with the first $i-1$ sides infinitesimal. Consider such a rectangle $[x_0, x_0 + t_1 e_1, \ldots, x_0 + t_{i-1} e_{i-1}, x_0 + t_i e_i, \ldots, x_0 + t_k e_k]$, where the $t_1, \ldots, t_{i-1}$ are in D. Consider this as a function of t_i alone, in other words, consider the function $g : R \to R$ given, for fixed x_0 and fixed t_js ($j \neq i$), by

$$g(t) := \Phi([x_0, x_0 + t_1 e_1, \ldots, x_0 + t_{i-1} e_{i-1}, x_0 + t e_i, \ldots, x_0 + t_k e_k]).$$

If $t = 0$, the input to Φ is a rectangle with i infinitesimal sides, and so $g(0) = 0$.

Let us spell out $g(t)$ in the matrix notation for affine maps;

$$g(t) = \Phi\left(\left\|\begin{array}{c|ccccc} x_{01} & t_1 & & & & \\ \vdots & & \ddots & & & \\ x_{0i} & & & t & & \\ \vdots & & & & \ddots & \\ x_{0k} & & & & & t_k \end{array}\right\|\right).$$

By the subdivision of rectangles exhibited in (A.8.6), it is now clear that $g(t+d) - g(t)$ equals

$$\Phi\left(\left\|\begin{array}{c|ccccc} x_{01} & t_1 & & & & \\ \vdots & & \ddots & & & \\ x_{0i}+t & & & d & & \\ \vdots & & & & \ddots & \\ x_{0k} & & & & & t_k \end{array}\right\|\right).$$

The input of Φ here is a rectangle with its first i sides infinitesimal, so by the induction assumption, the value of Φ on it is 0. Thus $g(t+d) = g(t)$ for all $d \in D$, i.e. $g' \equiv 0$. Since also $g(0) = 0$, we conclude by uniqueness of primitives that $g \equiv 0$. So Φ vanishes on all rectangles with basic vertex in x_0, but x_0 was arbitrary; so Φ vanishes on all rectangles. This proves the lemma. (This is essentially the argument given already in Kock *et al.*, 1979/1980, after Lemma 4.4 there, and reproduced in Kock, 1981/2006 and Lavendhomme, 1996.)

With these lemmas, the theorem follows.

The combinatorial differential forms that we consider presently are the cubical ones. The following is an extension of Theorem 3.3.2, which is the special case where $k = 1$.

Theorem 3.4.4 *Every k-dimensional observable Φ on a manifold M is of the form $\int \omega$ for a unique k-form ω. In particular, the process $\omega \mapsto \int \omega$ is a bijection between k-forms and k-dimensional observables.*

Proof. If such an ω exists, then by the second assertion in Theorem 3.3.6, we have for any infinitesimal k-parallelepipedum $[x_0, \ldots, x_k]$,

$$\omega[x_0, \ldots, x_k] = \Phi([x_0, \ldots, x_k]); \tag{3.4.1}$$

this proves that there is at most one k-form ω which gives rise to Φ by integration. Conversely, given a k-dimensional observable Φ; let us define ω by

(3.4.1). To see that the ω thus defined is indeed a differential form, we have to see that it vanishes on degenerate infinitesimal parallelepipeda. So suppose the parallelepipedum is degenerate by virtue of $x_i = x_0$. Then it is easy to see that this parallelepipedum γ subdivides in the ith direction into two copies of itself, and so it follows from the subdivision property for Φ that $\Phi(\gamma) = \Phi(\gamma) + \Phi(\gamma)$, whence $\Phi(\gamma) = 0$, so ω does get value 0 on this parallelepipedum.

The fact that for this form ω, $\int \omega = \Phi$ follows from Proposition 3.3.3 and from the uniqueness in Theorem 3.4.1.

Consider now a (cubical) k-form ω on a manifold M, and its coboundary (= exterior derivative) $d\omega$. Since $d\omega$ is a $(k+1)$-form, it defines a $(k+1)$-dimensional observable on M, $\gamma \mapsto \int_\gamma d\omega$, γ ranging over singular $(k+1)$-cubes. There is another functional on the set of singular $(k+1)$-cubes, namely $\gamma \mapsto \int_{\partial\gamma} \omega$, where $\partial\gamma$ is the k-chain in the (cubical) chain complex described in Section A.8. This is, in the notation from there, the coboundary $d \int \omega$ of the cochain $\int \omega$, and since $\int \omega$ is an observable, then so is its coboundary $d \int \omega$, by Proposition A.8.3.

Theorem 3.4.5 (Stokes' Theorem) *Let γ be a singular $(k+1)$-cube in a manifold M, and let ω be a k-form on M. Then*

$$\int_{\partial\gamma} \omega = \int_\gamma d\omega.$$

Proof. As functions of $\gamma \in S_{k+1}(M)$ (the set of singular $(k+1)$-cubes), both sides are observables (alternating and with subdivision property). On infinitesimal parallelepipeda, the two sides agree, because d for the cochain complex of differential forms was defined in terms of the ∂ in the chain complex of infinitesimal parallelepipeda. The result therefore follows from uniqueness of observables (Theorem 3.4.1).

Some aspects of the theory presented here may be summarized: the cubical complex of infinitesimal parallelepipeda in M is a subcomplex of the cubical complex of singular cubes. In the associated cochain complexes, the differential forms constitute a subcomplex of the first, and the observables constitute a subcomplex of the second cochain complex. The inclusion map of the complex of infinitesimal parallelepipeda into the complex of singular cubes induces a bijection between k-forms and k-observables. And these bijections, as k ranges, are compatible with the coboundary operators, so that the cochain complexes of cubical differential forms, and of observables, are isomorphic.

This line of reasoning and the proofs presented in this section run in parallel with those of Meloni and Rogora (1988) and Felix and Lavendhomme

(1990) (see also Lavendhomme, 1996, §4.5); however, these authors deal with a somewhat different synthetic notion of differential form; for the latter, it is the notion based on "marked microcubes", as described in Section 4.10.

Remark 3.4.6 The k-dimensional rectangles in R^k with all sides infinitesimal are not in general infinitesimal parallelepipeda; those rectangles P with infinitesimal sides that are at the same time infinitesimal parallelepipeda, are uninteresting from the viewpoint of differential forms, in the sense that Vol applied to such P gives 0. This follows from the fact that a diagonal $k \times k$ matrix in $\widetilde{D}(k,k)$ ($k \geq 2$) has determinant 0, see Exercise 1.2.20.

3.5 Wedge/cup product

Let W_1, W_2, and W_3 be vector spaces, and let

$$W_1 \times W_2 \xrightarrow{\;*\;} W_3$$

be a bilinear map. The most important special case is where $W_1 = W_2 = W_3 = R$, with $*$ being the multiplication map. Then there is a classical procedure from algebraic topology, called *cup product*: if S is a simplicial complex, and ω and θ are cochains on S (of degree k and l, respectively) with values in W_1 and W_2, respectively, one manufactures a cochain $\omega \cup_* \theta$ on S of degree $k+l$, with values in W_3. We shall describe this process for the case where S is the simplicial complex $M_{<\bullet>}$ of infinitesimal simplices in a manifold M. We shall be interested in the case where the W_i are KL vector spaces, and where ω and θ are simplicial differential forms.

In the present section, "differential form" means "simplicial differential form".

So let ω be a W_1-valued k-form on M, and let θ be a simplicial W_2-valued l-form on M. For any infinitesimal $k+l$-simplex $(x_0, x_1, \ldots, x_k, x_{k+1}, \ldots, x_{k+l})$, we define the W_3-valued $k+l$ cochain on $M_{<\bullet>}$ by the standard cup-product formula

$$\begin{aligned}(\omega \cup_* \theta)&(x_0, x_1, \ldots, x_k, x_{k+1}, \ldots, x_{k+l}) \\ &:= \omega(x_0, x_1, \ldots, x_k) * \theta(x_k, x_{k+1}, \ldots, x_{k+l}).\end{aligned} \tag{3.5.1}$$

Theorem 3.5.1 *If ω and θ are differential forms of degree k and l, as above, then the $k+l$-cochain given by the expression (3.5.1) is again a differential form.*

Proof. Note that x_k appears in both factors in (3.5.1); we first investigate what happens if the second occurrence of x_k is replaced by x_j ($j \leq k$): so consider for each $j = 0,\ldots,k$ the expression

$$\omega(x_0,\ldots,x_k) * \theta(x_j,x_{k+1},\ldots,x_{k+l}). \tag{3.5.2}$$

Lemma 3.5.2 *The expressions (3.5.1) and (3.5.2) have the same value, independently of the $j = 0,\ldots,k$ chosen. Furthermore, the expression vanishes if some x_r ($r \geq k+1$) equals some x_j ($j \leq k$).*

Proof. For the first assertion: we may assume that M is an open subset of a finite-dimensional vector space V, so that for any infinitesimal k-simplex $(y_0,y_1,\ldots,y_k)$ in M, we have $\omega(y_0,y_1,\ldots,y_k) = \Omega(y_0;y_1-y_0,\ldots,y_k-y_0)$ with $\Omega : M \times V^k \to W_1$ k-linear and alternating in the last k variables. Then

$$\begin{aligned}\omega(x_0,x_1,\ldots,x_k) &= \pm\omega(x_k,x_0,x_1,\ldots,x_{k-1}) \\ &= \pm\Omega(x_k;x_0-x_k,\ldots,x_{k-1}-x_k),\end{aligned}$$

so for $j<k$,

$$\begin{aligned}&\omega(x_0,\ldots,x_k) * \theta(x_j,x_{k+1},\ldots,x_{k+l}) \\ &\qquad = \pm\Omega(x_k;x_0-x_k,\ldots,x_{k-1}-x_k) * \theta(x_j,x_{k+1},\ldots,x_{k+l});\end{aligned}$$

because $\Omega(x_k;\ldots)$ is k-linear, and $*$ is bilinear, the occurrence of $x_j - x_k$ is linear, and by the Taylor principle, we may replace x_j in the θ-factor by x_k, proving the first assertion of the lemma. The second assertion is a formal consequence: if x_r for $r>k$ equals x_j for $j \leq k$, we use the "independence of j" already proved, so that the product equals $\omega(x_0,\ldots,x_k)\cdot\theta(x_j,x_{k+1},\ldots,x_r,\ldots)$, but now the θ factor is 0, due to the repeated occurrence of $x_j = x_r$.

It is now clear that the expression (3.5.1) does indeed satisfy the normalization condition required to deserve the name of differential form: assume $x_j = x_r$ for some $j \neq r$. It vanishes if $x_j = x_r$ for some $j,r \leq k$, because then the ω factor vanishes, and it vanishes if $x_j = x_r$ for some $r,j>k$, because then the θ factor vanishes; and it vanishes if $x_j = x_r$ with $j \leq k$ and $r>k$, by the lemma. So $\omega \cup_* \theta$ vanishes on a $k+l$-simplex if two of its vertices are equal. This proves the theorem.

When a function $f : M \to R$ is viewed as an R-valued 0-form, and ω is a W-valued k-form on M, the $0+k$-form $f \cup_. \omega$ is what we previously, and more naturally, have denoted $f\cdot\omega$.

It is clear that $\cup_*$ depends in a bilinear way on its two arguments.

Since the formula defining $\cup_*$ is a special case of the formula for cup products of simplicial cochains, we get some of the properties of $\cup_*$ by proofs that can be read out of books on algebraic topology; for instance, if $W_1 = W_2 = W_3$ and the bilinear map $* : W \times W \to W$ is associative, the cup product $\cup_*$ of simplicial differential forms is likewise associative.

But unlike the cup product in simplicial algebraic topology, the cup product here is graded-commutative: a sign change is introduced when commuting two odd-degree forms (in simplicial algebraic topology, we can assert this commutativity only "up to cohomology", see e.g. Hilton and Wylie, 1960, Theorem 4.1.8).

Proposition 3.5.3 *Let ω and θ be differential forms of degree k and l, respectively, with values in a commutative algebra A, whose underlying vector space is KL. Then*

$$\omega \cup \theta = (-1)^{k \cdot l} \theta \cup \omega.$$

Proof. Consider an infinitesimal $k+l$-simplex $(x_0, x_1, \ldots, x_{k+l})$. Then

$$(x_l, x_{l+1}, \ldots, x_{k+l}, x_0, x_1, \ldots, x_{l-1})$$

comes about by l cyclic permutations of the $k+l+1$-tuple $x_0, x_1, \ldots, x_k, \ldots, x_{k+l}$, so

$$\begin{aligned}(\omega \cup \theta)(x_0, x_1, \ldots, x_{k+l}) &= (-1)^{l \cdot (k+l)} (\omega \cup \theta)(x_l, \ldots, x_{k+l}, x_0, \ldots, x_{l-1}) \\ &= (-1)^{l \cdot (k+l)} \cdot \omega(x_l, \ldots, x_{k+l}) \cdot \theta(x_{k+l}, x_0, \ldots, x_{l-1}) \\ &= (-1)^{l \cdot (k+l)} \cdot \omega(x_l, \ldots, x_{k+l}) \cdot \theta(x_l, x_0, \ldots, x_{l-1})\end{aligned}$$

using Lemma 3.5.2; and performing one cyclic permutation in the inputs of θ, using that θ is alternating,

$$= (-1)^{l \cdot (k+l)} \cdot (-1)^l \omega(x_l, \ldots, x_{k+l}) \cdot \theta(x_0, \ldots, x_{l-1}, x_l);$$

the sign here equals $(-1)^{k \cdot l}$ (since $l \cdot (k+l) + l$ is congruent mod 2 to $k \cdot l$), and the rest of the expression is by commutativity of A equal to $\theta(x_0, \ldots, x_{l-1}, x_l) \cdot \omega(x_l, \ldots, x_{k+l}) = (\theta \cup \omega)(x_0, x_1, \ldots, x_{k+l})$, proving the proposition.

In particular, $\omega \cup \omega = 0$ if ω is a 1-form with values in a commutative algebra. On the other hand, for A a KL vector space with a non-commutative bilinear $* : A \times A \to A$, we have

Proposition 3.5.4 *Let ω be a 1-form on a manifold M, with values in A. Then for any infinitesimal 2-simplex (x, y, z) in M,*

$$\omega(x, y) * \omega(y, z) = -\omega(y, z) * \omega(x, y).$$

Hence also

$$\omega \wedge \omega = \tfrac{1}{2}[\omega, \omega],$$

where $\wedge$ refers to the cup product w.r.t. the multiplication $*$, and the square brackets refer to cup product with respect to the (bilinear) algebraic commutator map $A \times A \to A$ given by $(a,b) \mapsto a * b - b * a$.

Proof. For the first assertion, it suffices to consider the case where M is an open subset of a finite-dimensional vector space V, and $\omega(x,y) = \Omega(x; y-x)$ with $\Omega : M \times V \to A$, linear in the second variable. Then $\omega(x,y) = \Omega(x; y-x) = \Omega(y; y-x)$ by the Taylor principle, and $\omega(y,z) = \Omega(y; z-y)$. Consider the linear map $F : V \to A$ given by $\Omega(y; -)$. Then

$$V \times V \xrightarrow{F \times F} A \times A \xrightarrow{\quad * \quad} A$$

is bilinear, and thus behaves as if it were alternating on $\widetilde{D}(2,V)$. But $(y-x, z-y) \in \widetilde{D}(2,V)$ since (x,y,z) is an infinitesimal 2-simplex. For the second assertion, we have for any infinitesimal 2-simplex (x,y,z)

$$\begin{aligned}[\omega,\omega](x,y,z) &= [\omega(x,y), \omega(y,z)] \\ &= \omega(x,y) * \omega(y,z) - \omega(y,z) * \omega(x,y) \\ &= 2\omega(x,y) * \omega(y,z)\end{aligned}$$

by the first assertion of the proposition,

$$= 2(\omega \wedge \omega)(x,y,z).$$

The de Rham complex $\Omega^\bullet(M)$

We have now a purely combinatorial construction of a differential graded algebra. It deserves the name of (combinatorial) de Rham complex; for, it is essentially isomorphic to the classical de Rham complex, see Section 4.7 (except for some combinatorial factors like $k!$ or $k+1$).

We consider simplicial differential forms on M with values in R. The obvious linear structure on the set of simplicial k-forms (for each k) gives the set of simplicial forms on M the structure of a graded vector space $\Omega^\bullet(M)$. Furthermore, we have the coboundary operator d as described in (3.2.1), and also, because R has a bilinear multiplication $R \times R \to R$, we have the cup product, defined by (3.5.1) (with $*$ = the multiplication $R \times R \to R$). It deserves to be denoted $\wedge$, and we adopt this notation:

$$(\omega \wedge \theta)(x_0, \ldots, x_k, x_{k+1}, \ldots, x_{k+l}) := \omega(x_0, \ldots, x_k) \cdot \theta(x_k, x_{k+1}, \ldots, x_{k+l}); \tag{3.5.3}$$

it differs from the wedge product of the corresponding "classical" differential forms by a factor $(k+l)!/k!l!$, as will be (made meaningful and) proved in Chapter 4, in particular Theorem 4.7.3.

Together, the formulas describing these two structures are identical to the standard ones that describe, respectively, the coboundary and the cup product of cochains on a simplicial complex (here, the simplicial complex in question is $M_{<\bullet>}$). Therefore, the standard calculations from simplicial theory are valid, and give the first part in

Theorem 3.5.5 *(1) The graded vector space $\Omega^\bullet(M)$ carries the structure of a differential graded algebra; and the structure is contravariantly functorial in M. (2) The algebra structure is graded-commutative.*

For part (1), the meaning of these assertions, as well as their proofs, can be found in standard texts on algebraic topology, see e.g. Hilton and Wylie (1960, Chapter 4). Assertion (2) is Proposition 3.5.3.

The differential graded algebra $\Omega^\bullet(M)$ is essentially isomorphic to the standard de Rham complex of M, see Chapter 4. The fact that it appears as a sub-DGA of the standard singular cochain complex was proved in Kock (2000), but in a different way, using integration and Stokes' Theorem.

3.6 Involutive distributions and differential forms

An immediate consequence of Lemma 3.5.2 is that for two R-valued 1-forms ω and α on a manifold M, we have

$$(\omega \wedge \alpha)(x,y,z) = \omega(x,y) \cdot \alpha(x,z). \tag{3.6.1}$$

Note that it is the x which is repeated, rather than the y; this makes it more convenient for "localization at x", in a sense we shall explain now.

The notion of (simplicial) k-form may be localized, and we talk then about (simplicial) *k-cotangents* at x: this means a law ω which to each infinitesimal k-simplex $(x, x_1, \ldots, x_k)$ with first vertex x asssociates a number $\omega(x_1, \ldots, x_k) \in R$, subject to the requirement that the value is 0 if two of the vertices $x, x_1, \ldots, x_k$ are equal. So given a k-form ω on M, we get a k-cotangent ω_x for every $x \in M$. Given a p- and q-cotangent the same point x_0, we may form their wedge product by the formula (3.5.3) but with the second occurrence of x_k

replaced by x_0, and this will then be a $k+l$-cotangent at x_0. We are interested in the case $k = l = 1$.

In particular, if ω and α are 1-forms on M, and $x \in M$, then formula (3.6.1) tells us that

$$\omega_x \wedge \alpha_x = (\omega \wedge \alpha)_x.$$

Let $\omega_1, \ldots, \omega_q$ be a q-tuple of 1-forms on a manifold M (a "Pfaff system"). They define a pre-distribution $\approx$ by

$$x \approx y \quad \text{iff} \quad \omega_i(x,y) = 0 \text{ for } i = 1, \ldots, q.$$

We consider an arbitrary such system of differential 1-forms $\omega_1, \ldots, \omega_q$. Let I be the ideal in the de Rham complex which is *pointwise* generated by the ω_is in the following sense: a p-form θ is in I if for each $x \in M$, it is the case that θ_x can be written

$$\theta_x = \sum_{i=1}^{q} (\omega_i)_x \wedge \alpha_i \tag{3.6.2}$$

for suitable $p-1$-cotangents α_i at x. (We are not asserting or assuming that these cotangents α can be pieced together to a differential $p-1$-form.)

Proposition 3.6.1 *Assume $d\omega_i \in I$ for $i = 1, \ldots, q$. Then the pre-distribution $\approx$ defined by the ω_is is involutive.*

Proof. Given an infinitesimal 2-simplex (x,y,z) in M with $x \approx y$ and $x \approx z$, to prove $y \approx z$, we should prove that $\omega_i(y,z) = 0$ for every i. By assumption, we may for the given x and for each $i = 1, \ldots, q$ write

$$(d\omega_i)_x = \sum_{j=1}^{q} (\omega_j)_x \wedge \alpha_{ij}$$

for suitable 1-cotangents α_{ij} at x. Then on the one hand

$$d\omega_i(x,y,z) = \omega_i(x,y) - \omega_i(x,z) + \omega_i(y,z) = \omega_i(y,z),$$

the last equation since the first two terms in the middle are 0, by the assumption $x \approx y$ and $x \approx z$. On the other hand

$$d\omega_i(x,y,z) = \sum_j (\omega_j)_x(y) \cdot \alpha_{ij}(z) = 0,$$

the last equation since each $\omega_j(x,y)$ is 0 by $x \approx y$. By comparison, we conclude that $\omega_i(y,z) = 0$. Since this holds for all $i = 1, \ldots, q$, we conclude $y \approx z$.

We can now state the main theorem of this section, which contains the converse of this proposition. It shows that the simple combinatorial definition of "involutive" agrees with the standard "algebraic" one, i.e. the one in terms of the de Rham complex $\Omega^\bullet(M)$.

Theorem 3.6.2 *Let* $\omega_1,\ldots,\omega_q$ *be a system of differential 1-forms on* M*, and assume that the pre-distribution* $\approx$ *which it defines is a distribution of dimension* $\dim(M)-q$ *(the Pfaff system is of "maximal rank"). Then this distribution is involutive if and only if each* $d\omega_i$ *belongs to the ideal* I *pointwise generated by the* ω_i*s.*

Proof. The one implication is contained in Proposition 3.6.1. Conversely, assume that $\approx$ is involutive. Let $x\in M$ be given. Let θ_i be the 2-cotangent $(d\omega_i)_x$ at x. Then for any infinitesimal 2-simplex (x,y,z) with $x\approx y$ and $x\approx z$, we have, for each $i=1,\ldots,k$

$$\omega_i(y,z)=d\omega_i(x,y,z)=\theta_i(y,z),$$

the first equation sign because two of the three terms defining $(d\omega_i)(x,y)$ vanish by virtue of the two assumptions $x\approx y$ and $x\approx z$. Since $y\approx z$, by involutivity, we also have $\omega_i(y,z)=0$, so $\theta_i(y,z)=0$. Now take a coordinate chart around x, using a vector space V, and identify by KL the cotangents $(\omega_i)_x$ with linear maps $\Omega_i:V\to R$

$$(\omega_i)_x(y)=\Omega_i(y-x);$$

similarly $\Theta:V\times V\to R$,

$$\theta_i(y,z)=\Theta_i(y-x,z-x)$$

for a bilinear alternating map $\Theta_i:V\times V\to R$. The fact that $\theta_i(y,z)=0$ whenever x,y,z is an infinitesimal 2-simplex with $y\approx x$ and $z\approx x$ means that Θ_i vanishes on $(U\times U)\cap\widetilde{D}(2,V)=\widetilde{D}(2,U)$, where U denotes the meet of the null spaces of the Ω_i, and since Θ_i is bilinear alternating, it follows from Proposition 1.3.2 that Θ_i vanishes on $U\times U$. The "Nullstellensatz" (Appendix Proposition A.9.1) now implies that $\Theta_i=\sum_j\Omega_j\wedge\alpha_{ij}$ for suitable linear $\alpha_{ij}:V\to R$, which in turn gives the desired pointwise expression for $(d\omega_i)_x$.

3.7 Non-abelian theory of 1-forms

Group-valued forms

The notion of W-valued 1-form, and the notions of closed and exact 1-forms, as given in Definition 2.2.3 only make use of the additive group of W, not its vector space structure. However, for the crucial law (2.2.4), i.e. $\omega(x,y)=-\omega(y,x)$, the fact that W was assumed to be a KL vector space was used. We can generalize these notions in the sense that we may replace the commutative group $(W,+)$ by a not necessarily commutative group $(G,\cdot)$; we then let the multiplicative analogue of (2.2.4) be part of the definition (we return to cases where it can be *deduced* below, see (6.1.3)):

$$\omega(y,x) = \omega(x,y)^{-1}.$$

This idea of group-valued 1-forms seems also to be due to Bkouche and Joyal.

Thus, for a manifold M and a group $(G,\cdot)$ with unit e, we put

Definition 3.7.1 *A function $\omega : M_{(1)} \to G$ is called a (G-valued)* 1-form *on M if $\omega(x,x) = e$ for all $x \in M$ and if $\omega(x,y) = \omega(y,x)^{-1}$ for all $x \sim y$. A function $\theta : M_{<2>} \to G$ is called a (G-valued)* 2-form *if $\theta(x_0,x_1,x_2) = e$ if any two of the x_is are equal.*

The generalization to k-forms is evident for $k \geq 2$; and for $k = 0$: a 0-form is just a function $M \to G$. We consider G-valued forms in more detail in Section 6.1.

Just as for vector-space-valued simplicial forms, group-valued simplicial forms "pull back" along any map between manifolds: if $f : N \to M$ is such a map, and ω a G-valued k-form on M, we get a G-valued k-form $f^*\omega$ on N; the recipe is as before.

If ω is a G-valued 1-form, we get a G-valued 2-form $d.\omega$ by putting

$$d.\omega(x,y,z) := \omega(x,y)\cdot\omega(y,z)\cdot\omega(z,x), \tag{3.7.1}$$

for $(x,y,z) \in M_{<2>}$. The subscript "$\cdot$" is to remind us that it is the multiplication $\cdot$ of G that enters into the definition of this "coboundary-operator".

The concepts of *closed* and *exact* 1-forms, and the notion of *primitive* of a 1-form ramify in a *left* and a *right* version; formally, one gets one version from the other by replacing the group G with its opposite group G^{op}. The right version is visually the simplest, and we present it as our primary version. The subscripts "r" and "l" indicate "right" or "left", respectively. The above definition of $d.\omega$ should have been decorated with an "r" as well. Thus, (3.7.1) defines what should more completely be denoted $d_{\cdot,r}\omega(x,y,z)$.

Until further notice, everything is in the "right" version, and we omit the subscript r and the prefix "right".

A G-valued 1-form ω is called a *(right) closed* 1-form if $d.\omega$ is the "zero" 2-form, i.e. has constant value $e \in G$. Equivalently, if for any infinitesimal

2-simplex (x,y,z) in M, we have

$$\omega(x,y) \cdot \omega(y,z) = \omega(x,z). \qquad (r\text{-closed})$$

If $g : M \to G$ is a function, we get a G-valued 1-form $d.g$ on M by putting

$$d.g(x,y) := g(x)^{-1} \cdot g(y). \qquad (3.7.2)$$

Clearly, $d.g$ is closed; equivalently, for any function $g : M \to G$

$$d.(d.(g)) \equiv e \qquad (3.7.3)$$

(the "zero" form). Thus, we have $d. \circ d. =$"0".

A 1-form $\omega : M_{(1)} \to G$ is called *(right) exact* if $\omega = d_r g$ for some $g : M \to W$; and then g is called a *(right) primitive* of ω. So $g : M \to G$ is a primitive of ω if

$$\omega(x,y) = g(x)^{-1} \cdot g(y). \qquad (r\text{-primitive})$$

Let us for completeness write down the "left" versions of these notions: given a G-valued 1-form θ; we define the G-valued 2-form $d_{.,l}\theta$ by the formula

$$d_{.,l}\theta(x,y,z) := \theta(z,x) \cdot \theta(y,z) \cdot \theta(x,y).$$

Then θ is called *left closed* if $d_{.,l}\theta \equiv e$, equivalently, if $\theta(y,z) \cdot \theta(x,y) = \theta(x,z)$. If $h : M \to G$ is a function, we get a G-valued 1-form $d_{.,l}h$ on M by putting

$$d_{.,l}h(x,y) = h(y) \cdot h(x)^{-1}.$$

Clearly, $d_{.,l}h$ is left closed. A 1-form $\theta : M_{(1)} \to G$ is called *left exact* if $\theta = d_l h$ for some $h : M \to G$, and then h is a *left primitive* of θ. So h is a left primitive of θ if $\theta(x,y) = h(y) \cdot h(x)^{-1}$.

It is clear that if ω is right closed, then ω^{-1} is left closed, where $\omega^{-1} = \theta$ is the G-valued 1-form given by $\theta(x,y) := \omega(y,x)$ $(= \omega(x,y)^{-1})$. Also, if g is a right primitive of ω, then the function h given by $h(x) := g(x)^{-1}$ is a left primitive of $\theta = \omega^{-1}$.

Note that we have not here attempted to define when a G-valued 2-form θ should be called (right or left) closed. But see Section 6.2.

Example 3.7.2 If G is a group which is at the same time a manifold (so G is a Lie group), there is a canonical G-valued 0-form ω on G, namely the identity map $\mathrm{id} : G \to G$. The 1-form $\omega := d.(\mathrm{id})$, i.e.

$$\omega(x,y) = x^{-1} \cdot y;$$

it is closed, by "$d. \circ d. = 0$". This 1-form deserves the name (right) *Maurer–Cartan* form on G. The (trivial) fact that it is closed will appear to be the Maurer–Cartan equation, when translated into the language of Lie-algebra-valued forms, see Corollary 6.7.2.

For $(G,\cdot) = (R,+)$, the Maurer–Cartan form is dx,

$$dx(u,v) = -u + v = v - u.$$

Many questions in differential geometry can be reduced to the question: when are closed G-valued 1-forms on M exact? This will depend both on M and on G. For instance, it is the case (in suitable models for the axiomatic treatment) that if G is a (finite-dimensional) Lie group and M is simply connected, then closed G-valued 1-forms on M are exact.

As an example of such a reduction, we shall prove the following. We consider a manifold M modelled on a finite-dimensional vector space V, and we assume that closed $GL(V)$-valued 1-forms on M locally† are exact. (Here, $GL(V)$ is the group of linear automorphisms $V \to V$; it may by KL be identified with the group of invertible 0-preserving maps $D(V) \to D(V)$, and we shall make this identification.) If now $\kappa : D(V) \to \mathfrak{M}(x)$ is a frame at $x \in M$, and $g \in GL(V)$, we immediately get a new frame $\kappa \circ g : D(V) \to \mathfrak{M}(x)$ at x. The group structure in $GL(V)$ is $\circ$, composition (from right to left) of maps. Recall from Section 2.4 that a framing k on a manifold gives rise to an affine connection $\lambda = \lambda_k$. The transport law $\nabla(x,y) : \mathfrak{M}(y) \to \mathfrak{M}(x)$ for λ_k is $k_x \circ k_y^{-1}$.

Proposition 3.7.3 *A necessary and sufficient condition that an affine connection λ locally comes about from a framing k is that λ is flat.*

Proof. We already know from Proposition 2.4.1 that an affine connection which comes from a framing is flat. Assume conversely that λ is a flat connection. The flatness condition for λ is, in terms of the corresponding transport law ∇, that for any infinitesimal 2-simplex x,y,z in M

$$\nabla(x,z) = \nabla(x,y) \circ \nabla(y,z) \tag{3.7.4}$$

as maps $\mathfrak{M}(z) \to \mathfrak{M}(x)$. Now pick locally an auxiliary framing h on M, so $h_x : D(V) \to \mathfrak{M}(x)$ is a bijection taking 0 to x (for instance, take a coordinate chart from an open subset of M to V, and import the canonical framing). We can then locally construct a $GL(V)$-valued 1-form θ on M by putting

$$\theta(x,y) := h_x^{-1} \circ \nabla(x,y) \circ h_y : D(V) \to D(V). \tag{3.7.5}$$

† Here, 'locally' refers to the presupposed notion of "open inclusion".

Then an immediate calculation gives, using (3.7.4), that $\theta(x,y) \circ \theta(y,z) = \theta(x,z)$, which is to say that θ is a right closed ($GL(V)$-valued) 1-form. Hence by assumption, it has locally a right primitive $g : M \to GL(V)$, $\theta(x,y) = g(x)^{-1} \circ g(y)$. We define $k_x : D(V) \to \mathfrak{M}(x)$ by $k_x := h_x \circ g(x)^{-1}$. Then

$$\begin{aligned} k_x \circ k_y^{-1} &= h_x \circ g(x)^{-1} \circ g(y) \circ h_y^{-1} \\ &= h_x \circ \theta(x,y) \circ h_y^{-1} = \nabla(x,y), \end{aligned}$$

the last equality sign by rearranging (3.7.5).

This proposition is a special case of a similar result about cross-sections in principal bundles $P \to M$, and connections in the associated groupoids $PP^{-1} \rightrightarrows M$; see Section 5.6. The set of frames on M, as considered above, is in fact a principal $GL(V)$-bundle.

Combining Theorem 2.4.2 with Propositions 2.4.6 and 3.7.3, we conclude†

Theorem 3.7.4 . *Let M be a manifold modelled on a finite-dimensional vector space V. Assume that closed $GL(V)$-valued 1-forms and closed V-valued 1-forms on M locally are exact. Then any torsion-free and flat connection on M is locally integrable.*

(Conversely, a locally integrable connection is clearly torsion free and flat.)

Differential 1-forms with values in automorphism groups

Consider a constant bundle $M \times F \to M$, where M is a manifold. We consider the group $G = \mathrm{Diff}(F)$ of diffeomorphisms $F \to F$; for convenience, we let the group G act on F *from the right*, and denote the action by $\vdash$.

Proposition 3.7.5 *There are bijective correspondences between the following three kinds of data:*

(1) 1-forms on M with values in G;

(2) bundle connections in the bundle $M \times F \to M$;

(3) distributions in $M \times F$, transverse to the fibres.

The 1-form is closed iff the connection is flat. In this case, the distribution is involutive.

(Conversely, if the distribution is involutive, the connection is flat, under an extra mild hypothesis.)

† cf. Noll (1967).

Proof. The equivalence of (2) and (3) was proved in Section 2.6, even for non-constant bundles. We shall prove, for constant bundles $M \times F \to M$, the equivalence of (1) and (2); this is by pure logic: given a Diff(F)-valued 1-form ω on M, we get a bundle connection ∇ by putting (for $x \sim y$ in M and $a \in F$),

$$\nabla(y,x)(x,a) := (y, a \vdash \omega(x,y)).$$

Conversely, given a bundle connection ∇, then this formula describes ω completely in terms of ∇. Furthermore, for an infinitesimal 2-simplex x,y,z, the equation

$$\nabla(z,y)(\nabla(y,x)(x,a)) = \nabla(z,x)(a)$$

translates by the bijection immediately into

$$a \vdash \omega(x,y) \vdash \omega(y,z) = a \vdash \omega(x,z),$$

so the first equation holds for all such x,y,z,a iff the second one does so, proving the equivalence of "∇ flat" and "ω closed". The relationship between flatness of ∇ and involutiveness of the distribution was dealt with in Proposition 2.6.15; the "mild hypothesis" is recorded in (2.5.4).

There is no reason to expect that closed Diff(F)-valued forms are locally exact; the distribution transverse to the (vertical) fibres in $R \times R \to R$, given by the differential equation $y' = y^2$, should furnish a counterexample, but I haven't been able to prove this without further assumptions on R.

3.8 Differential forms with values in a vector bundle

Let $E \to M$ be a vector bundle; the fibres are assumed to be KL vector spaces. We have a notion of a simplicial differential k-form on M with values in $E \to M$; this means a law ω which to each infinitesimal k-simplex $(x_0, x_1, \ldots, x_k)$ in M associates an element $\omega(x_0, x_1, \ldots, x_k) \in E_{x_0}$, subject to the requirement that the value is 0 if $x_i = x_0$ for some $i = 1, \ldots, k$. If $E \to M$ is a product bundle $M \times W \to M$ with W a KL vector space, ω is of the form

$$\omega(x_0, x_1, \ldots, x_k) = (x_0, \overline{\omega}(x_0, \ldots, x_k))$$

for some $\overline{\omega} : M_{(k)} \to W$, where $\omega(x_0, x_1, \ldots, x_k) = 0$ if $x_i = x_0$ for some $i = 1, \ldots, k$. From Proposition 3.1.6 follows that this $\overline{\omega}$ is then a simplicial W-valued k-form on M, so that the notion of a bundle-valued simplicial k-form generalizes the notion of a (vector-space-valued) simplicial k-form.

We denote the space of differential k-forms on M with values in $E \to M$ by the symbol $\Omega^k(E \to M)$.

An example of a bundle-valued 1-form is the *solder form*: for any manifold M, it is a 1-form with values in the tangent bundle $T(M) \to M$, to be considered in Section 4.8.1.

Covariant derivative

Consider a k-form on M with values in a vector bundle $E \to M$, as above. One cannot immediately use the simplicial formula (3.2.1) for the exterior derivative $d\omega$; for, the first term in the standard sum (3.2.1) lives in E_{x_1}, whereas the remaining terms live in E_{x_0}. However, if there is given a linear bundle connection in E, $\nabla(y,x) : E_x \to E_y$ for $x \sim y$ in M, we can apply $\nabla(x_0,x_1)$ to the first term, and so we can consider the following sum in E_{x_0}:

$$(d^{\nabla}\omega)(x_0,x_1,\dots,x_{k+1}) \\ := \nabla(x_0,x_1)(\omega(x_1,\dots,x_{k+1})) + \sum_{i=1}^{k+1}(-1)^i\omega(x_0,x_1,\dots,\widehat{x_i},\dots,x_{k+1}); \tag{3.8.1}$$

this specializes to (3.2.1) for the case of a constant bundle $M \times W$, and with all $\nabla(y,x)$ given by the identity map of W; this is the *trivial connection* in the constant bundle $M \times W$.

We shall prove that $d^{\nabla}\omega$ is a bundle-valued $k+1$-form; so we should prove that we get value 0 on any $k+1$-simplex $(x_0,x_1,\dots,x_{k+1})$ where $x_i = x_0$ for some $i = 1,\dots,k+1$. We need the assumption that $E \to M$ is locally of the form $M \times W$. And without loss of generality, we may assume that M is an open subset of a finite-dimensional vector space V and that $\nabla(y,x)$ (for $x \sim y$) takes (x,w) to $(y, w + L(y-x,w))$ with $L : V \times W \to W$ bilinear. Then ω is given by some W-valued k-form $\overline{\omega}$ on M, as above. Consider e.g. the case where $x_{k+1} = x_0$. Then all terms in the sum (3.8.1), except possibly the first and the last, vanish; the two remaining terms are

$$\nabla(x_0,x_1)(\omega(x_1,\dots,x_k,x_{k+1})) \pm \omega(x_0,x_1,\dots,x_k).$$

Using that ω is alternating, and keeping track of signs, it then suffices to prove that

$$\nabla(x_0,x_1)(\omega(x_1,x_{k+1},\dots,x_k)) = -\omega(x_0,x_1,\dots,x_k).$$

Translated in terms of L and $\overline{\omega}$, this reads

$$\overline{\omega}(x_1,x_{k+1},\dots,x_k) + L(x_1 - x_0, \overline{\omega}(x_1,x_{k+1},\dots,x_k)) \\ = -\overline{\omega}(x_0,x_1,\dots,x_k).$$

But the L-term vanishes because it depends in a bilinear way on $x_1 - x_0 = x_1 - x_{k+1}$, and the result now follows because $\overline{\omega}$ is alternating (interchange the first two entries, and use $x_0 = x_{k+1}$).

We therefore have a (clearly linear) map $d^\nabla : \Omega^k(E \to M) \to \Omega^{k+1}(E \to M)$, called the *covariant derivative w.r.t.* ∇. We will not in general have $d^\nabla \circ d^\nabla = 0$; the curvature of ∇ enters, see Proposition 6.3.2 for a discussion in combinatorial terms.

There is a wedge product construction for bundle-valued forms, which we shall encounter in Section 6.3.

There is a similar theory of bundle-valued whisker and cubical forms which will not be developed here.

3.9 Crossed modules and non-abelian 2-forms

Recall that a *crossed module* $\mathcal{G}$ consists of two groups H, G, together with a group homomorphism $\partial : H \to G$, and an action (right action $\vdash$, say) of G on H by group homomorphisms, s.t.

(1) $\partial : H \to G$ is G-equivariant (takes the G-action $\vdash$ on H to the conjugation G-action on G),

$$\partial(h \vdash g) = g^{-1}.\partial(h).g$$

for all $h \in H$ and $g \in G$;

(2) the Peiffer identity

$$h^{-1}.k.h = k \vdash \partial(h)$$

holds for all h and k in H.

A homomorphism of crossed modules is a pair of group homomorphisms, compatible with the ∂s and the actions.

The notion of crossed module may seem somewhat ad hoc, but the category of crossed modules is equivalent to some other categories, whose descriptions are conceptually simpler: the category of group objects in the category of groupoids; the category of groupoid objects in the category of groups; the category of 2-groupoids with only one object (a 2-groupoid is a 2-category where all arrows and also all 2-cells are invertible); or the category of "edge-symmetric double groupoids with connections" (Brown and Higgins, 1981; Brown and Spencer, 1976). The latter description is particularly well suited for being lifted to higher dimensions, and for the theory of cubical differential forms, and higher connections, cf. Kock (2007, 2008); however, for the purpose of describing a theory of non-abelian 2-forms, the crossed module description is sufficient, and the one most readily adapted for concrete calculations. So we shall adopt this version (following in this respect Baez and Schreiber, 2006 and Schreiber and Waldorf, 2008); we shall consider differential forms in their simplicial manifestation.

Any group G gives rise canonically to two crossed modules, $INN(G)$ and $AUT(G)$; for $INN(G)$, $H = G$ and ∂ is the identity map, $\vdash$ is conjugation. And $AUT(G) = (G \to Aut(G))$ where $Aut(G)$ is the group of automorphisms of G and $\partial(g)$ is "conjugation by g".

Let $\mathcal{G} = (\partial : H \to G, \vdash)$ be a crossed module, and let M be a manifold. A *1-form on M with values in $\mathcal{G}$* is defined to be a 1-form on M with values in the group H, as in Section 3.7. If θ is such a form, $\theta : M_{(1)} \to H$, one gets also a G-valued 1-form ω, namely $\omega = \partial \circ \theta$. So one may, redundantly, rephrase the definition: a $\mathcal{G}$-valued 1-form is a *pair* of 1-forms (θ, ω), with values in the groups H and G, respectively, and with $\partial\theta = \omega$.

A 1-form with values in a group G may be identified with a 1-form with values in the crossed module $INN(G)$.

A *2-form on M with values in $\mathcal{G}$* is a pair (R, ω), where for any infinitesimal 2-simplex (x,y,z) in M, $R(x,y,z) \in H$, and for any infinitesimal 1-simplex (x,y), $\omega(x,y) \in G$, both subject to the normalization requirement that the value is the neutral element of H (resp. of G) if the simplex is degenerate, and such that

$$\partial(R(x,y,z)) = \omega(x,y).\omega(y,z).\omega(z,x). \tag{3.9.1}$$

Recall that under mild assumptions, $\omega(z,x) = \omega(x,z)^{-1}$. We shall assume this, and similarly for θ.

If $\theta = (\theta, \omega)$ is a $\mathcal{G}$-valued 1-form (so $\omega = \partial \circ \theta$), one gets a $\mathcal{G}$-valued 2-form $d\theta = (R, \omega)$ (same ω!) by defining R by the recipe

$$R(x,y,z) := \theta(x,y).\theta(y,z).\theta(z,x). \tag{3.9.2}$$

A *closed* $\mathcal{G}$-valued 2-form (R, ω) is one which satisfies the "Bianchi Identity" (see Section 5.2). For any infinitesimal 3-simplex (x,y,z,u), we have

$$(R(y,z,u) \vdash \omega(y,x)).R(xyu).R(xuz).R(xzy) = 1 \tag{3.9.3}$$

(1 denoting the unit element of the group H).

Proposition 3.9.1 *The coboundary of a $\mathcal{G}$-valued 1-form is a closed $\mathcal{G}$-valued 2-form.*

Proof. Let (θ, ω) be the given $\mathcal{G}$-valued 1-form, so θ is a H-valued 1-form, and $\omega = \partial \circ \theta$. Constructing R by the recipe (3.9.2), we are required to prove that for any infinitesimal 3-simplex (x, y, z, u), the following expression takes value 1:

$$\left[\left(\theta(y,z).\theta(z,u).\theta(u,y)\right) \vdash \omega(y,x)\right]$$

$$.\left[\left(\theta(x,y).\theta(y,u).\theta(u,x)\right).\left(\theta(x,u).\theta(u,z).\theta(z,x)\right).\left(\theta(x,z).\theta(z,y).\theta(y,x)\right)\right].$$

Since $\omega(y,x) = \partial(\theta(y,x))$, we may rewrite the first square bracket, using the Peiffer identity, and then we get

$$\left[\theta(x,y).\theta(y,z).\theta(z,u).\theta(u,y).\theta(y,x)\right].[\ldots],$$

where the last square bracket is unchanged (contains the same nine factors). Altogether, we have an expression which is a product of 14 factors in the group H, involving six elements and their inverses, $\theta(x,y), \ldots, \theta(z,u), \ldots$, and this product yields 1 by Ph. Hall's 14-letter identity,

$$(a^{-1}.b.c.d.a).(a^{-1}.d^{-1}.e).(e^{-1}.c^{-1}.f).(f^{-1}.b^{-1}.a) = 1 \qquad (3.9.4)$$

which holds for any six elements a, b, c, d, e, f in any group (and the proof of this identity is trivial). This proves the proposition.

4
The tangent bundle

The tangent bundle $T(M) \to M$ of a manifold M is traditionally the main vehicle for encoding the geometry of infinitesimals; a substantial part of existing literature on SDG deals with aspects of this, see e.g. Kock (1981/2006) and the references therein, notably the references for the second edition. The main tool for comparing the tangent bundle approach to the approach based on the (first-order) neighbour relation is what we call the log-exp bijection, which we introduce in Section 4.3 below.

4.1 Tangent vectors and vector fields

It is a classical conception in algebraic geometry (schemes) that the notion of tangent vectors may be represented by a scheme D, namely $D =$ the spectrum of the ring $k[\varepsilon] = k[Z]/(Z^2)$ of dual numbers, cf. e.g. Mumford (1965/1988, p. 338/238), who calls this D (in his notation I) "*a sort of disembodied tangent vector*", so that "*the set of all morphisms from D to M*"† is a "*sort of set-theoretic tangent bundle to M*".

In a seminal lecture in 1967, Lawvere proposed to axiomatize the object D, together with the category $\mathscr{E}$ of spaces in which it lives, and to exploit the assumed cartesian closedness of $\mathscr{E}$ (existence of function space objects) to comprehend the tangent vectors of a space M into a space M^D, which thus is not just a set, but a *space* (an object of $\mathscr{E}$). This was the seed that was to grow into SDG.

In the present text, D is, as in Section 1.2, taken to be the subspace of the ring R consisting of elements of square 0.‡

† I changed here Mumford's "I" into "D", and also his "X into "M".
‡ According to Lawvere, one should attempt to construct R out of D, rather than vice versa.

Definition 4.1.1 *A* tangent vector *at $x \in M$ is a map $\tau : D \to M$ with $\tau(0) = x$. The tangent bundle $T(M)$ is the space M^D.*

The space $T(M)$ of tangent vectors of M is thus the space M^D of maps from D to M; this space is a bundle $\pi : T(M) \to M$ over M; π associates to a tangent vector $\tau : D \to M$ its "base point" $\tau(0)$. This bundle is functorial in M; more precisely, to a map $f : M' \to M$, one gets a map $T(f) : T(M') \to T(M)$ making the square

$$\begin{array}{ccc} T(M') & \xrightarrow{T(f)} & T(M) \\ \Big\downarrow \pi & & \Big\downarrow \pi \\ M' & \xrightarrow[f]{} & M \end{array}$$

commutative. The map $T(f)$ takes $\tau : D \to M'$ to $f \circ \tau : D \to M$. It is sometimes denoted df. The functorality is just the standard (covariant) functorality of function space objects X^Y in the variable X. (Here, with $Y = D$.) If f is an open inclusion, or more generally, étale, the square is a pull back.

For $x \in M$, $T_x(M)$ denotes the fibre of $T(M) \to M$ over x; it is the space of those $\tau : D \to M$ with $\tau(0) = x$. Given $f : M \to N$, then the map $T(f)$: $T(M) \to T(N)$ restricts to a map $T_x(M) \to T_{f(x)}(N)$, denoted $T_x(f)$:

$$T_x(f)\tau = f \circ \tau : D \to N. \tag{4.1.1}$$

Since for any $d \in D$ and $t \in R$, we have $t \cdot d \in D$, we get an action of the multiplicative monoid of R on $T_x(M)$: if $\tau \in T_x(M)$, $t \cdot \tau \in T_x(M)$ is defined by

$$(t \cdot \tau)(d) := \tau(t \cdot d).$$

It is clear that if $f : M \to N$, then $T(f) : T(M) \to T(N)$ preserves this action; this follows by the associative law for composition of functions; consider

$$D \xrightarrow{\cdot t} D \xrightarrow{\tau} \mathfrak{M}(x) \xrightarrow{f} N.$$

Let now M be a manifold, so that $\mathfrak{M}(x) \subseteq M$ makes sense for $x \in M$. Then we don't need to have f defined on the whole of M in order to define $T_x(f)$: $T_x(M) \to T_{f(x)}(N)$, it suffices that f is defined on $\mathfrak{M}(x)$. In other words, it suffices that f is a 1-jet at x; for, if τ is a tangent at $x \in M$, then $\tau(d) \in \mathfrak{M}(x)$ for all $d \in D$, so $f(\tau(d))$ is defined, and (4.1.1) makes sense. We have in (2.1.8) described an action of the multiplicative monoid of R on $\mathfrak{M}(x)$; so for

$t \in R$, we have a map $t \cdot_x - : \mathfrak{M}(x) \to \mathfrak{M}(x) \subseteq M$. It follows from Proposition 2.7.5 that

$$T(t \cdot_x)(\tau) = t \cdot \tau.$$

4.2 Addition of tangent vectors

As we saw in the previous section, the fibres of $T(M) \to M$ have a natural algebraic structure: action by the multiplicative monoid of R. But for many spaces M (the "micro-linear" ones), the bundle $T(M) \to M$ has a richer natural algebraic structure: the fibres are vector spaces. We shall discuss the geometry (or better, the kinematics) of this addition structure on $T_x(M)$, provided M is a manifold.

This structure is best motivated in kinematic terms. Consider two rectilinear motions $\tau_1(t)$ and $\tau_2(t)$ in a vector space V, with the same initial value, i.e. with $\tau_1(0) = \tau_2(0)$; assume for simplicity that this initial value is $0 \in V$. The variable t is to be thought of as "time"; then by standard kinematics, the two motions can be superimposed into a third rectilinear motion τ, with

$$\tau(t) := \tau_1(t) + \tau_2(t),$$

likewise with initial value 0.

This conception generalizes in two ways. First, it is enough that the two motions take place in an *affine* space E (say physical space), provided the motions have the same initial value, $x \in E$, in which case the formula for the superimposed motion τ is given by

$$\tau(t) := \tau_1(t) - x + \tau_2(t). \tag{4.2.1}$$

Note that this is an affine combination, and it expresses the well-known parallelogram picture for superimposed motions (say a fish travelling according to τ_1 in a current floating according to τ_2). For the second generalization, one does not need to have the τ_is defined for all $t \in R$; it suffices that they are defined on some $U \subseteq R$ containing 0. Then $\tau_1 + \tau_2$ will be defined on the same U. In particular, U may be taken to be D, in which case rectilinearity is automatic: any $\tau : D \to E$ extends uniquely to a rectilinear (= affine) map $R \to E$, provided E satisfies the KL axiom (the KL axiom for an affine space means just this: the KL axiom for the associated vector space of translations; see Exercise 4.2.3 for a reformulation).

Since we here think of R as parametrizing *time*, and D is contained in R, a map $D \to M$ may be thought of as an (instantaneous) *motion* taking place in M, and sometimes we stress this aspect by talking about such maps $D \to M$ (i.e. tangent vectors) as *kinematic* entities.

Tangent vectors of an affine space E may be added, by the simple parallelogram law (4.2.1). More generally, one may form arbitrary *linear* combinations of tangent vectors τ_i at the same point x of an affine space:

$$(\sum_i t_i \cdot \tau_i)(d) := s \cdot x + \sum_i t_i \cdot \tau_i(d), \tag{4.2.2}$$

where $s \in R$ is chosen so as to make the right-hand side into an affine combination, i.e. by choosing $s = 1 - \sum_i t_i$.

Now let M be a manifold. If $\tau_i : D \to M$ $(i = 1, 2)$ have the same base point x, i.e. $\tau_1(0) = \tau_1(0) = x$, then the affine combination in (4.2.1) makes sense, since $x, \tau_1(d)$ and $\tau_2(d)$ form an infinitesimal 2-simplex, by (2.1.2). More generally, (4.2.2) makes sense.

Thus, for a manifold, we have the simple way of adding tangent vectors with common base point, more generally, of forming linear combinations of tangent vectors $\tau_1, \ldots, \tau_k$ with common base point, namely use the possibility of forming affine combinations of mutual neighbour points.

Let us, for future reference, record the formula for addition of tangent vectors: if τ_1 and τ_2 are tangent vectors with common base point x, then

$$(\tau_1 + \tau_2)(d) = \tau_1(d) + \tau_2(d) - x. \tag{4.2.3}$$

Similarly,

$$(\tau_1 - \tau_2)(d) = \tau_1(d) - \tau_2(d) + x. \tag{4.2.4}$$

There is another recipe, classical in SDG, for adding tangent vectors, which is more general, since it applies not just to manifolds, but to any microlinear space M (see Section A.4 in the Appendix); we recall this recipe (from e.g. Kock, 1981/2006, §I.7), in order to compare it with the one just given for manifolds (manifolds are always microlinear). Namely, given $\tau_i : D \to M$ $(i = 1, 2)$ with $\tau_1(0) = \tau_2(0)$ $(= x \in M$, say), then microlinearity of M implies that there exists a unique map $\tau_+ : D(2) \to D$ with $\tau_+(d, 0) = \tau_1(d)$ and $\tau_+(0, d) = \tau_2(d)$ for all $d \in D$; and then

$$\tau_1 + \tau_2 := \tau_+ \circ \Delta,$$

where $\Delta(d) := (d, d)$; equivalently $(\tau_1 + \tau_2)(d) = \tau_+(d, d)$, for all $d \in D$. In the case where M is a manifold, we can exhibit τ_+ explicitly, using affine combinations in M:

$$\tau_+(d_1, d_2) = \tau_1(d_1) - x + \tau_2(d_2). \tag{4.2.5}$$

The right-hand side is an affine combination of mutual neighbour points by Exercise 4.2.2 below. Putting $d_1 = d_2$, one recovers the definition (4.2.3).

For M a microlinear space, in particular for a manifold, the bundle $T(M) \to M$ is a *vector bundle* (the "tangent bundle"): the fibres $T_x(M)$ canonically have structure of "vector spaces" (meaning R-modules): the multiplication-by-scalars, and the addition, were described above. If M is an n-dimensional manifold, these vector spaces are isomorphic to R^n (although not canonically). This opens up the possibility of using notions from linear algebra, and this is one of the merits of the tangent-bundle (= kinematic) approach to infinitesimal geometry.

Exercise 4.2.1 Let $\tau : D \to M$ be a tangent vector, and let $t \in R$. Then $\tau(t \cdot d)$ equals the affine combination $(1-t)\cdot\tau(0)+t\cdot\tau(d)$. (*Hint*: It suffices to consider the case where M is an open subset of a vector space, so that one can work with principal parts of tangent vectors.)

Exercise 4.2.2 (Generalizing (2.1.2).) Let τ_1 and τ_2 be tangent vectors at M with common base point, $\tau_1(0) = \tau_2(0)$ ($= x$, say). If $(d_1, d_2) \in D(2)$, then $\tau_1(d_1) \sim \tau_2(d_2)$. (Also, $\tau_i(d_i) \sim x$.) (*Hint:* Same as in Exercise 4.2.1.)

Exercise 4.2.3 Prove that for an affine space A, with translation vector space V, the following conditions are equivalent: (1) V is KL; (2) any $D \to A$ extends uniquely to an affine map $R \to A$.

4.3 The log-exp bijection

Consider a manifold M. Since $T_x(M)$ is a finite-dimensional vector space, we may consider its first-order infinitesimal part $D(T_x(M))$, the set of $\tau \in T_x(M)$ with $\tau \sim 0$. The log-exp bijection which we shall describe, provides for each $x \in M$ a pair of maps

$$D(T_x(M)) \underset{\log_x}{\overset{\exp_x}{\rightleftarrows}} \mathfrak{M}(x)$$

which we shall prove are mutually inverse, in such a way that the zero vector in $T_x(M)$ corresponds to $x \in M$.

The map $\log_x$ is described in terms of affine combinations. If $y \in \mathfrak{M}(x)$, the affine combination $(1-t)x+ty$ makes sense for any $t \in R$, in particular for any $d \in D$, and we put

$$\log_x(y)(d) := (1-d)\cdot x + d \cdot y. \tag{4.3.1}$$

(Note that $\log_x(y)$ and $\log_y(x)$ cannot immediately be compared, since they live in different fibres of $T(M) \to M$.) For $d = 0$, we get x as value, so (4.3.1), as a function of $d \in D$ does provide a tangent vector $D \to M$ at x. Also if $y = x$, we get the constant function $D \to M$ with value x, and this function is the zero vector of $T_x(M)$. So $\log_x(x) = 0 \in T_x(M)$.

In other words, $\log_x$ is a 1-jet from $x \in M$ to $0 \in T_x(M)$. It is thus a map $\mathfrak{M}(x) \to \mathfrak{M}(0) = D(T_x(M))$; and it follows from Proposition 2.7.5 that it preserves the action by $(R, \cdot)$: if $y \sim x$ and $t \in R$,

$$\log_x(t \cdot_x y) = t \cdot \log_x(y); \tag{4.3.2}$$

here, $t \cdot_x y$ denotes the action of $(R, \cdot)$ on $\mathfrak{M}(x)$, i.e. $t \cdot_x y$ is the affine combination $(1-t) \cdot x + t \cdot y$ of the neighbour points x and y, cf. (2.1.8).

The definition of $\log_x$ can also be phrased:

$$(\log_x(y))(d) = d \cdot_x y$$

for $y \sim x$ and $d \in D$. In particular, for instance,

$$\log_x(2y - x) = 2 \cdot \log_x(y). \tag{4.3.3}$$

We note the following naturality property of log: if $f : M \to N$ is a map between manifolds, then

$$\log_{f(x)}(f(y)) = f \circ \log_x(y) \quad (= T(f)(\log_x(y)));$$

this follows because any map between manifolds preserves affine combinations of mutual neighbours (Theorem 2.1.4); in particular, it preserves the ones which define log.

We may construe the log-exp bijection in more global terms, as a map of bundles over the given manifold M. The subsets $D(T_x(M)) \subseteq T_x(M)$ together define a subbundle of the tangent bundle $T(M) \to M$, and we shall denote this subbundle $D(T(M)) \to M$. We also have the bundle $M_{(1)} \to M$ (the first neighbourhood of the diagonal), $(x, y) \mapsto x$ for $x \sim y$. The family of mutually inverse maps $\exp_x$ and $\log_x$ now comes about from an isomorphism of these bundles over M,

$$D(T(M)) \underset{\log}{\overset{\exp}{\rightleftarrows}} M_{(1)}.$$

The maps exhibited here are in fact natural in M; this follows from the naturality property of log.

The map exp in this sense was described already in Dubuc and Kock (1984); the corresponding log was described in Kock (2003).

Since M and N are manifolds, any 1-jet from a point x in M to N extends to a map $U \to N$, for some open subset $U \subseteq M$ containing x. So the naturality applies also to 1-jets: for a 1-jet $f : \mathfrak{M}(x) \to N$, we have commutativity of

$$\begin{array}{ccc} \mathfrak{M}(x) & \xrightarrow{f} & \mathfrak{M}(f(x)) \\ \Big\downarrow{\scriptstyle \log_x} & & \Big\downarrow{\scriptstyle \log_{f(x)}} \\ T_x(M) & \xrightarrow[T_x(f)]{} & T_{f(x)}(N) \end{array} \tag{4.3.4}$$

Let V be a finite-dimensional vector space.

Proposition 4.3.1 *If M is an open subset of V, the principal part of the tangent vector* $\log_x(y) \in T_x(M)$ *is* $y - x$.

Proof. We have $\log_x(y)(d) = (1-d) \cdot x + d \cdot y$; since we now are in a vector space, this affine combination can be rewritten as the linear combination $x + d \cdot (y - x)$. As a function of $d \in D$, this is a tangent vector at x with principal part $y - x$.

To construct an inverse $\exp_x : D(T_x(M)) \to \mathfrak{M}(x)$ for $\log_x : \mathfrak{M}(x) \to D(T_x(M))$, we first assume that M is an open subset of a finite-dimensional vector space V. This implies, by KL for V, that any $\tau : D \to M$, tangent vector at $x \in M$, is of the form $\tau(d) = x + d \cdot v$ for a unique $v = \gamma(\tau) \in V$, the "principal part" of τ. This establishes a linear bijection $T_x(M) \cong V$, and under this bijection, $D(T_x(M))$ corresponds to $D(V)$. So assume now that $\tau \in D(T_x(M))$, so $\gamma(\tau) \in D(V)$. For such τ, we put

$$\exp_x(\tau) = x + \gamma(\tau);$$

since $\gamma(\tau) \sim 0$, $x + \gamma(\tau) \sim x$, so $\exp_x(\tau) \in \mathfrak{M}(x)$, so $\exp_x : D(T_x(M)) \to \mathfrak{M}(x)$, and it clearly takes $0 \in T_x(M)$ into x. (Note that $\mathfrak{M}(x)$, as formed in V, equals $\mathfrak{M}(x)$ as formed in M, since we assumed that $M \subseteq V$ was open.)

We then note that $\exp_x(\log_x(y)) = x + (y - x) = y$ since the principal part of $\log_x(y)$ is $y - x$, by Proposition 4.3.1. Also

$$\begin{aligned} \log_x(\exp_x(\tau))(d) &= (1-d) \cdot x + d \cdot \exp_x(\tau) \\ &= (1-d) \cdot x + d \cdot (x + \gamma(\tau)) = x + d \cdot \gamma(\tau) = \tau(d). \end{aligned}$$

This proves that for M open in V, the processes $\log_x$ and $\exp_x$ are mutual inverse $D(T_x(M)) \cong \mathfrak{M}(x)$. The result for general manifold M now follows from the naturality (4.3.4) of log already established. Thus we have proved

Theorem 4.3.2 *For M a manifold and $x \in M$, $\log_x : \mathfrak{M}(x) \to D(T_x(M))$ and $\exp_x : D(T_x(M)) \to \mathfrak{M}(x)$ are mutually inverse maps.*

Recall that for $t \in R$ and $\tau \in T_x(M)$, $t \cdot \tau \in T_x(M)$ is the tangent vector given by $(t \cdot \tau)(d) := \tau(t \cdot d)$. If $t = d \in D$, $d \cdot \tau \in D(T_x(M))$, so $\exp_x$ may be applied to it. It is easy to establish that for the case where $t = d \in D$,

$$\exp_x(d \cdot \tau) = \tau(d). \tag{4.3.5}$$

It suffices to do this in a coordinatized situation, with M an open subset of a vector space V. If $\tau \in T_x(M)$ has principal part $v \in V$, then $d \cdot \tau$ has principal part $d \cdot v$, and so

$$\exp_x(d \cdot \tau) = x + d \cdot v = \tau(d).$$

Applying $\log_x$ to both sides of (4.3.5) yields another useful relation:

$$d \cdot \tau = \log_x(\tau(d)) \tag{4.3.6}$$

for $d \in D$, $\tau \in T_x(M)$.

Exercise 4.3.3 Prove that $\log_x$ is compatible with the multiplicative action by R, and also that $\log_x(y - x + z) = \log_x(y) + \log_x(z)$ if x, y, z form an infinitesimal 2-simplex.

The log-exp bijection established by the theorem has an important consequence:

Theorem 4.3.4 *For M and N manifolds, and for $x \in M$, $y \in N$, there is a canonical bijective correspondence between 1-jets $\mathfrak{M}(x) \to \mathfrak{M}(y)$ from x to y, and linear maps $T_x(M) \to T_y(N)$.*

Proof/Construction. Given a 1-jet $f : \mathfrak{M}(x) \to \mathfrak{M}(y)$ ($y = f(x)$, $x \in M, y \in N$), then the map $T_x(f) : T_x(M) \to T_y(N)$ is homogeneous, hence it is linear, by Theorem 1.4.1. Conversely, given a linear $l : T_x(M) \to T_y(N)$, it restricts to a zero-preserving $D(T_x(M)) \to D(T_y(N))$, by the functorality of the D-construction. By the bijections of Theorem 4.3.2, there is a unique 1-jet $f : \mathfrak{M}(x) \to \mathfrak{M}(y)$ such that $\log_y \circ f = l \circ \log_x$. Now both maps l and $T_x(f)$ are linear. To see that $l = T_x(f)$, it therefore suffices, by KL, to see that these two maps agree on $D(T_x(M))$. But by construction of l, l satisfies $\log_y \circ f = l \circ \log_x$, and by naturality (4.3.4) of log, $T_x(f)$ satisfies the same equation. From the fact that $\log_x$ and $\log_y$ are bijections, it now follows that l and $T_x(f)$ agree on $D(T_x(M))$. This proves the theorem.

Corollary 4.3.5 *Let $x \in M$, with M a manifold, Then there is a canonical bijective correspondence between cotangents (Definition 2.7.3) $\mathfrak{M}(x) \to R$, and linear maps $T_x M \to R$, "classical cotangents".*

Corollary 4.3.6 *Given a manifold M and a finite-dimensional vector space V, then there is a canonical bijective correspondence between V-framings of M, in the sense of Section 2.2,*

$$M \times D(V) \to M_{(1)},$$

and "classical" framings

$$M \times V \to T(M).$$

Here, we use $T_0(V) \cong V$, by KL for V.

Affine connections as connections in the tangent bundle

We consider an affine connection λ on a manifold M. It may be construed as a bundle connection ∇ in the bundle $M_{(1)} \to M$, see (2.5.5); so to $x \sim y$ in M, λ gives rise to a transport law $\nabla(y,x) : \mathfrak{M}(x) \to \mathfrak{M}(y)$, i.e. a 1-jet from x to y. By the above Theorem 4.3.4, the information contained in such a 1-jet is the same as a linear $\overline{\nabla}(y,x) : T_x(M) \to T_y(M)$, and $\overline{\nabla}(y,x)$ is invertible since $\nabla(y,x)$ is so. Thus we see that an affine connection λ on M may be encoded as a bundle connection $\overline{\nabla}$ in the tangent bundle $T(M) \to M$, with the property that each of its transport maps $\overline{\nabla}(y,x)$ is a *linear* isomorphism. In other words, affine connections on M may be identified with linear bundle connections in the tangent bundle $T(M) \to M$. For completeness, let us describe explicitly how ∇ and $\overline{\nabla}$ are related: for $x \sim y$ and $\tau \in T_x(M)$, $\overline{\nabla}(y,x)(\tau) \in T_y(M)$ is the tangent vector whose value at $d \in D$ is given by

$$\big(\overline{\nabla}(y,x)(\tau)\big)(d) := \nabla(y,x)(\tau(d)) = \lambda(x,y,\tau(d));$$

and for $x \sim y$, $x \sim z$, $\nabla(y,x)(z)$ $(= \lambda(x,y,z))$ is defined by the formula

$$\nabla(y,x)(z) = \exp_y\big(\overline{\nabla}(y,x)(\log_x z)\big).$$

Let us summarize:

Proposition 4.3.7 *There is a bijective correspondence between affine connections on M, and linear bundle connections in $T(M) \to M$.*

Note that the lack of symmetry between y and z in the (combinatorial) notion of affine connection $\lambda(x,y,z)$ is, in terms of the corresponding linear bundle connection, pinpointed as follows: what we have called the *active* aspect y is still combinatorial ($x \sim y$), whereas the *passive* aspect, i.e. the neighbours z of x that are being transported, are replaced by tangent vectors at x.

In the first synthetic formulations of the notion of affine connections (as in Kock and Reyes, 1979a, reproduced in Kock, 1981/2006, §I.7 and Lavendhomme, 1996, §5.1), both the active and passive aspects are replaced by tangent vectors.

Generally, bundle connections can be formulated so as to conceive the active aspect in kinematic terms (tangent vectors), cf. Nishimura (1998) for a synthetic treatment; this notion of bundle connections is more general than the combinatorial one, since the base space of the bundle need not be assumed to be a manifold; only microlinearity is assumed.

It is worthwhile to record the expression of $\overline{\nabla}(y,x) : T_xM \to T_yM$ in a coordinatized situation $M \subseteq V$, where λ may be expressed by Christoffel symbols $\Gamma(x;u,v)$ as in (2.3.12). Let $\tau \in T_xM$ be a tangent vector with principal part $a \in V$, $\tau(d) = x + d \cdot a$. Then for $y \sim x$,

$$\begin{aligned}\overline{\nabla}(y,x)(\tau)(d) &= \lambda(x,y,\tau(d)) = \lambda(x,y,x+d\cdot a)\\ &= y - x + (x + d\cdot a) + \Gamma(x;y-x,d\cdot a)\\ &= y + d\cdot(a + \Gamma(x;y-x,a))\end{aligned}$$

which has principal part $a + \Gamma(x;y-x,a)$. So identifying tangent vectors with their principal parts (when their base point is understood from the context)

$$\overline{\nabla}(y,x)(a) = a + \Gamma(x;y-x,a). \tag{4.3.7}$$

The notion of symmetric affine connection can be reformulated in terms of "second-order exponential map" (which in turn is a geometric variant of the notion of *spray*, which is a kinematic notion).

Definition 4.3.8 *A* second-order exponential map *on a manifold M is a map* $\exp^{(2)} : D_2(T(M)) \to M_{(2)}$ *extending the exponential map* $D(T(M)) \to M_{(1)}$.

We postpone the discussion of this until the chapter on metric notions (Chapter 8), which is where we have applications for second-order exponential maps.

4.4 Tangent vectors as differential operators

Let M be a manifold, and $\tau : D \to M$ a tangent vector at $x \in M$. Then for all $d \in D$, $\tau(d) \sim x$, in other words, τ factors through $\mathfrak{M}(x) \subseteq M$. Let W be a KL vector space (the most important case being $W = R$), and consider a W-valued

1-jet at x, i.e. a map $f : \mathfrak{M}(x) \to W$. Consider the composite

$$D \xrightarrow{\tau} \mathfrak{M}(x) \xrightarrow{f} W.$$

As a map $D \to W$, it has a principal part, which we denote $D_\tau f$. Thus, the equation characterizing $D_\tau f$ is

$$f(\tau(d)) = f(x) + d \cdot D_\tau f.$$

It is the *derivative* of f in the *direction* of τ, or *along* τ.

As a function of f and τ, $D(f,\tau) := D_\tau f$ defines a map

$$D : TM \times_M J^1(M,W) \to W.$$

Proposition 4.4.1 *$D_\tau f$ depends in a bilinear way on (τ, f).*

Proof. It is easy to see that the dependence on f is linear, for fixed τ. Now consider a fixed $f : \mathfrak{M}(x) \to W$; then $\tau \mapsto D_\tau f$, as a function of τ, defines a map $T_x(M) \to W$. To prove that this map is linear, we use Theorem 1.4.1: it suffices (since $T_x(M)$ is a finite-dimensional vector space) to prove the homogeneity condition $D_{t\cdot\tau} f = t \cdot D_\tau f$ for all $t \in R$. To see this for a given t, it suffices to see that for all $d \in D$,

$$d \cdot D_{t\cdot\tau} f = d \cdot t \cdot D_\tau f.$$

The left-hand side here is by definition $f((t\cdot\tau)(d)) - f(x)$. Recall from Exercise 4.2.1 that $(t\cdot\tau)(d)$ (for τ a tangent vector at x) may be expressed in terms of the affine combination $t \cdot (\tau(d)) + (1-t) \cdot x$. Thus

$$\begin{aligned} d \cdot D_{t\cdot\tau} f &= f\big((t\cdot\tau)(d)\big) - f(x) = f\big(t \cdot \tau(d) + (1-t)\cdot x\big) - f(x) \\ &= t \cdot f(\tau(d)) + (1-t) \cdot f(x) - f(x) \end{aligned}$$

(using the fact that f preserves affine combinations of mutual neighbour points, cf. Theorem 2.1.4)

$$= t \cdot f(\tau(d)) - t \cdot f(x) = t \cdot \big(f(\tau(d)) - f(x)\big) = t \cdot d \cdot D_\tau f,$$

which is the right-hand side of the desired equation. This proves the proposition.

Exercise 4.4.2 Prove the Leibniz rule for D_τ:

$$D_\tau(f \cdot g) = D_\tau f \cdot g + f \cdot D_\tau g,$$

where f and g are R-valued 1-jets at $x \in M$; g may even be replaced by a W-valued 1-jet, with W a KL vector space. Even more generally, there is such a law $D_\tau(f * g) = D_\tau(f) * g + f * D_\tau g$ for a bilinear $* : W_1 \times W_2 \to W_3$, with the W_is KL vector spaces.

The construction $D_\tau f$ immediately globalizes into the notion of derivative of a function along a vector *field*. A *vector field* on M is a law X which to each $x \in M$ associates a tangent vector $X(x) \in T_x(M)$; equivalently, X is a cross-section of $T(M) \to M$. Let X be a vector field on M, and let f be a function $M \to W$. Then we get a new function $D_X f : M \to W$ (sometimes denoted $X(f)$), namely

$$(D_X f)(x) := D_{X(x)} j_x f,$$

the derivative of (the 1-jet at x of) f along the field vector $X(x)$ of X at x. Thus $D_X f$ is characterized by validity, for all $d \in D$, of

$$f(X(x)(d)) = f(x) + d \cdot (D_X f)(x). \tag{4.4.1}$$

This is standard in SDG, cf. Kock (1981/2006, §I.10), Lavendhomme (1996, §3.3.1).

Exercise 4.4.3 Prove the Leibniz rule for D_X,

$$D_X(f \cdot g) = D_X f \cdot g + f \cdot D_X g.$$

Here f and g are functions $M \to R$; g may even be replaced by a function $g : M \to W$ (with W a KL vector space). Similarly for the more general situation described in Exercise 4.4.2.

4.5 Cotangents, and the cotangent bundle

In this section, we consider the *cotangent bundle* $T^*M \to M$ of a manifold M; its fibre over $x \in M$ are the *cotangents* or *cotangent vectors* at $x \in M$; the notion of cotangent may be defined in a combinatorial or in a classical way, but the bundles (vector bundles, in fact) obtained by these definitions are canonically isomorphic, and we shall use the same notation for them. The notions may be seen as the point-localized versions of the notions of combinatorial, respectively, classical 1-form.

The notion of combinatorial cotangent was introduced already in Definition 2.7.3: a cotangent at $x \in M$ is a 1-jet from $x \in M$ to $0 \in R$. The cotangents at x clearly form an R-module, by pointwise addition and multiplication by scalars. It is actually finite dimensional. This may be seen in many ways, for instance by virtue of the comparison with the classical cotangents which we shall define now.

For an n-dimensional manifold M, each tangent space T_xM (for $x \in M$) is locally an n-dimensional vector space. Therefore, its dual $(T_xM)^*$ is an n-dimensional vector space as well, whose elements we call *classical cotangents* at x. From local cartesian closedness of the category in which we work, it follows that there is a *bundle* over M whose fibre over x is $(T_xM)^*$. This is a finite-dimensional vector bundle over M, called the (classical) *cotangent bundle*, and it is denoted $T(M)^* \to M$. Thus $(T(M)^*)_x = (T_xM)^*$. A classical cotangent ψ at x is thus a linear map $T_xM \to R$. One also writes T_x^*M for $(T^*M)_x = (T_xM)^*$, and similarly one writes $T_x^{**}M$ for $(T_xM)^{**}$.

There is a natural bijective correspondence between classical and combinatorial cotangents at $x \in M$ (cf. Corollary 4.3.5): given a combinatorial cotangent ω at x, the classical $\overline{\omega} : T_xM \to R$ corresponding to it is determined by the condition for all $d \in D$,

$$d \cdot \overline{\omega}(\tau) = \omega(\tau(d)) \tag{4.5.1}$$

for $\tau \in T_xM$. Conversely, a classical cotangent $\overline{\omega}$ at x determines a combinatorial cotangent ω by putting, for $y \sim x$,

$$\omega(y) = \overline{\omega}(\log_x y). \tag{4.5.2}$$

Both the combinatorial and the classical notion of cotangent may be generalized into W-valued cotangents with W a KL vector space. The special case considered above is then the case where $W = R$.

Thus, a combinatorial W-valued cotangent at $x \in M$ is a 1-jet from $x \in M$ to $0 \in W$; and a classical W-valued cotangent at x in M is a linear $T_xM \to W$. The correspondence between combinatorial and classical cotangents provided above also applies to W-valued cotangents; we get an isomorphism from the vector bundle of combinatorial W-valued cotangents to the vector bundle of classical W-valued cotangents, provided by the same formula (4.5.1) as above.

We may denote the bundle of W-valued cotangents by something like $T^*(M,W)$; but consideration of the vector space of (classical) W-valued cotangents at $x \in M$ in terms of coordinates around x shows that the canonical linear map

$$T_x^*M \otimes W \to T^*(M,W) \tag{4.5.3}$$

is an isomorphism.

Note that a section of the bundle of combinatorial W-valued cotangents is the same as a combinatorial W-valued 1-form: if for each $x \in M$ we are given a combinatorial cotangent ω_x at x, we get a combinatorial 1-form ω by putting $\omega(x,y) := \omega_x(y)$.

Consider a function $f : M \to W$. We have by Proposition 4.4.1 that $D_\tau f \in W$ depends linearly on $\tau \in T_x M$, and so defines a classical W-valued cotangent at x, denoted $df(x; -)$, and it deserves the name: the *differential* of f at x. For $W = R$, the map $x \mapsto df(x; -)$ is the *differential* of f; thus the differential $df : M \to T^*M$ is a section of the bundle of classical (R-valued) cotangents.

If M is an open subset of a finite-dimensional vector space V, the tangent spaces $T_x M$ may be identified with V, via principal-part formation, and in this case, df gets identified with a map $df : M \times V \to R$, linear in the second variable. We leave to the reader to see that this is consistent with the use of the term "differential", and with the notation for it, as given in Section 1.4.

The combinatorial cotangents corresponding to the classical cotangents $df(x; -)$ essentially make up the combinatorial 1-form df considered in Definition 2.2.3.

4.6 The differential operator of a linear connection

We consider a vector bundle $\pi : E \to M$ whose fibres are KL vector spaces. Recall that a *linear* connection in such a bundle is a bundle connection ∇ such that each transport law $\nabla(y,x) : E_x \to E_y$ is linear.

We shall associate to ∇ a differential operator $\tilde{\nabla}$, generalizing the construction of $D_X f$ in the previous section (which is the special case where $E = M \times W$ and ∇ the trivial connection).

Let τ be a tangent vector at $x \in M$, and let ζ be a section 1-jet of E at x, i.e. a map $\mathfrak{M}(x) \to E$ with $\pi(\zeta(x_1)) = x_1$ for all $x_1 \in \mathfrak{M}(x)$. Then we can construct an element $\tilde{\nabla}_\tau(\zeta) \in E_x$ as follows. For $d \in D$, we consider the element in E_x given as $\nabla(x, \tau(d))(\zeta(\tau(d)))$. As a function of $d \in D$, it defines a map $D \to E_x$, and $\tilde{\nabla}_\tau(\zeta)$ is by definition the principal part of this map. Thus, $\tilde{\nabla}_\tau(\zeta)$ is characterized by the validity in E_x of

$$\nabla(x, \tau(d))(\zeta(\tau(d))) = \zeta(x) + d \cdot \tilde{\nabla}_\tau(\zeta)$$

for all $d \in D$. If $f : \mathfrak{M}(x) \to R$, we get out of ζ a new section 1-jet $f \cdot \zeta$ at x, by putting $(f \cdot \zeta)(x_1) = f(x_1) \cdot \zeta(x_1)$, for $x_1 \in \mathfrak{M}(x)$. (The multiplication here is multiplication of vectors by scalars in the vector space E_{x_1}.) Note that $\tilde{\nabla}$ itself is a map

$$\tilde{\nabla} : T_x M \times (J^1 E)_x \to E_x,$$

for each $x \in M$.

Proposition 4.6.1 *For $f : \mathfrak{M}(x) \to R$ and τ and ζ as above, we have*

$$\tilde{\nabla}_\tau(f \cdot \zeta) = f(x) \cdot \tilde{\nabla}_\tau(\zeta) + D_\tau f \cdot \zeta(x).$$

Proof. By definition, $\tilde{\nabla}_\tau(f \cdot \zeta)$ is the principal part of the function of $d \in D$ given by the expression $\nabla(x, \tau(d))\big((f \cdot \zeta)(\tau(d))\big)$; now

$$\begin{aligned}
&\nabla(x, \tau(d))\big((f \cdot \zeta)(\tau(d))\big)\\
&\quad = \nabla(x, \tau(d))\big(f(\tau(d)) \cdot \zeta(\tau(d))\big)\\
&\quad = \nabla(x, \tau(d))\big((f(x) + d \cdot D_\tau f) \cdot \zeta(\tau(d))\big)\\
&\quad = \nabla(x, \tau(d))\big(f(x) \cdot \zeta(\tau(d)) + d \cdot D_\tau f \cdot \zeta(\tau(d))\big)\\
&\quad = f(x) \cdot \nabla(x, \tau(d))(\zeta(\tau(d))) + d \cdot D_\tau f \cdot \nabla(x, \tau(d))(\zeta(\tau(d)));
\end{aligned}$$

using that $\nabla(x, \tau(d))$ is linear. Now, because of the factor d on the last term, we may use Taylor principle and replace the other occurrences in this term by 0, so we get

$$\begin{aligned}
&= f(x) \cdot \nabla(x, \tau(d))(\zeta(\tau(d))) + d \cdot D_\tau f \cdot \zeta(x);\\
&= f(x) \cdot \big(\zeta(x) + d \cdot \tilde{\nabla}_\tau \zeta\big) + d \cdot D_\tau f \cdot \zeta(x)\\
&= f(x) \cdot \zeta(x) + d \cdot \big[f(x) \cdot \tilde{\nabla}_\tau \zeta + D_\tau f \cdot \zeta(x)\big].
\end{aligned}$$

So the square bracket is that principal part which defines $\tilde{\nabla}_\tau(f \cdot \zeta)$, which thus equals $f(x) \cdot \tilde{\nabla}_\tau \zeta + D_\tau f \cdot \zeta(x)$. This proves the proposition.

Exercise 4.6.2 Prove that $\tilde{\nabla}_\tau(\zeta)$ depends in a linear way on τ as well as on ζ. (Use Theorem 1.4.1.) This generalizes Proposition 4.4.1.

The proposition immediately globalizes: if Y is a vector field on M, and $Z : M \to E$ is a section, we have for each $x \in M$ a tangent vector $\tau := Y(x)$ and a 1-jet section ζ, namely the restriction $j_x(Z)$ of Z to $\mathfrak{M}(x)$, and then $\tilde{\nabla}_\tau \zeta \in E_x$; as x ranges, we thus get a new section of E, which we may denote $\tilde{\nabla}_Y Z$. Its value $(\tilde{\nabla}_Y Z)(x)$ at x is characterized by

$$\nabla(x, Y(x)(d))\big(Z(Y(x)(d))\big) = Z(x) + d \cdot (\tilde{\nabla}_Y Z)(x).$$

Globalizing Proposition 4.6.1, we then immediately get, for a linear connection ∇ in the vector bundle $E \to M$ and the associated operator $\tilde{\nabla}$:

Theorem 4.6.3 (Koszul's Law) *Let Y be a vector field on M and Z a cross-section of $E \to M$. Let $f : M \to R$ be a function. Then*

$$\tilde{\nabla}_Y(f \cdot Z) = f \cdot \tilde{\nabla}_Y Z + D_Y f \cdot Z.$$

Recall that a vector field on M is a cross-section of the tangent bundle $T(M) \to M$. Since an affine connection on a manifold M may be encoded as a linear bundle connection on this bundle (Proposition 4.3.7), it follows that an affine connection gives rise to a differential operator $\tilde{\nabla}$ which to a pair of tangent vector fields Y, Z on M provides a new vector field $\tilde{\nabla}_Y Z$, and that Koszul's Law, as stated in the theorem, holds.

In most classical texts on differential geometry, the notion of linear connection in general, and in particular the notion of affine connection, is *defined* in terms of these differential operators. An alternative formulation in classical terms, more geometric in nature, is the description in terms of distributions transverse to the fibres; this description is the immediate counterpart of the synthetic description given at the end of Section 2.5.

4.7 Classical differential forms

We give here a comparison between the theory of classical differential forms, on the one hand, and the theory of combinatorial differential forms (simplicial, combinatorial, and whisker) on the other. We use the term "classical differential form" for the form notion where the inputs of the k-forms on M are k-tuples of tangent vectors on M, but we conceive tangent vectors in their synthetic manifestation: as maps $D \to M$. In this sense, the exposition is still entirely in the context of SDG, and this synthetic theory of "classical" differential forms is essentially expounded in the standard treatises on SDG (Kock, 1981/2006, §I.14.1; Lavendhomme, 1996, §4.1.2); a (not quite classical) variant of the classical notion is where the inputs of k-forms on M are maps $D^k \to M$, "microcubes", or "marked microcubes" (Kock, 1981/2006, §I.14.2; Lavendhomme, 1996, §4.1.1; Moerdijk and Reyes, 1991, §IV.1). (We shall return to the microcubes, with or without marks, below.)

We consider the case where the values of the forms are in a KL vector space W – usually just R. Recall that if M is a manifold, a *classical* differential k-form $\overline{\omega}$ on M with values in W is a law which to each k-tuple of tangent vectors $(\tau_1, \ldots, \tau_k)$ with the same base point associates an element

$$\overline{\omega}(\tau_1, \ldots, \tau_k) \in W,$$

subject to two requirements: it is k-linear, and it is alternating. The latter requirement refers to the action of the symmetric group $\mathfrak{S}_k$ in k letters (not in $k+1$ letters, as for simplicial differential forms).

It is a consequence of Theorem 1.4.1 that the k-linearity (linearity in each of the k inputs τ_i) may be replaced by the weaker requirement of homogeneity in each of these arguments,

$$\overline{\omega}(\tau_1, \ldots, t \cdot \tau_i, \ldots, \tau_k) = t \cdot \overline{\omega}(\tau_1, \ldots, \tau_i, \ldots, \tau_k)$$

for each $i = 1, \ldots, k$ and $t \in R$.

The log-exp construction provides a way of passing from a classical differential k-form $\overline{\omega}$ to a simplicial k-form ω: given $\overline{\omega}$, we define the simplicial k-form ω by the explicit formula

$$\omega(x_0, x_1, \ldots, x_k) := \overline{\omega}(\log_{x_0}(x_1), \ldots, \log_{x_0}(x_k)). \qquad (4.7.1)$$

Note that all the $\log_{x_0}(x_i)$ are tangent vectors at the same point of M, namely at x_0, so that $\overline{\omega}$ may indeed be applied to the k-tuple of them.

If $x_i = x_0$ for some $i \geq 1$, then $\log_{x_0}(x_i)$ is the zero tangent vector at x_0, and so by multilinearity of $\overline{\omega}$, the right-hand side gives the value $0 \in W$. So the simplicial k-cochain ω described satisfies the normalization condition of Proposition 3.1.6, and thus is a simplicial differential form. (Actually, (4.7.1) defines also a whisker k-form, since $\overline{\omega}$ is alternating.)

If the value vector space W is finite dimensional, a simplicial differential form always take infinitesimal values, i.e. takes values in $D(W)$, whereas a classical form may take any value, infinitesimal or finite, in W. Note that the tangent vectors $\log_{x_0}(x_i)$ occurring in the formula (4.7.1) are infinitesimal (belong to $D(T_{x_0})$), and that the right-hand side of (4.7.1) (therefore) is infinitesimal.

Theorem 4.7.1 *The correspondence $\overline{\omega} \mapsto \omega$ between classical and simplicial differential W-valued forms is bijective.*

Proof. Note that the correspondence itself was described in coordinate-free terms. Therefore, it is sufficient to establish the bijectivity of the correspondence in a coordinatized situation, i.e. with M an open subset of a finite-dimensional vector space V. Since for $x \in M$, $T_x(M)$ may be identified with V, by associating to a tangent vector at x its principal part, we see that a classical differential W-valued k-form $\overline{\omega}$ for each x provides a map $\overline{\Omega}(x; -, \ldots, -) : V^k \to W$, which is k-linear and alternating. Also, we have that a simplicial differential k-form for each x provides a map $\Omega(x; -, \ldots, -) : \widetilde{D}(k, V) \to W$

with the normalization property that its value vanishes if one of the k arguments is 0. We know already by the KL property (cf. Section 1.3) that there is a bijective correspondence between these two kinds of data, obtained simply by restricting $\overline{\Omega}(x; -, \ldots, -) : V^k \to W$ to $\widetilde{D}(k,V) \subseteq V^k$. We have to see that this correspondence obtained via the coordinatization by V agrees with the one constructed using log; this follows from Proposition 4.3.1.

It is possible to describe directly, in a coordinate-free way, the classical k-form $\overline{\omega}$ corresponding to a simplicial k-form ω, i.e. to describe $\overline{\omega}(\tau_1, \ldots, \tau_k)$ for an arbitrary k-tuple of tangent vectors, when ω is given; this is not simple, but may be done along the line of reasoning in (3.1.5). However, if ω is given as a *whisker* form, rather than just as a *simplicial* form, it is easy: if τ_i $(i = 1, \ldots, k)$ are tangent vectors at the same point, say x, then $\overline{\omega}(\tau_1, \ldots, \tau_k)$ is the unique element in V such that for $(d_1, \ldots, d_k) \in D^k$ we have

$$d_1 \cdot \ldots \cdot d_k \cdot \overline{\omega}(\tau_1, \ldots, \tau_k) = \omega(\tau_1(d_1), \ldots, \tau_k(d_k)).$$

Note that $\tau_i(d_i) \sim x$, but not necessarily $\sim \tau_j(d_j)$, so that $(x, \tau_1(d_1), \ldots, \tau_k(d_k))$ is an infinitesimal k-whisker, but not necessarily an infinitesimal k-simplex.

The construction of (3.1.5) provides the transition from a simplicial form to the corresponding whisker form, so the combination of these two constructions provides the passage from simplicial to classical forms (modulo a combinatorial factor).

We shall prove that the coboundary operator for combinatorial forms matches the classical coboundary operator for classical differential forms; this was stated in Kock (2000, p. 259), and a proof was sketched (using integration, and the validity of the classical Stokes' Theorem). We shall here present a direct proof, by working in coordinates, i.e with a chart from an abstract finite-dimensional vector space V. We denote, temporarily, the coboundary operator for classical differential forms by $\overline{d}$, to distinguish it from the coboundary d for simplicial forms, and from the coboundary for cubical forms, which we temporarily denote by $\tilde{d}$. The forms considered are supposed to take values in a KL vector space W.

Theorem 4.7.2 *Let $\overline{\omega}$ be a classical k-form on M. Let ω be the corresponding simplicial k-form. Then $\overline{d}(\overline{\omega})$ corresponds to $k+1$ times the simplicial form $d\omega$. Also, let $\tilde{\omega}$ be the cubical k-form corresponding to $\overline{\omega}$; then $\overline{d}(\overline{\omega})$ corresponds to $\tilde{d}(\tilde{\omega})$.*

Proof. In view of the correspondence between simplicial and cubical coboundary expressed in Theorem 3.2.4, it suffices to prove the latter of the two assertions of the theorem. We already calculated the function $M \times V^{k+1} \to W$

($k+1$-linear alternating in the last $k+1$ arguments) corresponding to $\tilde{d}(\tilde{\omega})$; this is the function expressed in (3.2.4). This, however, is also the classical coordinate calculation (or definition) of the function $M \times V^{k+1} \to W$, for the exterior derivative of the classical form $\overline{\omega}$, see e.g. Madsen and Tornehave (1997, Definition 3.2).

We shall next prove that the cup product for simplicial forms (modulo a combinatorial factor) matches the wedge (exterior) product of classical differential forms. (This was proved also at the end of Kock (2000), except there the combinatorial factors were swept under the carpet by building them into the correspondence $\omega \leftrightarrow \overline{\omega}$.) We assume given KL vector spaces W_1, W_2, and W_2, and a bilinear map $* : W_1 \times W_2 \to W_3$; ω and $\overline{\omega}$ are supposed to take values in W_1, and θ and $\overline{\theta}$ in W_2; the cup and wedge products to be compared then take values in W_3. We omit the symbol $*$ on $\cup$ and $\wedge$

Theorem 4.7.3 *Let $\overline{\omega}$ and $\overline{\theta}$ be classical k- and l-forms, respectively, on a manifold M, and let ω and θ be the corresponding simplicial forms. Then to the classical $k+l$-form $\overline{\omega} \wedge \overline{\theta}$ corresponds the simplicial form $(k,l) \cdot (\omega \cup \theta)$, where (k,l) denotes the integer $(k+l)!/k!l!$.*

Proof. Recall that (one of) the standard versions of the formula for $\wedge$ involves an alternating sum ranging over the set of *k,l-shuffles* σ; a k,l-shuffle σ is a permutation of the numbers $1,2,\ldots,k+l$ with $\sigma(1) < \ldots < \sigma(k)$ and with $\sigma(k+1) < \ldots < \sigma(k+l)$. We analyse now the combinatorial form corresponding to $\overline{\omega} \wedge \overline{\theta}$; consider an infinitesimal $k+l$-simplex $(x_0, x_1, \ldots, x_{k+l})$. Then

$$\begin{aligned}
&(\overline{\omega} \wedge \overline{\theta})(\log_{x_0}(x_1), \ldots, \log_{x_0}(x_{k+l})) \\
&\quad = \sum_{\sigma} \operatorname{sign}(\sigma) \cdot \overline{\omega}(\log_{x_0}(x_{\sigma(1)}), \ldots, \log_{x_0}(x_{\sigma(k)})) \\
&\qquad * \overline{\theta}(\log_{x_0}(x_{\sigma(k+1)}), \ldots, \log_{x_0}(x_{\sigma(k+l)}))
\end{aligned}$$

where σ ranges over the set of $k+l$-shuffles

$$\begin{aligned}
&= \sum_{\sigma} \operatorname{sign}(\sigma) \cdot \omega(x_0, x_{\sigma(1)}, \ldots, x_{\sigma(k)}) * \theta(x_0, x_{\sigma(k+1)}, \ldots, x_{\sigma(k+l)}) \\
&= \sum_{\sigma} \operatorname{sign}(\sigma) \cdot (\omega \cup \theta)(x_0, x_{\sigma(1)}, \ldots, x_{\sigma(k+l)})
\end{aligned}$$

using Lemma 3.5.2 to exchange the x_0 in the θ factor by $x_{\sigma(k)}$. Now $\omega \cup \theta$ is alternating (being a simplicial form), and so all the terms here are equal, as σ ranges; in particular they are all equal to $(\omega \cup \theta)(x_0, x_1, \ldots x_{k+l})$. Since there are (k,l) shuffles σ, we conclude

$$(\overline{\omega}\wedge\overline{\theta})\big(\log_{x_0}(x_1),\ldots,\log_{x_0}(x_{k+l})\big)=(k,l)\cdot(\omega\cup\theta)\big(x_0,x_1,\ldots,x_{k+l}\big),$$

and this proves the theorem.

Classical geometric distributions

A classical (geometric) distribution on a manifold M consists in giving for each $x\in M$ a linear subspace $S(x)\subseteq T_x(M)$, with suitable regularity conditions. Such data gives rise to a pre-distribution $\approx$, in the sense of Section 2.6: for $x\sim y$ we put $x\approx y$ if $\log_x(y)\in S(x)$. This is clearly a reflexive relation; to see that it is symmetric, it suffices to consider a coordinatized situation, i.e. we may assume that M is an open subset of a finite-dimensional vector space V. In this case, $\log_x y=y-x$ by Proposition 4.3.1. Also, we may assume that $S(x)$ is the kernel of a linear $\sigma: T_x(M)\to W$ for some finite-dimensional vector space W (this being part of the regularity condition which we did not specify fully). Identifying each $T_x(M)$ with V via principal-part formation, we thus have a map $\sigma: M\times V\to W$, linear in its second variable, such that $\tau\in S(x)$ iff $\sigma(x;\tau)=0$. Assume now that $\log_x(y)\in S(x)$, so $\sigma(x;y-x)=0$. To prove that $\log_y(x)\in S(y)$ amounts to proving $\sigma(y;x-y)=0$. We have

$$0=\sigma(x;y-x)=\sigma(y;y-x)=-\sigma(y;x-y),$$

the middle equality by the Taylor principle (using $x\sim y$), and the last equality by linearity of σ in the second variable. This proves that $\approx$ is a symmetric relation.

We note that $\mathfrak{M}_{\approx}(x)\subseteq\mathfrak{M}(x)$ sits in a pull-back diagram

$$\begin{array}{ccccc}
\mathfrak{M}_{\approx}(x) & \longrightarrow & D(S(x)) & \longrightarrow & S(x)\\
\downarrow & & \downarrow & & \downarrow\\
\mathfrak{M}(x) & \xrightarrow[\log_x]{\cong} & D(T_x(M)) & \longrightarrow & T_x(M)
\end{array}$$

(the right-hand square being a pull back by Proposition 1.2.4). It follows from this that $\approx$ is actually a distribution, not just a pre-distribution.

Suppose now that there are given q classical 1-forms ω_i. Let linear subspaces $S(x)$ be given as the joint zero set of the ω_is. Then the corresponding combinatorial distribution $\approx$ is given, as at the end of Section 3.5, in terms of those combinatorial 1-forms that correspond to the ω_is. The condition for $\approx$ to be involutive, in terms of the exterior algebra $\Omega^\bullet(M)$ (Theorem 3.6.2), then transfers immediately to the classical condition for involutiveness of the classical distribution given by the $S(x)$s. For, the combinatorial and classical de Rham algebras are isomorphic (modulo some scalar factors), by Theorem 4.7.2 and Theorem 4.7.3.

4.8 Differential forms with values in $TM \to M$

In Section 3.8, we discussed simplicial differential forms with values in a vector bundle on a manifold M.

We have available a particular vector bundle on M, namely the tangent bundle $TM \to M$. There is an almost tautological combinatorial 1-form θ on M with values in this bundle, the *solder form* (terminology from Kobayashi and Nomizu, 1963). In our context, it is really just the log map: recall that given $x \sim y$ in M, we have a particular tangent vector $\log_x(y)$ at x, given by $\log_x(y)(d) = (1-d)\cdot x + d\cdot y$. This construction may be interpreted as a simplicial 1-form θ on M with values in $TM \to M$,

$$\theta(x,y) := \log_x(y). \tag{4.8.1}$$

Another example of a $TM \to M$-valued simplicial differential form – now a 2-form – is obtained by applying log to the "intrinsic torsion" (2.3.11) $b_x(y,z)$ of an affine connection λ on M; recall $b_x(y,z) = \lambda(\lambda(x,y,z),y,z)$ for an infinitesimal 2-simplex x,y,z in M. Because we get x as value if $y = x$ or $z = x$, $\log_x(b_x(y,z)) \in T_x(M)$ is 0 if $y = x$ or $z = x$, so we have a $TM \to M$-valued simplicial 2-form $\log b$ given by

$$(\log b)(x,y,z) := \log_x(b_x(y,z)).$$

(One may define a notion of *whisker* differential form with values in a vector bundle, in analogy with the *simplicial* vector-bundle-valued forms, cf. Section 3.8; the form $\log b$ is defined on 2-whiskers, not just on 2-simplices, so is an example of such a $TM \to M$-valued 2-form in the whisker sense.)

Now given an affine connection λ on M; it may be re-interpreted as a linear bundle connection ∇ in the tangent bundle $TM \to M$, and therefore gives rise to covariant exterior derivatives d^∇ of TM-valued differential forms. We also write d^λ for this exterior derivative.

Theorem 4.8.1 *The covariant exterior derivative of the solder form θ agrees with* log *of the intrinsic torsion b of λ except for a factor 2:*

$$2 \cdot d^\lambda(\theta) = \log b.$$

The $TM \to M$-valued simplicial 2-form $d^\lambda(\theta)$, we call the *torsion form* of the affine connection λ.

Proof. Consider an infinitesimal 2-simplex (x,y,z). We calculate $d^\lambda(\theta)(x,y,z) \in T_xM$. We have

$$d^\lambda(\theta)(x,y,z) = \theta(x,y) - \theta(x,z) + (\nabla(x,y) \dashv \theta(y,z)).$$

Here ∇ denotes the transport law associated to λ, thus $\nabla(x,y) \dashv u = \lambda(y,x,u)$ for $u \sim y$. Let us calculate the value of this tangent vector at $d \in D$; recall the simple way (4.2.3) of adding tangent vectors, using affine combinations of mutual neighbour points; and recall (4.8.1) that the solder form was defined in terms of log, which in turn (4.3.1) was defined in terms of such affine combinations. Expressing the right-hand side here in terms of affine combinations, we get

$$\begin{aligned} &d^\lambda(\theta)(x,y,z)(d) \\ &\quad = \big[(1-d)\cdot x + d\cdot y\big] - \big[(1-d)\cdot x + d\cdot z\big] + \big[\nabla(x,y)((1-d)\cdot y + d\cdot z)\big]; \end{aligned}$$

now the map $\nabla(x,y) : \mathfrak{M}(y) \to \mathfrak{M}(x)$ preserves the multiplicative action by scalars (Proposition 2.3.7), so that we may continue the equation

$$= \big[(1-d)\cdot x + d\cdot y\big] - \big[(1-d)\cdot x + d\cdot z\big] + \big[(1-d)\cdot x + d\cdot \nabla(x,y)(z)\big].$$

This affine combination we can rewrite by simple arithmetic, and continue the equation

$$= (1-d)\cdot x + d\cdot \big[y - z + \lambda(y,x,z)\big] = (\log_x(y - z + \nabla(x,y)(z))(d).$$

So we have proved (writing again $\lambda(y,x,z)$ for $\nabla(x,y)(z)$)

$$d^\lambda(\theta)(x,y,z) = \log_x(y - z + \lambda(y,x,z)). \tag{4.8.2}$$

On the other hand, by Proposition 2.3.5, we have the first equality sign in

$$\log_x(b_x(y,z)) = \log_x(2[y - z + \lambda(y,x,z)] - x) = 2\log_x(y - z + \lambda(y,x,z)) \tag{4.8.3}$$

(the last equality sign is one of the rules for log, cf. (4.3.3)). Comparing (4.8.2) and (4.8.3) gives the result.

Corollary 4.8.2 *An affine connection λ is torsion free if and only if $d^\lambda(\theta) = 0$, where θ is the solder form.*

In a coordinatized situation, with M an open subset of a finite-dimensional vector space V, differential forms with values in $T(M) \to M$ may be identified with V-valued differential forms on M; in this case, (4.8.2) and the Taylor principle gives that the coordinate expression for the V-valued 2-form $d^{\lambda}(\theta)$ is given by the Christoffel symbol of λ,

$$d^{\lambda}(\theta)(x,y,z) = \Gamma(x; x-y, z-x).$$

There is a classical calculus of tangent-bundle-valued differential forms, involving the Frölicher–Nijenhuis bracket, Lie derivative, and contractions. Such calculus has been dealt with in synthetic terms by Nishimura (2008a), for differential forms in terms of "microcubes" (in the sense of Section 4.10 below).

4.9 Lie bracket of vector fields

Recall that a *vector field* on a manifold M is a cross-section of the tangent bundle $T(M) \to M$. Seeing the total space of $T(M)$ as the function space M^D, and the structural map $T(M) \to M$ as "evaluation at $0 \in D$", one can by sheer logic (cartesian closed categories) get three equivalent manifestations of this notion; these equivalences go back to Lawvere's 1967 lecture:

a map $X : M \to M^D$ such that $X(m)(0) = m$ for all $m \in M$

a map $X : M \times D \to M$ with $X(m,0) = m$ for all $m \in M$

a map $X : D \to M^M$ with $X(0) =$ the identity map of M.

The first of these manifestations is just saying that X is a cross-section of $T(M) \to M$; the second has the advantage of not mentioning any function space objects; and the third gives each $X(d)$ as a map $M \to M$, an *infinitesimal transformation*, a terminology and viewpoint which goes back to Sophus Lie. The infinitesimal transformation $X(d) : M \to M$ is often denoted X_d. It has X_{-d} for its inverse, see Kock (1981/2006, Corollary I.8.2).

This viewpoint of infinitesimal transformations leads classically to the idea that the group-theoretic *commutator* of infinitesimal transformations for two vector fields X, Y forms, "in the limit", the infinitesimal transformations of a new vector field $[X,Y]$, cf. e.g. Nelson (1967, §2.4). The rendering of this idea in the synthetic context (due to Reyes and Wraith, 1978) is well expounded in the literature, cf. e.g. Kock (1981/2006), Lavendhomme (1996, §3.2.2), and we shall not repeat it in full here. The basic picture is the following, reproduced from Kock (1981/2006):

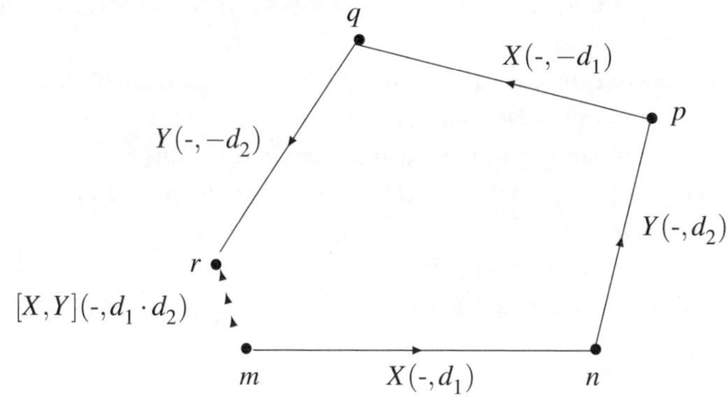

(4.9.1)

with $r = [X,Y](m,d_1 \cdot d_2)$. The defining equation for the Lie bracket may thus be written

$$[X,Y]_{d_1 \cdot d_2} = \{X_{d_1}, Y_{d_2}\},$$

where curly brackets denote group-theoretic commutator.

The pentagon itself makes sense for general microlinear spaces M, not just for manifolds. For manifolds, the following assertions make sense.

Points in this pentagon connected by a line segment are (first-order) neighbours, and also $m \sim r$, as well as $m \sim q$ (see Proposition 4.9.1 below); but m and p cannot be asserted to be neighbours in general, nor can n and q, see Exercise 4.10.4.

There are two other points which it is natural to put into the picture above, namely $q' := Y(m,d_2)$ and $p' := X(q',d_1)$. They lie approximately 1 cm southeast of q and p, respectively.

For fixed m, we may consider the points n, $p, \ldots, p'$ as functions of $(d_1,d_2) \in D \times D$ and write $n = n(d_1,d_2)$, $\ldots$, $p' = p'(d_1,d_2)$. This makes sense for a general microlinear space M, not just for a manifold; but for M a manifold, we can make some further assertions. Recall the "strong difference construction" $\tau' \dot{-} \tau$, described in Remark 2.1.9, or in the Appendix, Section A.4 (condition ML 4).

Proposition 4.9.1 *(1) For fixed (d_1,d_2), the points m, q and q' form an infinitesimal 2-simplex; and $r = m - q' + q$. (2) For $(d_1,d_2) \in D(2) \subseteq D \times D$,*

$$p(d_1,d_2) = p'(d_1,d_2) \text{ and } p' \dot{-} p = [X,Y](m,-).$$

Proof. For (1): Note that the second assertion in (1) makes invariant sense, because of the first: $m - q' + q$ is an affine combination of mutual neighbour points. For (2): Note that the second assertion in (2) makes sense because of the first: the strong difference of two maps $D \times D \to M$ which agree on $D(2)$ is defined, and is a tangent vector.

To prove these assertions, it suffices to consider the coordinatized case, i.e. with M an open subset of some finite-dimensional vector space V, so that the vector fields X and Y are given by their principal-part functions ξ and η, respectively; thus, $\xi : M \to V$ with $X(x,d) = x + d \cdot \xi(x)$ for $x \in M$ and $d \in D$; similarly for Y and η.

Then we can calculate the points in the figure, and also q' and p', in terms of ξ and η and their directional derivatives. This is a completely standard calculation:

$$\begin{aligned} n &= X(m,d_1) = m + d_1 \cdot \xi(m) \\ p &= Y(n,d_2) = m + d_1 \cdot \xi(m) + d_2 \cdot \eta(m + d_1 \cdot \xi(m)) \\ &= m + d_1 \cdot \xi(m) + d_2 \cdot (\eta(m) + d_1 \cdot d\eta(m;\xi(m))) \end{aligned}$$

by a Taylor expansion of η in the direction of $\xi(m)$;

$$\begin{aligned} q = X(p,-d_1) &= m + d_1 \cdot \xi(m) + d_2 \cdot (\eta(m) + d_1 \cdot d\eta(m;\xi(m))) \\ &\quad - d_1 \cdot \xi\left[m + d_1 \cdot \xi(m) + d_2 \cdot (\eta(m) + d_1 \cdot d\eta(m;\xi(m)))\right]. \end{aligned}$$

In the second line, we can put $d_1 = 0$ inside the square bracket, by the Taylor principle; using $\xi[m + d_2 \cdot \eta(m)] = \xi(m) + d_2 \cdot d\xi(m;\eta(m)))$ (Taylor expansion), the expression simplifies, and we end up with

$$= m + d_2 \cdot \eta(m) + d_1 \cdot d_2 \cdot (d\eta(m;\xi(m)) - d\xi(m;\eta(m))).$$

Using this expression for q, and the Taylor principle again, we arrive similarly at

$$r = Y(-d_2,q) = m + d_1 \cdot d_2 \cdot (d\eta(m;\xi(m)) - d\xi(m;\eta(m))).$$

Also,

$$q' = Y(m,d_2) = m + d_2 \cdot \eta(m).$$

Finally

$$p' = q' + d_1 \cdot \xi(q') = m + d_2 \cdot \eta(m) + d_1 \cdot \xi(m + d_2 \cdot \eta(m))$$

and Taylor expanding the last term, we thus get

$$p' = m + d_2 \cdot \eta(m) + d_1 \cdot \xi(m) + d_1 \cdot d_2 \cdot d\xi(m;\eta(m)).$$

The coordinate expressions for q and q' reveal immediately that m, q, and q' form an infinitesimal 2-simplex, and combining it with the coordinate expression for r, we see by pure additive calculation the validity of statement (1). For (2): we also see from the coordinate expressions derived that if $d_1 \cdot d_2 = 0$, then p and p' agree, proving that p and p' agree on $D(2) \subseteq D \times D$. The last assertion in (2) now again is by pure additive calculation. This proves the proposition.

The proposition provides us with an alternative way to construct the Lie bracket of two vector fields (cf. Kock and Lavendhomme, 1984; Kolar, 1982; White, 1982) in terms of strong difference:

$$[X,Y] = p' \dot{-} p.$$

Note that to construct the points p and p', we only need to know the value of the vector fields X and Y in the points m, n, and q', which all are 1-neighbours of m; in other words, to construct $p' \dot{-} p$, we only need to know the 1-jets at m of the vector fields X and Y. (In contrast, the construction of q in the figure (4.9.1) requires knowledge of $q = X(p,-)$, and p is not in general a 1-neighbour of m, only a 2-neighbour; so we need the 2-jet of X at m to construct $[X,Y](m,-)$ via the "pentagon" construction.)

For given $m \in M$, we have, by the strong-difference construction, constructed the Lie bracket formation as a map

$$(J^1TM)_m \times (J^1TM)_m \to T_mM.$$

So the Lie bracket is exhibited as a map of vector bundles

$$J^1TM \times_M J^1TM \to TM. \tag{4.9.2}$$

So $TM \to M$ is not only a vector bundle, but with the map (4.9.2) as structure, it is an *algebroid* (a notion we shall study in more generality in Chapter 5).

A further way of constructing the Lie bracket of vector fields is in terms of the Lie derivative construction, cf. Section 5.4 below. This construction likewise exhibits the Lie bracket as a bundle map (4.9.2).

Left-invariant vector fields on a Lie group

Let G be a Lie group (or just a microlinear group), i.e. a manifold (or just a microlinear space) equipped with a group structure. There is a classical correspondence between the set of left-invariant vector fields on G, and the space T_eG, the tangent space at the neutral element $e \in G$. In the present context, this correspondence is rendered as follows: to $\xi \in T_eG$, we associate the vector field $X : G \times D \to G$ given by $X(g,d) := g \cdot \xi(d)$. This vector field is left invariant in the sense that $X(h \cdot g, d) = h \cdot X(g,d)$. If left-invariant vector fields X and Y on G are given by $\xi \in T_eG$ and $\eta \in T_eG$, respectively, then an immediate calculation gives that, for $m \in G$,

$$(Y_{d_2}^{-1} \circ X_{d_1}^{-1} \circ Y_{d_2} \circ X_{d_1})(m) = m \cdot \xi(d_1) \cdot \eta(d_2) \cdot \xi(d_1)^{-1} \cdot \eta(d_2)^{-1},$$

from which we see that

$$[X,Y](m, d_1 \cdot d_2) = m \cdot \{\xi(d_1), \eta(d_2)\},$$

where $\{\xi(d_1), \eta(d_2)\}$ denotes the group-theoretic commutator of the elements $\xi(d_1)$ and $\eta(d_2)$ in G.

Thus, identifying $T_e G$ with the space of left-invariant vector fields, we see that the Lie bracket of two such vector fields is again left invariant, and that the Lie bracket on $T_e G$ induced by the identification may be described

$$[\xi, \eta](d_1 \cdot d_2) = \{\xi(d_1), \eta(d_2)\}. \tag{4.9.3}$$

4.10 Further aspects of the tangent bundle

Geometric distributions and vector fields

There are two ways to express when a (classical) geometric distribution is involutive; one is in terms of a set of differential 1-forms defining the distribution, the other is in terms of vector fields "subordinate" to the distribution. We dealt with the differential-form formulation in Section 3.6, proving the essential equivalence of this formulation with the combinatorial notion of involutive (modulo the comparison of exterior derivative and wedge product, see Theorem 4.7.3).

We shall state the formulation of "involutive" in terms of vector fields, and prove the implication from "combinatorial involutive" to "vector field involutive" (but not the converse implication; the converse would require *choice* or construction of suitable subordinate vector fields, and such techniques are not readily available in the synthetic context).

Given a combinatorial distribution $\approx$ on a manifold M, then a vector field X is called *subordinate* to $\approx$ if $X(m,d) \approx m$ for any $m \in M$ and $d \in D$ (note that automatically $X(m,d) \sim m$).

Theorem 4.10.1 *Let $\approx$ be an involutive distribution on M. Then if X and Y are vector fields subordinate to $\approx$, then also $[X,Y]$ is subordinate to $\approx$.*

Remark 4.10.2 This is almost immediate if we use the Frobenius Integrability Theorem; for, it is easy to see that a vector field is subordinate to $\approx$ precisely when each field vector $X(m,-)$ is a tangent vector to the leaf through m. So if X and Y are subordinate to $\approx$, they restrict to vector fields on each leaf, and the Lie bracket of vector fields on the leaf is again a vector field on the leaf.

We shall, however, give a proof that does not depend on the Frobenius Integrability Theorem. We first prove the special case where the two vector fields in question have a certain property (which in turn implies that all five points in the "commutator pentagon" (4.9.1) are mutual neighbours); more precisely

Lemma 4.10.3 *Consider two vector fields X and Y on M with the property that for any m in M, and d_1, d_2 in D, we have $X(m,d_1) \sim Y(m,d_2)$. Then if X and Y are subordinate to an involutive distribution $\approx$, then so is $[X,Y]$.*

Proof. We first see that all five points in the diagram (4.9.1) are mutual neighbours; for instance, m and p are neighbours, because $p = Y(n,d_2)$, $m = X(n,-d_1)$. Now $m \approx n$ because X is subordinate to $\approx$; similarly $n \approx p$ because Y is subordinate to $\approx$. From $m \sim p$ and the assumed involutivity of $\approx$, we now conclude $m \approx p$. Similarly $p \approx r$. But $m \sim r$ holds (even without any assumptions on X and Y), so we conclude, again by involutivity, that $m \approx r$, i.e. $m \approx [X,Y](m, d_1 \cdot d_2)$ for all d_1, d_2 in D. This is not quite to say that $m \approx [X,Y](m,d)$ for all $d \in D$, since we cannot assert that the multiplication map $D \times D \to D$ is surjective. However, the assumption that $\mathfrak{M}_{\approx}(m)$ is a linear subset of $\mathfrak{M}(m)$ implies that it can be carved out of $\mathfrak{M}(m)$ as the zero set of functions with values in R, and "R perceives the multiplication map to be surjective" (cf. Section 1.3). We conclude that $m \approx [X,Y](m,d)$ for all $d \in D$. So $[X,Y]$ is subordinate to $\approx$. This proves the lemma.

We can now prove the theorem. Given two vector fields X and Y subordinate to $\approx$. Since $\approx$ is a distribution (not just a pre-distribution), it follows that any linear combination of these vector fields is again subordinate to $\approx$. So for an arbitrary 2×2 matrix d_{ij}, the pair of vector fields X' and Y',

$$X' = d_{11}X + d_{12}Y, \quad Y' = d_{21}X + d_{22}Y$$

is subordinate to $\approx$. Furthermore, since the Lie bracket is bilinear and alternating, it follows that

$$[X',Y'] = \det(\underline{\underline{d}}) \cdot [X,Y].$$

If now furthermore $\underline{\underline{d}} \in \widetilde{D}(2,2)$, we conclude from the ideal properties of the $\widetilde{D}$s that X', Y' satisfy the special assumptions in the lemma. From the lemma, we therefore conclude that $[X',Y']$ is subordinate to $\approx$, and hence that $\det(\underline{\underline{d}}) \cdot [X,Y]$ is subordinate to $\approx$. Again, as in the proof of Lemma 4.10.3 and using the third cancellation principle from Section 1.3, we deduce that $[X,Y]$ is subordinate to $\approx$. This proves the theorem.

Exercise 4.10.4 The relation between X and Y assumed in this lemma: $X(a,d_1) \sim Y(a,d_2)$ is very strong; it need not obtain even between X and

X: $X(a,d_1)$ need not be $\sim X(a,d_2)$. Consider for instance the vector field X on R given by $X(a,d) = a + d$. One does not in general have $a + d_1 \sim a + d_2$ for all $(d_1,d_2) \in D \times D$. (This will be the case precisely when $(d_1,d_2) \in D(2)$.)

Contraction of a combinatorial differential form against a vector field

If X is a vector field on a manifold M, and $\overline{\omega}$ a classical differential k-form on M, there is an obvious way to construct a $k-1$-form $X \rightharpoonup \overline{\omega}$, namely

$$(X \rightharpoonup \overline{\omega})(\tau_1, \ldots, \tau_{k-1}) := \overline{\omega}(X(x_0), \tau_1, \ldots, \tau_{k-1}),$$

where the τ_is are tangent vectors at the same point $x_0 \in M$ and $X(x_0)$ is the field vector of X at this point.

This construction cannot immediately be paraphrased for simplicial k-forms on M; one would like to have $X \rightharpoonup \omega$ characterized by

$$d \cdot \left[(X \rightharpoonup \omega)(x_0, \ldots, x_{k-1})\right] := \omega(x_0, X(x_0,d), x_1, \ldots, x_{k-1}) \qquad (4.10.1)$$

for all $d \in D$ and for $x_0, \ldots, x_{k-1}$ an infinitesimal $k-1$-simplex. However, there is no reason why $X(x_0,d)$ and x_i (for $i > 0$) should be 1-neighbours, so the $k+1$-tuple given as input to ω cannot be asserted to be an infinitesimal k-simplex. So to describe the contraction of ω against X, one needs to invoke the bijective correspondence between simplicial forms and whisker forms. For, since $X(x_0,d) \sim x_0$, the input for ω in (4.10.1) is clearly a k-whisker if $(x_0, \ldots, x_{k-1})$ is a $k-1$-whisker.

Microcubes and marked microcubes

One of the aims of the various synthetic theories of differential forms is to have the notion of exterior derivative in geometric form, even prior to Stokes' Theorem (which classically is a description of the relationship between coboundary and exterior derivative, but in a form which makes a theory of integration a necessary preliminary).

In the theory of combinatorial forms, in the simplicial or cubical manifestation, the exterior derivative is the same as for simplicial and cubical cochains in algebraic topology, so is immediately of geometric nature, and defined in terms of the geometric *faces* of certain figures (infinitesimal simplices, respectively infinitesimal parallelepipeda). The whisker manifestation of combinatorial differential forms does not admit such geometrically evident coboundary formula, and the same applies to classical differential forms: neither an infinitesimal

k-whisker, nor a k-tuple of tangent vectors with the same base point, have natural faces.

In SDG, the earliest descriptions of exterior derivative in geometric terms replaced the idea of a k-tuple of tangent vectors (with same base point) with a richer kind of geometric figure, the *marked microcube*.† A k-dimensional *microcube* in a space M is a map $\tau : D^k \to M$; such a microcube gives rise to a k-tuple of tangent vectors, namely by restricting τ to the k "axes" of D^k, the ith axis $D \to D^k$ being given by the embedding $d \mapsto (0,\dots,d,\dots,0)$ (with d in the ith position). Such microcubes (as k ranges) do not quite form a cubical complex; the faces δ_i^0 may be defined, but the faces δ_i^1 not, because D "does not have a natural endpoint".

Therefore, one is led to consider a still richer kind of geometric figure, namely the *marked* microcube; a *marked* k-dimensional microcube in M consists of a microcube $\tau : D^k \to M$ *together with* an element $\underline{d} = (d_1,\dots,d_k) \in D^k$. A marked k-dimensional microcube does have $2k$ faces, as required in a cubical complex: $\delta_i^1(\tau,\underline{d})$ is the pair consisting of $\tau(-,-,\dots,d_i,-,\dots)$ together with $(d_1,\dots,\widehat{d_i},\dots,d_k) \in D^{k-1}$; and $\delta_i^0(\tau,\underline{d})$ is the pair consisting of $\tau(-,-,\dots,0,-,\dots,-)$ together with the same $k-1$-tuple. Then the marked microcubes do form a cubical complex.

The theory of microcubes, and marked microcubes, as a basis for a geometric theory of differential forms, has the advantage that it works for any M, not necessarily a manifold. But the input data for differential forms, in this manifestation, in some sense is too rich: a microcube contains information which is not first order. This richness of the input in turn means that more equations have to be imposed in order for a function, defined on microcubes, or on marked microcubes, to qualify as a differential form (see Proposition I.14.4 in Kock, 1981/2006). More objectively, under mild assumptions, any differential n-form gives the same value on two microcubes $D^n \to M$ which agree on $D(n) \subseteq D^n$ (see Kock, 1981/2006, Exercise I.14.2 for a sketch).

Let us also note that the bundle of microcubes on M is not a first-order bundle (in the terminology of Section 5.3), i.e. it does not carry a canonical action by the groupoid of 1-jets of M (except when $k = 1$). Similarly for marked microcubes.

Microcubes are used for other purposes than differential forms in White (1982) in the context of Riemannian geometry, and in several articles by Nishimura, including Nishimura (1997, 2008a,b).

† Terminology from Lavendhomme (1996); in Kock (1981/2006, §I.14), they are called infinitesimal singular rectangles; in Moerdijk and Reyes (1991, §IV.1), they are called infinitesimal cubes.

5

Groupoids

"The concept of groupoid is one of the means by which the twentieth century reclaims the original application of the group concept. The modern, rigorous concept of group is far too restrictive for the range of geometrical applications envisaged in the work of Lie. There have arisen the concepts of Lie pseudogroup, of differentiable and of Lie groupoid, and of principal bundle ..."†

We discuss in this chapter, from a synthetic viewpoint, some differential-geometric aspects of the theory of groupoids, principal bundles, and connections in groupoids (including principal connections); in most of the theory presented, the space of objects of the groupoid is assumed to be a manifold. For some of the theory, it is also assumed that the space of arrows is a manifold.

5.1 Groupoids

A groupoid is a category‡ where all arrows are invertible. We are interested in *small* groupoids; we here understand this as a groupoid internal to the category $\mathscr{E}$ of spaces; thus a (small§) groupoid $\mathbb{G}$ consists of a space G_1 of *arrows*, and a space G_0 of *objects*. To each arrow f is associated two objects, the domain $d_0(f)$ and the codomain $d_1(f)$ of the arrow, and this is exhibited

† From the introduction to Mackenzie (1987).

‡ The basic algebraic structure is thus that of *composition* of arrows. There comes the inevitable question whether to compose from the left to the right (diagrammatic order), or from the right to the left (function composition notation). Our experience is that it is not advisable to make the choice once and for all; for categories or groupoids whose arrows are actual *maps* or *functions*, the right-to-left notation is preferable, whereas for formulae and calculations in abstract categories/groupoids, the diagrammatic order is better. In the present exposition, we use both conventions, and state which of them is in use in a given formula, if this is not clear from the context. As an aid, we usually denote left-to-right composition by a dot, thus $f.g$ or $f \cdot g$ denotes "first f, then g"; for right-to-left composition, the same composition is denoted by the circle: "$g \circ f$". Plain concatenation "fg" or "gf" is used in both kinds of notation.

§ We usually omit the word "small".

graphically by

$$x \xrightarrow{\quad f \quad} y$$

with $x = d_0(f)$ and $y = d_1(f)$. Arrows can be *composed*, subject to the usual book-keeping conditions, and the composition is associative, has units, as well as inverses; thus

$$x \xrightarrow{\quad f \quad} y \xrightarrow{\quad f^{-1} \quad} x$$

compose to give the identity arrow id_x or $i(x)$ at x. We often display a groupoid $\mathbb{G}$, somewhat incompletely, with the hieroglyph $G_1 \rightrightarrows G_0$, the two maps displayed being "formation of domain" and "formation of codomain", respectively. Formation of inverse is a map $t : G_1 \to G_1$; it is an involution, and it interchanges d_0 and d_1. Formation of identities is a map $i : G_0 \to G_1$. The space G_1 we sometimes call the *total space* of the groupoid.

A groupoid with only one object is a group (more precisely, the space of arrows of it is a group). Just like groups encode the notion of *global* symmetry, groupoids are encoding *local*, or even infinitesimal, symmetries, and in this role, they appear already implicitly in the work of S. Lie, in his work on differential equations, contact transformations, etc.; their role in differential geometry was made more explicit through the work of C. Ehresmann.

A groupoid is in particular a category. A *homomorphism* of groupoids is the same as a functor. So if $\mathbb{G} = (G_1 \rightrightarrows G_0)$ and $\mathbb{H} = (H_1 \rightrightarrows H_0)$ are groupoids, a homomorphism $F : \mathbb{G} \to \mathbb{H}$ consists of maps $F_0 : G_0 \to H_0$, $F_1 : G_1 \to H_1$, commuting with the book-keeping maps (domain- and codomain-formation), and commuting with composition, and with the formation of identity arrows (it then automatically commutes with inversion).

We sometimes say that such F is a homomorphism *over* F_0, and if F_0 is the identity map of $G_0 = H_0 = M$, we say that it is a homomorphism *over* M.

We are mainly interested in the case of groupoids $\Phi \rightrightarrows M$ where the space M is a manifold, but we don't usually assume that the space Φ is a manifold. In Example 5.1.2 below, $\mathfrak{S}(E \to M) \rightrightarrows M$ will not in general have a manifold as total space, even when E and M are manifolds.

Example 5.1.1 Given a space M, there are two "extreme" groupoids with M as space of objects, namely the "discrete" groupoid $M \rightrightarrows M$ (both the displayed maps $M \to M$ being the identity map of M), and the "codiscrete" groupoid $M \times M \rightrightarrows M$ (the displayed maps being the two projections). In the discrete groupoid on M, the only arrows are the identity arrows id_x; in the codiscrete case, for every pair of objects x, y, there is exactly one arrow from x to y.

If $\mathbb{G} = (G_1 \rightrightarrows M)$ is a groupoid, there is exactly one homomorphism over M from the discrete groupoid on M to $\mathbb{G}$, and there is exactly one homomorphism over M from $\mathbb{G}$ to the codiscrete groupoid on M.

Example 5.1.2 Let $\pi : E \to M$ be a map (a bundle over M). We have a groupoid $\mathfrak{S}(E \to M)$ with M as space of objects, and where an arrow from $x \in M$ to $y \in M$ is a bijective map $E_x \to E_y$. (For the existence of such as an object in the category $\mathscr{E}$, one needs local cartesian closedness of $\mathscr{E}$.) For $M = 1$, i.e. for a space E, this is the standard "symmetric group" $\mathfrak{S}(E)$ of permutations of E. The groupoid $\mathfrak{S}(E \to M)$ deserves the name "symmetric groupoid on $E \to M$". (In topos-theoretic terms, this is the groupoid of invertible arrows in the internal category $Full(\pi)$, as described in Johnstone, 1977, Example 2.3.8.)

In case $\pi : E \to M$ is a *vector* bundle (i.e. each individual E_x is equipped with the structure of a vector space), we have a subgroupoid $GL(E \to M)$ of $\mathfrak{S}(E \to M)$, namely an arrow from x to y is a *linear* bijection $E_x \to E_y$. For $M = 1$, i.e. for a vector space E, this is the standard "general linear group" $GL(E)$. The groupoid $GL(E \to M)$ deserves the name "general linear groupoid on $E \to M$".

In case $\pi : E \to M$ is a bundle of *pointed* spaces (i.e. each individual E_x is equipped with a base point p_x or equivalently, there is given a cross-section p of π), we have a subgroupoid $\mathfrak{S}_*(E \to M)$ of $\mathfrak{S}(E \to M)$, namely an arrow from x to y is a base-point-preserving bijection $E_x \to E_y$.

One has similar groupoids for bundles where the fibres have some other kind of algebraic structure, e.g. group bundles. For a bundle $G \to M$ of groups, one might write $Iso(G \to M) \rightrightarrows M$, for the groupoid of group isomorphisms between the fibres; for a single group G (= a bundle of groups over 1), one would traditionally rather write $Aut(G)$ ("*auto*morphism group" of G), but when more than one group is involved, "iso" is more adequate than "auto".

Example 5.1.3 Let M be a manifold, and k a non-negative integer. Then we have the "k-jet groupoid $\Pi^{(k)}(M) \rightrightarrows M$", where an arrow from $x \in M$ to $y \in M$ is an invertible k-jet from x to y, i.e. a bijective map $\mathfrak{M}_k(x) \to \mathfrak{M}_k(y)$ taking x to y. (There are also groupoids whose arrows are non-holonomous jets, cf. Kock, 1985.) The groupoid $\Pi^{(k)}(M) \rightrightarrows M$ can be seen as arising from the bundle $M_{(k)} \to M$ of pointed sets: $(\Pi^{(k)}(M) \rightrightarrows M) = \mathfrak{S}_*(M_{(k)} \to M)$.

Note that there are evident "restriction" functors $\Pi^{(l)}(M) \to \Pi^{(k)}(M)$ for $l \geq k$; for, $\mathfrak{M}_k(x) \subseteq \mathfrak{M}_l(x)$, and a bijective base-point-preserving $\mathfrak{M}_l(x) \to \mathfrak{M}_l(y)$ restricts to a bijective base-point-preserving $\mathfrak{M}_k(x) \to \mathfrak{M}_k(y)$.

There is an isomorphism of groupoids over M

$$(\Pi^{(1)}(M) \rightrightarrows M) \cong GL(T(M) \to M); \tag{5.1.1}$$

this follows from Theorem 4.3.4.

Example 5.1.4 Let $S \subseteq M \times M$ be an equivalence relation on a space M. It may be viewed as a groupoid $S \rightrightarrows M$, with $(x,y) \in S$ being considered as an arrow from x to y. The transitive law for S gives the composition, the symmetry for S gives inversion, and reflexivity gives the units.

Example 5.1.5 Let G be a group. Then for any M, there is a groupoid over M, whose space of arrows is $M \times M \times G$; domain and codomain formation are the first two projections, and composition is given by

$$(x,y,a) \cdot (y,z,b) := (x,z,a \cdot b),$$

where $\cdot$ denotes the multiplication in G. A groupoid of this form, we call a *constant* groupoid.

Exercise 5.1.6 If M is an open subset of a finite-dimensional vector space V, $\Pi^{(1)}(M)$ is isomorphic to $M \times M \times GL(V) \rightrightarrows M$. (*Hint:* Use Proposition 1.3.1.)

Example 5.1.7 A group bundle $E \to M$ (i.e. each fibre E_x is equipped with a group structure) may be viewed as a groupoid over M, with the special property that there only exist arrows $x \to y$ when $x = y$.

Example 5.1.8 Given a groupoid $\Phi \rightrightarrows M$, we get a group bundle $gauge(\Phi) \to M$, the *gauge group bundle* (terminology from Mackenzie, 1995) of $\Phi \rightrightarrows M$, whose fibre over $x \in M$ is the group $\Phi(x,x)$ of endo-arrows of Φ at x. If $f : x \to y$ is an arrow in Φ, we get a group homomorphism from $\Phi(x,x) \to \Phi(y,y)$ "by conjugation by f", i.e. it is given by (in right-to-left notation for composition)

$$\phi \mapsto f \circ \phi \circ f^{-1}$$

for $\phi \in \Phi(x,x)$.

We have a subgroupoid $Gauge(\Phi)$ of $\mathfrak{S}(gauge(\Phi) \to M)$; $Gauge(\Phi)(x,y)$ consists of those maps $\Phi(x,x) \to \Phi(y,y)$ that happen to be group homomorphisms.

We have a morphism ad of groupoids over M, $ad : \Phi \to Gauge(\Phi)$; ad takes $f : x \to y$ in Φ into $\phi \mapsto f \circ \phi \circ f^{-1}$ for $\phi \in \Phi(x,x)$ (using right-to-left composition).

5.2 Connections in groupoids

Graphs

We first introduce a notion of reflexive symmetric graph, of which both groupoids, the first neighbourhood of the diagonal of a manifold, and also the path space of a manifold (see Example 5.2.3 below), are examples.

A *reflexive symmetric graph* **X** is a pair of sets (X_1, X_0), together with four maps: $d_0 : X_1 \to X_0$, $d_1 : X_1 \to X_0$, $i : X_0 \to X_1$, and $t : X_1 \to X_1$. The elements of X_0 are the *vertices* of the graph, the elements of X_1 the *edges*; for $u \in X_1$, $d_0(u)$ (resp. $d_1(u)$) is the *source* (resp. the *target*) vertex of u. The symmetry is a *structure*, namely a map $t : X_1 \to X_1$; it is assumed to be an involution ($t \circ t = \text{id}$), and to interchange source and target, i.e. $d_0 \circ t = d_1$ and $d_1 \circ t = d_0$. Also, reflexivity of the graph is a structure, given by the map $i : X_0 \to X_1$. We assume $d_0 \circ i = d_1 \circ i = \text{id}$ and $t \circ i = i$. The morphisms $\xi : X \to X'$ in the category of reflexive symmetric graphs are pairs of maps $\xi_0 : X_0 \to X_0'$, $\xi_1 : X_1 \to X_1'$ which commute with the four structural maps in an evident sense.

We will often denote a graph **X**, as above, with the hieroglyph $X_1 \rightrightarrows X_0$. When we say "graph" and "graph-morphism" in the following, it will be understood that we mean "reflexive symmetric graph (-morphism)".

(To whom it may concern: the category of reflexive symmetric graphs may be seen as symmetric simplicial sets, truncated in dimension 1; or also, as the category of symmetrical cubical sets, truncated in dimension 1; or, as the category of presheaves on the category $\{\mathbf{1}, \mathbf{2}\}$ consisting of all maps between one- and two-point sets.)

Example 5.2.1 The reflexive symmetric graph arising from a groupoid (the underlying graph of the groupoid) has the objects of the groupoid as its vertices, the arrows for its edges, t is inversion in the groupoid $t(u) = u^{-1}$, and $i(x)$ is the identity arrow id_x at the object x. We also call the graph thus obtained the *underlying graph* of the groupoid.

The functor from groupoids to graphs thus described is faithful, but not full. In particular, a map of reflexive symmetric graphs between the underlying graphs of groupoids preserves identities and inversion (by definition), but *does not* necessarily preserve composition. This is related to the notion of *curvature* or *flatness* of connections, as will be considered below.

The following example is the main structure in the present book.

Example 5.2.2 If M is a manifold, we can see the first neighbourhood of the diagonal of M as a reflexive symmetric graph $M_{(1)} \rightrightarrows M$. Its vertices are the elements of M and its edges are (ordered) pairs (x,y) of neighbour points $x \sim y$, $d_0((x,y)) = x$, $d_1((x,y)) = y$. Since the 1-neighbour relation is symmetric, we have an involution t given by $t(x,y) = (y,x)$, and i is given by $i(x) = (x,x)$.

The functor from manifolds to graphs thus described is full and faithful.

The kth neighbourhood of the diagonal of M is likewise a graph. Also, the "non-holonomous monads" considered in Section 2.7 can be seen as arising from graphs.

For a graph arising as the first neighbourhood of the diagonal of a manifold M, the map $(d_0,d_1) : M_1 \to M_0 \times M_0$ is mono; and the existence of the involution t is therefore a *property* rather than an added structure. For the underlying graph of a groupoid, neither of these simplifications obtain.

Example 5.2.3 "The" fundamental graph $P(M)$ of paths of a manifold: there are several candidates. We are not asserting any composition structure (unlike in the fundamental groupoid), but only graph structure, on $P(M)$, and do not bother about concatenation of paths. The simplest version of $P(M)$ has for its total space the space of maps $R \to M$, and for the two structural maps $P(M) \rightrightarrows M$ evaluation at 0 and 1, respectively. The symmetry comes about by precomposing by the affine map $R \to R$ consisting of reflection in $1/2$. The reflexivity structure j comes about by precomposing with the constant map $R \to 1$, equivalently $j(x)$ is the map $R \to M$ with constant value x. This graph is identical to the 1-skeleton $S_{[1]}(M), S_{[0]}(M)$ of the complex of singular cubes, as considered in the Appendix. As expounded there, it carries some further structure, "subdivision", which we shall exploit in Section 5.8.

There also exists a graph of *piecewise* paths, see Section 5.8.

Example 5.2.4 Let M be a manifold. Then there is a canonical morphism of reflexive symmetric graphs

$$[-,-] : M_{(1)} \to P(M) \tag{5.2.1}$$

defined using affine combinations of neighbour points, as in Section 2.1; explicitly, if $x \sim y$ in M, we have the path $[x,y] : R \to M$ given by $t \mapsto (1-t)\cdot x + t\cdot y$. This construction takes the "edge" (x,y) in the graph $M_{(1)}$ to the "edge" $[x,y] : R \to M$ in the graph $P(M)$. Both edges have domain x and codomain y, and it is also easy to see that the reflexivity and symmetry structures of these graphs are preserved by the $[-,-]$-construction.

Base change (full image)

For groupoids as well as for graphs, one has the notion of *full image*: if $\Phi = (\Phi \rightrightarrows M)$ is a groupoid, and $f : N \to M$ is a map, there is a groupoid $f^*(\Phi \rightrightarrows M)$ over N, where an arrow in $f^*(\Phi)$ from u to v $(u, v \in N)$ is by definition an arrow in Φ from $f(u)$ to $f(v)$; and similarly for graphs (using the terms "vertex" and "edge", rather than "object" and "arrow"). For the groupoid case, there is an evident composition of arrows in $f^*(\Phi)$, making it into a groupoid; this is the *full image of* $\Phi \rightrightarrows M$ *under* f. (Thus, groupoids in fact form a fibred category over the category of manifolds $M, N, \ldots$)

The groupoid $M \times M \times G$ considered in Example 5.1.5 is a special case: it is the full image of the groupoid $G \rightrightarrows 1$ along the unique map $M \to 1$.

A functor $\xi : (\Psi \rightrightarrows N) \to (\Phi \rightrightarrows M)$ given by $\xi_0 : N \to M$ and $\xi_1 : \Psi \to \Phi$ may be identified with a functor over N, $(\Psi \rightrightarrows N) \to \xi_0^*(\Phi \rightrightarrows M)$.

Connections

We can now describe the notion of *connection in a groupoid*, as it may be rendered in the language of SDG, cf. Kock (1980, Remark 6.4; 1989, 1996).

More precisely, we are discussing the notion of "*infinitesimal connection*" or even more pedantically, "*first-order infinitesimal connection*".

We consider a groupoid $\Phi \rightrightarrows M$ whose space of objects is a manifold. We shall compose from left to right in Φ. Recall the graph $M_{(1)} \rightrightarrows M$ associated to a manifold M ("first neighbourhood of the diagonal").

Definition 5.2.5 *A connection* ∇ *in a groupoid* $\Phi \rightrightarrows M$ *is a morphism of reflexive symmetric graphs from* $M_{(1)}$ *to (the underlying graph of)* Φ.

In other words, if $(x, y) \in M_{(1)}$ (i.e. if $x \sim y$), $\nabla(x, y)$ is an arrow $x \to y$ in Φ; and the following laws hold:

$$\nabla(x, x) = \mathrm{id}_x \tag{5.2.2}$$

for all $x \in M$, and, for all $x \sim y$,

$$\nabla(y, x) = (\nabla(x, y))^{-1}. \tag{5.2.3}$$

In the context of SDG, it often happens that (5.2.3) is a consequence of (5.2.2). Thus, (2.5.2) can be reinterpreted as asserting (5.2.3) for a certain type of groupoid.

Example 5.2.6 Let λ be an affine connection on a manifold M. It gives rise to a connection in the groupoid $\Pi^{(1)}(M) \rightrightarrows M$ of invertible 1-jets; in fact, for $x \sim y$

in M, the transport law $\nabla(y,x) : \mathfrak{M}(x) \to \mathfrak{M}(y)$ (cf. (2.3.14) is an invertible 1-jet from x to y (cf. Propositions 2.3.2 and 2.3.3), thus an arrow $x \to y$ in the groupoid. It is an easy exercise to see that this correspondence between affine connections and groupoid-theoretic connections in $\Pi^{(1)}(M) \rightrightarrows M$ is in fact a bijection. Also, since $\Pi^{(1)}(M) \rightrightarrows M$ is isomorphic to $GL(TM \to M) \rightrightarrows M$ (cf. (5.1.1)), affine connections may also be construed as groupoid-theoretic connections in $GL(TM \to M) \rightrightarrows M$.

This groupoid-theoretic connection notion subsumes the notion of bundle connection: a bundle connection in a bundle $E \to M$ can be construed as a connection in the groupoid $\mathfrak{S}(E \to M)$; to say that a bundle connection in a vector bundle $E \to M$ is linear is to say that it is a connection in the groupoid $GL(E \to M)$.

The notion of flatness of a bundle-connection is (via the comparison of Proposition 5.2.10 below) a special case of the following. Let $\Phi \rightrightarrows M$ be a groupoid, with M a manifold.

Definition 5.2.7 *A connection* ∇ *in* $\Phi \rightrightarrows M$ *is called* flat *or* curvature free *if*

$$\nabla(x,y) \cdot \nabla(y,z) = \nabla(x,z) \tag{5.2.4}$$

whenever $x \sim y$, $y \sim z$, *and* $x \sim z$.

Definition 5.2.8 *Let* ∇ *be a connection in the groupoid* $\Phi \rightrightarrows M$. *The* curvature *of* ∇ *is the law* $R = R_\nabla$, *which to an infinitesimal 2-simplex* (x,y,z) *in* M *associates the arrow*

$$R(x,y,z) := \nabla(x,y) \cdot \nabla(y,z) \cdot \nabla(z,x) \in \Phi(x,x).$$

Thus (since $\nabla(x,z)$ and $\nabla(z,x)$ are mutually inverse), $R(x,y,z)$ is an identity arrow in case two of the three "vertices" are equal; and a connection ∇ is flat iff all values of R are identity arrows in Φ. The curvature R will be an example of a (simplicial) 2-form with values in the group bundle $gauge(\Phi) \to M$, in a sense which will be made more explicit in Chapter 6.

Affine connections may be seen as connections in the groupoid $\mathfrak{S}_*(M_{(1)} \to M)$.

Let $F : \Phi \to \Psi$ be a morphism (functor) between groupoids over M. Then F is also a morphism of reflexive symmetric graphs over M, and therefore from a connection $\nabla : M_{(1)} \to \Phi$ in Φ, we get by composition a connection $F \circ \nabla$ in Ψ.

This applies in particular to the morphism of groupoids over M

$$ad : \Phi \to Gauge(\Phi) \subseteq \mathfrak{G}(gauge(\Phi))$$

considered in Example 5.1.8. Thus, if ∇ is a connection in a groupoid $\Phi \rightrightarrows M$, we get a connection $ad \circ \nabla$ in the groupoid $Gauge(\Phi) \rightrightarrows M$. This connection we denote just $ad\nabla$:

$$(ad\nabla)(x,y) = \text{ conjugation by } \nabla(x,y) = [\phi \mapsto \nabla(y,x) \cdot \phi \cdot \nabla(x,y)].$$

If $f : N \to M$ is a map between manifolds, and ∇ is a connection in a groupoid $\Phi \rightrightarrows M$, then since f preserves $\sim$, it is easy to see that we get a connection $f^*(\nabla)$ in the groupoid $f^*(\Phi) \rightrightarrows N$ (= the full image of Φ along f), namely

$$f^*(\nabla)(n_1, n_2) := \nabla(f(n_1), f(n_2))$$

for $n_1 \sim n_2$ in N. We may call this the *pull back of the connection* ∇ *along* f.

Given a group bundle $G \to M$ with M a manifold, there is a group $\Omega^1(G \to M)$ consisting of "1-forms with values in $G \to M$", meaning maps $\omega : M_{(1)} \to G$ with $\omega(x,y) \in G_x$, and $\omega(x,x) = e_x$ (= the unit element in the group G_x). (Group-bundle-valued differential forms will be studied in more detail in Chapter 6.)

Proposition 5.2.9 *The space C of connections in $\Phi \rightrightarrows M$ carries a canonical left action by the group $G = \Omega^1(gauge(\Phi))$ of 1-forms with values in the gauge group bundle of $\Phi \rightrightarrows M$, and with this action, it is a translation space over G.*

(The sense of the term "translation space" is the evident one: given two connections in Φ, there is a unique $g \in G$ which takes the one connection to the other, by the left action described.)

Proof. Given a connection ∇ in Φ, and a 1-form ω with values in the gauge group bundle, we get the connection $\omega \cdot \nabla$ in Φ by putting

$$(\omega \cdot \nabla)(x,y) := \omega(x,y) \cdot \nabla(x,y). \tag{5.2.5}$$

(Note that $\omega(x,y)$ is an endo-arrow at x, so that the composition here does make sense.) Given two connections ∇ and Γ in Φ, their "difference" $\Gamma \cdot \nabla^{-1}$ is the gauge-group-bundle-valued 1-form given by

$$(\Gamma \cdot \nabla^{-1})(x,y) := \Gamma(x,y) \cdot \nabla(y,x) = \Gamma(x,y) \cdot (\nabla(x,y))^{-1}. \tag{5.2.6}$$

Then clearly

$$(\Gamma \cdot \nabla^{-1}) \cdot \nabla = \Gamma,$$

and it is unique with this property. The verifications are trivial.

Proposition 5.2.10 *Let $E \to M$ be a bundle. Then there is a natural bijective correspondence between bundle connections in it, and connections in the groupoid $\mathfrak{S}(E \to M)$.*

Proof/Construction. Given a bundle connection ∇, we define a groupoid connection $\hat{\nabla}$ as follows: for $x \sim y$ in M, the arrow $\hat{\nabla}(x,y) \in \mathfrak{S}(E \to M)(x,y)$ is the map

$$e \mapsto \nabla(y,x)(e),$$

for $e \in E_x$; conversely, given a groupoid connection ∇ in $\mathfrak{S}(E \to M)$, we define a bundle connection $\check{\nabla}$ by putting

$$\check{\nabla}(y,x)(e) := \text{ the value of } \nabla(x,y) \text{ on } e.$$

It is trivial to check that the two processes are inverse of each other. The interchange of the order in which x and y occur is due to the fact that we see the $\nabla(y,x)$ of a bundle connection as a *function* $E_x \to E_y$ and that we use standard functional notation for its action on elements, i.e. write it on the left of the argument. For the groupoid theory, we compose diagrammatically (left to right).

In case $E \to M$ is a bundle with some fibrewise algebraic structure, there is similarly a bijective correspondence between bundle connections that preserve this structure, and connections in the appropriate subgroupoid of $\mathfrak{S}(E \to M)$.

We shall adopt a notational shortcut in the statement and proof of the following result, by writing xy for $\nabla(x,y) : x \to y$. Also, we omit commas; upper right indices denote conjugation. Then we have (cf. Kock, 1996)

Theorem 5.2.11 (Combinatorial Bianchi Identity) *Let ∇ be a connection in a groupoid Φ, and let R be its curvature. Then for any infinitesimal 3-simplex (x,y,z,u),*

$$\mathrm{id}_x = R(yzu)^{(yx)}.R(xyu).R(xuz).R(xzy).$$

We shall below (Theorem 6.4.1) interpret the expression here as the *coboundary* or *covariant derivative* $d^{\nabla}(R)$ of R in a "complex" of group-bundle-valued forms. This is still a purely "combinatorial" gadget, but we shall later specialize to the case of the general linear groupoid of a vector bundle with connection, and see that our formulation contains the classical Bianchi Identity.

Proof. With the streamlined notation mentioned, the identity to be proved is

$$xy.(yz.zu.uy).yx.(xy.yu.ux).(xu.uz.zx).(xz.zy.yx) = \mathrm{id}_x;$$

the proof is now simply repeated cancellation: first remove all parentheses, then keep cancelling anything that occurs, or is created, which is of the form $yx.xy$, etc., using (5.2.3); one ends up with nothing, i.e. the identity arrow at x. (This is essentially Ph. Hall's "14-letter identity" from group theory.)

The piece of drawable geometry which is the core of the theorem is the following: consider a tetrahedron. *Then the four triangular boundary loops compose (taken in a suitable order) to a loop which is null-homotopic inside the 1-skeleton of the tetrahedron* (the loop opposite the first vertex should be "conjugated" back to the first vertex by an edge, in order to be composable with the other three loops). This is the Homotopy Addition Lemma, in one of its forms, cf. e.g. Brown and Higgins (1981) (notably Proposition 2).

Note that the theorem applies to any (non-commutative) tetrahedron of arrows in an arbitrary groupoid. There also exist cubical versions, deriving from a combinatorial identity for the arrows of a (non-commutative) cube in an arbitrary groupoid, cf. Nishimura (2008b).

Remark 5.2.12 A connection in a groupoid $\Phi \rightrightarrows M$ may be seen as a cross-section in a certain bundle over M, the bundle of *connection elements*. If $x \in M$, a connection element in Φ at x is a section 1-jet δ of the bundle $d_1 : \Phi \to M$ with $\delta(x) = \mathrm{id}_x \in \Phi$, which furthermore satisfies $d_0(\delta(x_1)) = x$ for all x_1. Thus, δ associates to each x_1 with $x \sim x_1$ an arrow in ϕ from x to x_1, and to x itself it associates id_x.

If we comprehend the connection elements for all $x \in M$, we get a bundle over M, the bundle of connection elements in Φ. A cross-section of this bundle thus associates to each $x \in M$ a connection element δ_x at x, and it defines a connection ∇ in Φ by putting

$$\nabla(x,y) = \delta_x(y) : x \to y$$

for $x \sim y$, and conversely, a connection ∇ in Φ defines a section: for $x \in M$, $\delta_x : \mathfrak{M}(x) \to \Phi$ is given by the same equation, now read from the right to the left.

5.3 Actions of groupoids on bundles

When discussing groupoids abstractly, we continue to compose from the left to the right. Therefore also, we consider actions of groupoids $G \rightrightarrows M$ on bundles $\pi : E \to M$ as *right* actions. Actions are supposed to be associative and unitary. Usually, we just write the action by concatenation, just as the composition in a groupoid is written by concatenation; however, sometimes it is clarifying to have separate symbols for an action, and for the composition; we choose $\vdash$ for the action, and a dot "$\cdot$" for groupoid composition. With this notation, the associative law for the action reads (assuming the relevant book-keeping conditions, $\pi(a) = d_0(g)$, $d_1(g) = d_0(h)$):

$$(a \vdash g) \vdash h = a \vdash (g \cdot h).$$

(Mnemotechnical device for $a \vdash g$: think of $\vdash$ as a hammer; $g \in G$ *acts on* (hammers on) $a \in E$.) Similarly, the unitary law reads

$$a \vdash 1 = a$$

where 1 is the identity arrow at $\pi(a) \in M$. *Left* actions may similarly be denoted $\dashv$. The "hammer" symbol is supposed to bind stronger than other connectives, thus $a + h \dashv b$ means $a + (h \dashv b)$.

An action of a groupoid $\Phi \rightrightarrows M$ on a bundle $E \to M$ may be re-interpreted as a morphism (functor) of groupoids over M from $\Phi \rightrightarrows M$ to $\mathfrak{S}(E \to M)$, and vice versa. The proof is essentially as that of Proposition 5.2.10.

If a bundle $E \to M$ carries some fibrewise algebraic structure, it makes sense to say that an action on $E \to M$ by a groupoid $\Phi \rightrightarrows M$ *preserves* this structure: this is just to say that for each $g \in \Phi$ from x to y, say, the map $a \mapsto a \vdash g$ from E_x to E_y is a homomorphism of the kind of algebraic structure considered. This notion may also be encoded by describing a certain subgroupoid of $\mathfrak{S}(E \to M)$; thus, if $E \to M$ is a vector bundle, a $\Phi \rightrightarrows M$-action by *linear* maps may be encoded as a groupoid homomorphism $\Phi \to GL(E \to M) \subseteq \mathfrak{S}(E \to M)$.

Example 5.3.1 Recall from Example 5.1.8 the gauge group bundle $gauge(\Phi) \to M$ derived from a groupoid $\Phi \rightrightarrows M$. The groupoid Φ acts on the right on $gauge(\Phi) \to M$, by group isomorphisms, using conjugation:

$$a \vdash g := g^{-1} \cdot a \cdot g \tag{5.3.1}$$

where $a : x \to x$; so $a \in (gauge(\Phi))_x$, and $g : x \to y$. This is a re-interpretation of the groupoid homomorphism $ad : \Phi \to Gauge(\Phi)$ of Example 5.1.8.

One calls it the *adjoint action* of a groupoid $\Phi \rightrightarrows M$ on its gauge group bundle; this action clearly preserves the fibrewise group structure. The groupoid $\Phi \rightrightarrows M$ also acts on the left on $gauge(\Phi)$,

$$g \dashv a := g \cdot a \cdot g^{-1}. \tag{5.3.2}$$

The discrete groupoid $M \rightrightarrows M$ on M acts, in a unique way, on any bundle over M. For, the only arrows in the category are identity arrows, and their action is determined by the unitary law.

On the other hand, an action by the codiscrete groupoid $M \times M \rightrightarrows M$ on a bundle $\pi : E \to M$ is a very strong kind of structure: it amounts to a *trivialization* of the bundle, i.e. it implies that the bundle is isomorphic to a product bundle $M \times F \to M$ for some F. Namely, take F to be the orbit space of the action, so there is a canonical surjection $p : E \to F$. Define a morphism $\phi : E \to M \times F$ by sending $e \in E$ to $(\pi(e), p(e))$. To construct an inverse ψ for ϕ, consider $(x, f) \in M \times F$. Since p is surjective, we may pick some $e' \in E$ with $p(e') = f$. Let $x' := \pi(e')$. Then we put $\psi(x, f) := e' \vdash (x', x)$. This does not depend on the choice of e'; for if e'' is another choice, $p(e'') = p(e') = f$, so e'' and e' are in the same orbit for the action, i.e. there exists an arrow $g : x'' \to x'$ with $e'' \vdash g = e'$. But there is only one arrow $x'' \to x'$ in $M \times M \rightrightarrows M$, namely (x'', x'). Then

$$e'' \vdash (x'', x) = e'' \vdash (x'', x') \vdash (x', x) = e' \vdash (x', x).$$

Note that for any $x \in X$, E_x is mapped bijectively to F by p, so in this sense F "is" the fibre of $E \to M$.

Note also that a trivialization of a bundle $E \to M$ does not necessarily preserve a fibrewise algebraic structure that may be present in the bundle. Thus, a vector bundle may conceivably have trivial underlying bundle, without being trivial as a vector bundle.

A *kth-order natural structure* on a bundle $E \to M$ is an action of $\Pi^{(k)}(M) \rightrightarrows M$ on E.

A zero-order natural structure is thus an action by $\Pi^{(0)}(M)$. Since this groupoid is isomorphic to the codiscrete groupoid $M \times M \rightrightarrows M$, a zero-order natural structure on $E \to M$ amounts to a trivialization of the bundle, $E \cong M \times F$. This is a very strong kind of structure.

Generally, an action by $\Pi^{(k)}(M)$ for a *low* k provides a stronger structure than an action for a *high* k. This is particularly clear when viewing actions on a bundle $E \to M$ as functors into $\mathfrak{S}(E \to M)$. For, if $l \geq k$, we have a canonical (restriction) functor

$$\Pi^{(l)}(M) \to \Pi^{(k)}(M)$$

over M; so viewing an action by $\Pi^{(k)}(M)$ on $E \to M$ as a functor $\Pi^{(k)}(M) \to \mathfrak{S}(E \to M)$, one gets an $\Pi^{(l)}(M)$ action by precomposing with the restriction functor.

Traditionally, one says then that a vector bundle is *tensorial* or is a *tensor bundle* if it is equipped with a *first*-order natural structure (assumed to be compatible with the fibrewise linear structure).

Examples of first-order bundles:

$M_{(1)} \to M$ carries a canonical first-order natural structure. This is almost tautological: an arrow in $\Pi^{(1)}(M)$ from $x \in M$ to $y \in M$ is a bijection $f : \mathfrak{M}_1(x) \to \mathfrak{M}_1(y)$ taking x to y. The fibre of $M_{(1)} \to M$ over x is $\mathfrak{M}_1(x)$, so if $x' \sim_1 x$, $f(x')$ makes sense and belongs to $\mathfrak{M}_1(y)$, and this defines the action. It is better, however, to write $x'f$ rather than $f(x')$, since we want composition in $\Pi^{(1)}(M)$ to be from left to right. So, explicitly:

$$x' \vdash f := x'f \tag{5.3.3}$$

for $x' \sim_1 x$ and f an invertible 1-jet from x to y. (Here, we identify the fibre of $M_{(1)}$ over x with $\mathfrak{M}_1(x)$.)

The bundles $M_{<k>} \to M$, $Wh^k(M) \to M$, and $M_{[k]} \to M$ all carry first-order structure, by the similar formula, e.g.

$$(x_0, x_1, \ldots, x_k) \vdash f := (y_0, x_1 f, \ldots, x_k f) \tag{5.3.4}$$

for f a 1-jet from x_0 to y_0, and for $(x_0, x_1, \ldots, x_k)$ an element in the fibre over x_0 of $M_{<k>} \to M$ or of $Wh^k(M) \to M$; for the case of the bundle $M_{[k]} \to M$, the parentheses in (5.3.4) should be replaced by square brackets.

An even more tautological example is the bundle of frames $D(n) \to M$ on a manifold; if $k : D(n) \to \mathfrak{M}_1(x)$ is a frame at $x \in M$, and f a 1-jet, as above, we define $k \vdash f := k.f$, the composite (from left to right) of the maps k and f.

Examples of tensorial bundles:

$T(M) \to M$ carries a canonical tensor bundle structure; it is essentially described in the context of Theorem 4.3.4: given a 1-jet $f : \mathfrak{M}_1(x) \to \mathfrak{M}_1(y)$, as in the previous series of examples, and given $\tau \in T_x(M)$, since for any $d \in D$, we have $\tau(d) \in \mathfrak{M}_1(x)$, $f(\tau(d))$ makes sense. The definition of $\tau \vdash f$ then reads

$$(\tau \vdash f)(d) := f(\tau(d)). \tag{5.3.5}$$

The fact that this action consists of *linear* maps follows from Theorem 4.3.4.

The bundle of k-tuples of common-base-point tangent vectors carries, by the same formulas, a natural first-order structure; it is linear w.r.t. each of the k vector bundle structures which are present. (The exact formulation of this k-fold linearity is somewhat complicated to formulate in abstract terms, so we shall not be more precise.)

Note that the bundle in the last example consists of the inputs of classical differential k-forms. On the other hand, the bundle of inputs of differential forms in the microcube formulations does not (for $k \geq 2$) carry first-order natural structure; rather:

Example 5.3.2 The bundle of k-dimensional microcubes on M, or of k-dimensional marked microcubes, carries kth-order natural structure. This follows because if $\tau : D^k \to M$ is a microcube at x, we have for all $(d_1, \ldots, d_k) \in D^k$ that $\tau(d_1, \ldots, d_k) \in \mathfrak{M}_k(x)$, so that a k-jet $f : \mathfrak{M}_k(x) \to \mathfrak{M}_k(y)$ may be applied to it. Then

$$(\tau \vdash f)(d_1, \ldots, d_k) := f(\tau(d_1, \ldots, d_k)). \tag{5.3.6}$$

Similarly for marked microcubes. But unlike the bundle of infinitesimal parallelepipeda, the bundle of marked microcubes does not carry first-order structure (unless $k = 1$).

In the examples presented so far, the invertibility of the acting jets did not play a role. It does so, however, in the following important example of a tensor bundle:

Example 5.3.3 The cotangent bundle $T^*(M)$ as considered in Section 4.5 is a tensor bundle. The description of the action of invertible 1-jets $f : x \to y$ in M on cotangents in their combinatorial manifestation is particularly simple: if j is a cotangent at $x \in M$, i.e. a 1-jet from $x \in M$ to $0 \in R$, $j \vdash f$ is just the composite $j \circ f^{-1}$. On the other hand, if $\overline{j} : T_x \to R$ is a classical cotangent at x, we get a classical cotangent $\overline{j} \vdash f$ as the composite

$$T_y(M) \xrightarrow{T_y(f^{-1})} T_x(M) \xrightarrow{\quad \overline{j} \quad} R.$$

Proposition 5.3.4 *Assume that the bundle $E \to M$ is equipped with an lth-order natural structure. Then the bundle $J^k(E) \to M$ may canonically be equipped with a $k+l$-order natural structure.*

Proof/Construction. Given a $k+l$-jet $f : x \to y$ and $\sigma \in (J^k E)_x$. We should construct $\sigma \vdash f \in (J^k E)_y$. So to $y' \sim_k y$, we should construct $(\sigma \vdash f)(y')$. Since $f : \mathfrak{M}_{k+l}(x) \to \mathfrak{M}_{k+l}(y)$ is bijective, there is a unique $x' \sim_k x$ with $f(x') = y'$, and hence we have the element $\sigma(x') \in E_{x'}$. Also, since $\mathfrak{M}_l(x') \subseteq \mathfrak{M}_{k+l}(x)$, the $k+l$-jet f restricts to an l-jet $\hat{f} : x' \to y'$. Since $E \to M$ by assumption carries an lth-order structure, we have the action $\vdash \hat{f} : E_{x'} \to E_{y'}$. We put

$$(\sigma \vdash f)(y') := (\sigma(x')) \vdash \hat{f}.$$

Notice that nothing is involved in this construction except for lambda calculus.

5.4 Lie derivative

Given a tensor bundle $E \to M$ and a vector field X on M, there is a classical construction L_X, which to a global section f of $E \to M$ associates a new global section $L_X f$, its *Lie derivative* of f along X. When $E \to M$ is the constant bundle $M \times W \to M$ (with W a KL vector space), this construction was already considered in Section 4.4 with notation $D_X f$ (provided we identify sections of $M \times W \to M$ with functions $M \to W$); the defining equation was (4.4.1).

We shall describe the construction of general Lie derivative in groupoid-theoretic terms. As for $D_X f$, we obtain the more general construction of $L_X f$, from a construction on individual jets.

We consider a vector bundle $E \to M$ equipped with an action (left action, say) by the 1-jet groupoid $\Pi^{(1)}(M)$ (actually, it suffices that the graph of near-identities in this groupoid acts, see the Remark 5.4.1 below). The fibres of $E \to M$ are supposed to be KL vector spaces.

For any vector field X on M, we shall define $L_X : J^1(E)_x \to E_x$, for any $x \in M$. Recall the infinitesimal transformations $X_d : M \to M$ of the vector field (Section 4.9).

Consider a 1-jet section f at x, $f : \mathfrak{M}_1(x) \to E$. Let $d \in D$; we cannot immediately compare $f(X_d(x))$ and $f(x)$, since they live in different fibres of $E \to M$. But we can transport $f(X_d(x))$ back to the x-fibre by means of the action of inverse of the 1-jet at x of X_d. This leads to the following recipe.

For each $d \in D$, we form in E_x the difference

$$[(j_1^x X_d)^{-1} \dashv f(X_d(x))] - f(x). \tag{5.4.1}$$

(Here, '$\dashv$' denotes the left action.) Note that $X_d : M \to M$ is an invertible map, so its 1-jet $j_1^x X_d$ at x is an invertible map $\mathfrak{M}_1(x) \to \mathfrak{M}_1(X_d(x))$, i.e. an arrow $x \to X_d(x)$ in $\Pi^{(1)}(M)$; the inverse of this arrow then acts on $f(X_d(x)) \in E_{X_d(x)}$ to bring it back to E_x. If $d = 0$, the difference (5.4.1) is 0, so by KL for E_x, it may be written, as a function of $d \in D$, in the form $d \cdot L_X(f)$ for a unique $L_X(f) \in E_x$, and this element is the Lie derivative of f along X. So the characterizing equation for the Lie derivative $L_X f$, for an individual jet f, is that for all $d \in D$,

$$d \cdot L_X f = [(j_1^x X_d)^{-1} \dashv f(X_d(x))] - f(x).$$

Here, f is a 1-jet of a section at x of $E \to M$. If $f : M \to E$ is an everywhere defined section, we get another section $L_X f$ by sending $x \in M$ to $L_X f_x$, where f_x is the 1-jet of f at x, i.e. the restriction of f to $\mathfrak{M}_1(x)$.

We note that to define $L_X f$, for f a section 1-jet at x, we don't need to know the whole infinitesimal transformation $X_d : M \to M$, only its 1-jet at x; equivalently, we don't need to know the whole X as a section of $T(M) \to M$; we only need a 1-jet section of $T(M) \to M$ at x, i.e. we need to know an element of $J^1(T(M))$.

The construction of $L_X f$ thus provides a map of bundles

$$L : J^1TM \times_M J^1E \to E.$$

The construction itself does not depend on the fact that the action $\dashv$ of $\Pi^{(1)}M \rightrightarrows M$ on $E \to M$ is by fibrewise linear maps; however, linearity of the action implies, as is easily seen (using Theorem 1.4.1) that $L_X f$, for fixed X, depends in a linear way on f. It is also easily seen to be linear in X, for fixed f.

The classical Lie derivative for global sections f of $E \to M$ along a vector field on M comes about by noting that a vector field on M, i.e. a global section X of $TM \to M$, prolongs to a global section $X^{(1)}$ of $J^1TM \to M$, cf. Section 2.7. Then the classical $L_X f$ comes about from the above L as the composite

$$M \xrightarrow{(X^{(1)}, f)} J^1TM \times_M E \xrightarrow{L} E.$$

Remark 5.4.1 The arrow $j_1^x X_d$ used in (5.4.1) is a *near-identity* arrow in the groupoid in question (meaning that it is a 1-neighbour of an identity arrow), since for $d = 0$, it is an identity arrow. Therefore, to define Lie derivatives of 1-jet sections of $E \to M$, it suffices to have a unitary action on $E \to M$ by the (reflexive symmetric) *graph* consisting of the near-identities of $\Pi_1(M)$.

The Lie bracket $[X, Y]$ of vector fields is a special case of Lie derivatives. The tangent bundle $TM \to M$ of a manifold is a tensor bundle (see Section 5.3). Therefore, the Lie derivative construction $L : J^1TM \times_M J^1E \to E$, as a bundle map, applies with $E = TM$, so we have a Lie derivative map

$$L : J^1TM \times_M J^1TM \to TM.$$

In particular, given two vector fields, i.e. global sections $TM \to M$, say X and Y, we may prolong these to sections $X^{(1)}$ and $Y^{(1)}$ of $J^1TM \to M$; jointly, they provide a section $(X^{(1)}, Y^{(1)})$ of $J^1TM \times_M J^1TM$, which we may compose by L.

Proposition 5.4.2 *We have*

$$L \circ (X^{(1)}, Y^{(1)}) = [X,Y] : M \to TM,$$

in standard short notation, $L_X Y = [X,Y]$.

Proof. Let $m \in M$; we denote the 1-jet at m of X by X, and similarly for Y. Then by definition of Lie derivative, we have for any $d_1 \in D$ that

$$d_1 \cdot L_X Y = ((j_1^x X_{d_1})^{-1} \dashv Y(X_{d_1}(m))) - Y(m), \qquad (5.4.2)$$

where the subtraction is to be performed in $T_m M$. Now subtraction of tangent vectors at a fixed $m \in M$ can be performed by using affine combinations of mutual neighbour points in M, see (4.2.4). In particular, the value of (5.4.2) on $d_2 \in D$ is given by the affine combination

$$((j_1^x X_{d_1})^{-1} \dashv Y(X_{d_1}(m)))(d_2) - Y(m)(d_2) + m,$$

which by the tautological construction of the action $\dashv$ unravels into

$$X_{d_1}^{-1}(Y_{d_2}(X_{d_1}(m))) - Y_{d_2}(m) + m.$$

Referring to the notation of (4.9.1), this is the affine combination

$$q - q' + m.$$

But $q - q' + m = [X,Y](m, d_1 \cdot d_2)$, by Proposition 4.9.1. So we have proved the first equality sign in

$$(d_1 \cdot L_X Y)(d_2) = [X,Y](m, d_1 \cdot d_2) = (d_1 \cdot [X,Y])(m, d_2)$$

(omitting the fixed $m \in M$ from the notation). Cancelling the universally quantified d_1 yields $(L_X Y)(d_2) = [X,Y](d_2)$. Since this holds for all d_2, $L_X Y = [X,Y]$ (at the given point m).

Note the advantage of the $L_X Y$-description over the "pentagon" (=group-theoretic commutator) description; the description of the latter does not reveal that the field vector of $[X,Y]$ at m only depends on the 1-jets of X and Y at m, since the point p in the pentagon only can be asserted to be a second-order neighbour of m, so that the pentagon-style construction $q := X(p, -d_1)$ requires knowledge of the 2-jet of X at m.

5.5 Deplacements in groupoids

The deplacement bundles to be introduced now are, for suitable groupoids, their *Lie algebroids*. The Lie algebra $T_e(G)$ of a Lie group G is a special case. Deplacements in groupoids were considered by Ehresmann (1950); their comprehension into an algebroid is due to Pradines (1967).

Definition 5.5.1 *A* deplacement *in a groupoid* $\Phi \rightrightarrows M$ *is a map* $\overline{X} : D \to \Phi$ *such that all* $\overline{X}(d)$ *have the same domain* $x \in M$*, and such that* $\overline{X}(0)$ *is the identity arrow* id_x*.*

We say then that $\overline{X}$ is a deplacement *at* x. Thus, a deplacement looks like this:

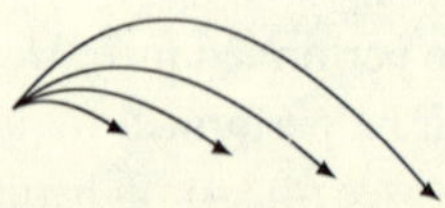

A tangent vector τ to M may be seen as a deplacement $\overline{\tau}$ in the groupoid $\Pi^{(0)}M \rightrightarrows M$, i.e. in the codiscrete groupoid $M \times M \rightrightarrows M$, by putting $\overline{\tau}(d) := (\tau(0), \tau(d))$.

The deplacements in $\Phi \rightrightarrows M$ at $x \in M$ form, as x ranges, a bundle over M. We denote it $\mathscr{A}(\Phi) \to M$ (notation from Nishimura, 2007), and call it the *deplacement bundle*. Its total space is clearly a subspace of the total space $T(\Phi)$ of tangent vectors to Φ. In fact, a deplacement $\overline{X}$ at x may be seen as a tangent vector at $\mathrm{id}_x \in \Phi$ which is d_0-vertical, in the sense that $d_0 \circ \overline{X}$ is the zero tangent at x. If Φ is a manifold (or even just microlinear), linear combinations of deplacements at x (as formed in $T_{\mathrm{id}_x}(\Phi)$) again are deplacements, so $\mathscr{A}(\Phi) \to M$ is a vector bundle.

Any deplacement $\overline{X}$ has an underlying tangent vector $X : D \to M$, namely the map which takes $d \in D$ to the codomain $\partial_1(\overline{X}(d))$ of the arrow $\overline{X}(d)$; it is called the *anchor* of $\overline{X}$. (We also say that $\overline{X}$ is a deplacement *along* X.) Thus there is a map of bundles over M, the *anchor* map $\mathscr{A}(\Phi) \to TM$. If Φ and M are manifolds (or just microlinear spaces), the anchor map is fibrewise linear.

We have more generally a notion of D'-deplacement in $\Phi \rightrightarrows M$ for any pointed object D'. We leave it as an easy exercise for the reader to see that a D'-deplacement in $\Phi \rightrightarrows M$ is tantamount to a functor from the groupoid $D' \times D' \rightrightarrows D'$ to $\Phi \rightrightarrows M$.

We note that for a constant bundle $E = (M \times F) \to M$, a deplacement $\overline{X}$ in the groupoid $\mathfrak{S}(E \to M) \rightrightarrows M$ at $x \in M$ amounts to a pair (τ, σ) with $\tau \in T_x(M)$ and $\sigma \in T_{\mathrm{id}}(\mathfrak{S}(F))$; then τ is the anchor of $\overline{X}$.

Recall that for a vector bundle $E \to M$, we have the groupoid $GL(E) = GL(E \to M)$ of linear isomorphisms between the fibres. Then we have an exact sequence of vector spaces

$$0 \to [E_x, E_x] \to \mathscr{A}(GL(E))_x \to T_x M \to 0,$$

where square brackets denote "space of linear maps".

Example 5.5.2 Recall that for a manifold M, we have the groupoid $\Pi^{(1)}M \rightrightarrows M$, where an arrow $x \to y$ is an invertible 1-jet from x to y, i.e. an invertible $\mathfrak{M}(x) \to \mathfrak{M}(y)$ with $x \mapsto y$. Given a deplacement $\overline{X}$ at $x \in M$ in this groupoid, we construct a 1-jet section $\hat{X}$ of $TM \to M$ by putting, for $y \sim x$,

$$\hat{X}(y)(d) := \overline{X}(d)(y).$$

The anchor X of this deplacement is $\hat{X}(x) \in T_x(M)$. Conversely, given a 1-jet section $\hat{X}$ at x of $TM \to M$, and given $d \in D$, we get a 1-jet $\overline{X}(d)$ from x to $X(x)(d)$ by putting, for $y \sim x$,

$$\overline{X}(d)(y) := \hat{X}(y)(d).$$

It is not immediate that this $\overline{X}(d)$ is an *invertible* 1-jet; this can be verified by working in coordinates.

It is clear that these two processes $\overline{X} \leftrightarrow \hat{X}$ are mutually inverse; thus: *1-jets of vector fields on M are tantamount to deplacements in* $\Pi^{(1)}M \rightrightarrows M$.

Combining with Theorem 4.3.4 (in the explicit form of (5.1.1)) we therefore also have: *1-jets of vector fields on M are tantamount to deplacements in* $GL(TM \to M)$.

The Lie algebroid of a groupoid

We consider now a groupoid $\Phi \rightrightarrows M$, where M is a manifold and where Φ is microlinear. Then we have the vector bundle of deplacements in Φ, namely $\mathscr{A}(\Phi) \to M$, as described above. We shall describe a "Lie bracket"

$$J^1\mathscr{A}(\Phi) \times_M J^1\mathscr{A}(\Phi) \to \mathscr{A}(\Phi)$$

providing $\mathscr{A}(\Phi) \to M$ with the structure of what is called a *Lie algebroid* (Mackenzie, 1987; Pradines, 1967).

For simplicity, we first consider the case of the Lie bracket of two deplacement *fields* $\overline{X}$ and $\overline{Y}$, i.e. global sections of $\mathscr{A}(\Phi) \to M$, not just section 1-jets. (This is anyway the classical formulation, cf. Mackenzie, 1987, Definition III.2.1.) The Lie bracket of vector fields on M will be a special case, namely by considering the groupoid $M \times M \rightrightarrows M$.

A deplacement field $\overline{X}$ in $\Phi \rightrightarrows M$ thus provides for each $m \in M$ and $d \in D$ an arrow $\overline{X}(m,d) : m \to X(m,d)$, where $X(m,-)$ is the anchor of the deplacement

$\overline{X}(m,-)$. Then X itself is an ordinary vector field on M, called the anchor of the deplacement field.

Such a deplacement field $\overline{X}$ gives rise to a vector field $\tilde{X}$ on Φ, namely

$$\tilde{X}(\phi,d) := \phi \cdot \overline{X}(n,d),$$

where n is the codomain of the arrow ϕ (we compose from left to right). One may characterize vector fields on Φ which arise in this way as the left-invariant vertical vector fields on Φ, with suitable understanding of the words; if Φ is microlinear, the Lie bracket of vector fields on Φ makes sense, and it can be proved that the Lie bracket of the vertical left-invariant vector field is again vertical left invariant, thus inducing a Lie bracket for any two deplacement fields. We shall give another description of the bracket which is more "compact" in the sense of not involving the whole of Φ.

Let us note the following property, which holds under the assumption of microlinearity of Φ: let $\overline{X}$ be a deplacement field in Φ, with anchor X. Then for $m \in M$ and $d \in D$,

$$(\overline{X}(m,d))^{-1} = \overline{X}(n,-d) \tag{5.5.1}$$

where $n = X(m,d)$. This can be seen by considering the vector field $\tilde{X}$ on Φ, as described above. From the general theory of microlinear spaces, we know that the infinitesimal transformation $\tilde{X}_{-d}$ is inverse of $\tilde{X}_d$. In particular

$$\mathrm{id}_m = \tilde{X}_{-d}(\tilde{X}_d(\mathrm{id}_m)) = \overline{X}(m,d) \cdot \overline{X}(n,-d)$$

where n is $X(m,d)$.

The anchors X and Y of $\overline{X}$ and $\overline{Y}$ are then vector fields on M, and we may (for $m \in M$ and $(d_1,d_2) \in D \times D$) form the pentagon (4.9.1), with $n := X(m,d_1)$, $p = Y(n,d_2)$, etc.; but now we replace the straight-line segments connecting m to n, n to p, etc., by *arrows* in the groupoid Φ; thus the line segment from m to n is replaced by the arrow $\overline{X}(m,d_1) : m \to X(m,d_1) = n$; the line segment from n to p is replaced by the arrow $\overline{Y}(n,d_2) : n \to Y(n,d_2) = p$; ...; the line segment from q to r is replaced by $\overline{Y}(q,-d_2) : q \to Y(q,-d_2) = r$. These four arrows are composable in Φ, and the composite is an arrow $m \to r$. If either d_1 or d_2 is 0, we get the value id_m for the composite arrow, which therefore by microlinearity of Φ is of the form $t(d_1 \cdot d_2)$ for some unique $t : D \to \Phi$ with $t(0) = \mathrm{id}_m$, and this t, we declare to be $[\overline{X},\overline{Y}]$. This construction is clearly a generalization of the "pentagon" (=group-theoretic commutator) construction of Lie bracket of vector fields.

Let us also give the construction of $[\overline{X},\overline{Y}]$ in terms of strong difference, cf. Nishimura (2007). Besides the arrows $\overline{X}(m,d_1) : m \to n$ and $\overline{Y}(n,d_2) : n \to p$, we have the arrows $\overline{Y}(m,d_2) : m \to q'$ and $\overline{X}(q',d_1) : q' \to p'$. To construct these four arrows, we need only to know the 1-jets of $\overline{X}$ and $\overline{Y}$ at m. Also, if $(d_1,d_2) \in D(2) \subseteq D \times D$, $p = p'$ and $q = q'$, as for 1-jets of vector fields. Therefore, for $(d_1,d_2) \in D(2)$, we can make the following composite arrow $m \to m$ in Φ:

$$m \xrightarrow{\overline{X}(m,d_1)} n \xrightarrow{\overline{Y}(n,d_2)} p \xrightarrow{\overline{X}(p,-d_1)} q \xrightarrow{\overline{Y}(q,-d_2)} m. \qquad (5.5.2)$$

We claim that this has constant value id_m on $D(2)$. It suffices by microlinearity of Φ to see that we get value id_m if $d_1 = 0$ and also if $d_2 = 0$. But this is clear: if for instance $d_2 = 0$, $n = p$ and $m = q$, so the composite is

$$m \xrightarrow{\overline{X}(m,d_1)} n \xrightarrow{\mathrm{id}_n} n \xrightarrow{(\overline{X}(n,-d_1))} m \xrightarrow{\mathrm{id}_m} m.$$

It follows from (5.5.1) that for $(d_1,d_2) \in D(2)$, the two composites

$$m \xrightarrow{\overline{X}(m,d_1)} n \xrightarrow{\overline{Y}(n,d_2)} p \qquad (5.5.3)$$

and

$$m \xrightarrow{\overline{Y}(m,d_2)} q' \xrightarrow{\overline{X}(q',d_1)} p' \qquad (5.5.4)$$

agree as arrows $m \to p = p'$; it follows that we may form the strong difference of the functions $D \times D \to \Phi$ described by (5.5.4) and (5.5.3), and this will provide us with the tangent vector $t : D \to \Phi$, and again, we declare $[\overline{X},\overline{Y}](m,-)$ to be this t. Note that the construction only uses the values of $\overline{X}$ and $\overline{Y}$ on neighbour points n and q' of m; thus, the construction defines, for each $m \in M$, a map

$$J^1(\mathscr{A}(\Phi))_m \times J^1(\mathscr{A}(\Phi))_m \to \mathscr{A}(\Phi)_m.$$

Jointly, we get a bundle map

$$J^1(\mathscr{A}(\Phi)) \times_M J^1(\mathscr{A}(\Phi)) \to \mathscr{A}(\Phi),$$

which provides the vector bundle $\mathscr{A}(\Phi) \to M$ with algebroid structure.

5.6 Principal bundles

A main example to have in mind is the *frame bundle* of a manifold, as considered in Section 2.2. We shall return to this motivating example.

Recall that a groupoid $\Phi \rightrightarrows M$ is called *transitive* if for any $x, y \in M$, there is at least one arrow in Φ from x to y. We compose from left to right in the present section.

One way of formulating an abstract algebraic notion of "principal bundle" is by making it subordinate to the notion of groupoid, more precisely to the notion of *principal* groupoid (cf. Kock, 2007b):

Definition 5.6.1 *A* principal *groupoid is a transitive groupoid* $\Phi \rightrightarrows N$ *whose space* N *of objects is given as* $M + \{*\}$.

So the space of objects is decomposed into a special point $*$ and "the rest" M. In the applications, M will be a manifold. The set of arrows with codomain $*$ and with domain in M organizes itself into a bundle P over M, with the structural map $P \to M$ given by domain formation. It is a surjective map, by the assumed transitivity of Φ. It is the *principal bundle* associated to the principal groupoid $\Phi \rightrightarrows M + \{*\}$.

The group $G := \Phi(*, *)$ acts from the right on $P \to M$ by postcomposition in Φ. Let us temporarily denote this action $\vdash$. The action is fibrewise, and, in each fibre, it is free and transitive: if $p, q \in P_x$ with $x \in M$, there is a unique $g \in G$ with $p \vdash g = q$, namely $g = p^{-1} \cdot q$.

For the case of the frame bundle on an n-dimensional manifold M, the groupoid Φ in question may be described as follows. Consider the n-dimensional manifold $M + R^n$, and the groupoid $\Pi^{(1)}(M + R^n)$ of invertible 1-jets between points of this manifold. Now take Φ to be the full subgroupoid of this groupoid given by the subset of $M + \{0\} \subseteq M + R^n$ (here, 0 denotes the 0 of R^n). Thus $P \to M$ in this case is the set of invertible 1-jets from points x of M to $0 \in R^n$. Such a jet is exactly what we in Section 2.2 called (the inverse of) a *frame* at x. Therefore, an element $u \in P_x$ may be called an *(abstract) (co-) frame* at $x \in M$. In the following, we omit the phrase "co-".†

Since elements in P_x should be thought of as (abstract) frames, a cross-section k of $P \to M$ may be thought of as an (abstract) framing.

Associated to the principal groupoid $\Phi \rightrightarrows (M + \{*\})$ is the full subgroupoid of Φ given by the subset $M \subseteq M + \{*\}$. It deserves the notation $PP^{-1} \rightrightarrows M$; for, every arrow $x \to y$ in it may be written $p \cdot q^{-1}$ with $p \in P_x = \Phi(x, *)$ and $q \in P_y = \Phi(y, *)$. For similar reasons, G deserves the notation $P^{-1}P$. The groupoid $PP^{-1} \rightrightarrows M$ acts on the left on the bundle $P \to M$, the group $P^{-1}P (= G)$ acts, fibrewise, from the right; the actions commute with each other – this is just the associative law for the composition in the groupoid Φ.

Remark 5.6.2 The principal groupoid $\Phi \to M + \{*\}$ may be reconstructed from $P \to M$ together with some suitable algebraic structure on it; classically,

† It is anyway a matter of convention whether a map or its inverse deserves the name "frame" or "coframe"; like also left-to-right or right-to-left is conventional, and we have changed the conventions compared to Section 2.2 on framings.

this extra algebraic structure consists in the group G and a free fibrewise transitive action on P by G; an alternative approach, cf. Kock (2007b), is to give the extra algebraic structure in terms of a certain partially defined ternary operation, providing $P \to M$ with a *pregroupoid* structure, namely (for the case of a principal groupoid) $p \cdot q^{-1} \cdot r$ (where q and r are in the same fibre, i.e. have the same domain).

This "reconstruction" fact makes it meaningful to talk not only about the principal bundle associated to a principal groupoid $\Phi \rightrightarrows M + \{*\}$, but also conversely; in Kock (2007b), we called $\Phi \rightrightarrows M + \{*\}$ the *enveloping groupoid* of the principal bundle P. In particular, we may talk about the groupoid $PP^{-1} \rightrightarrows M$ associated to the principal bundle $P \to M$ (Ehresmann), and also we may talk about the group $G = P^{-1}P$ of the principal bundle (classically, G is given prior to the bundle).

An advantage of the principal groupoid viewpoint is that both the right action on $P \to M$ by $G = P^{-1}P$ and the left action by $PP^{-1} \rightrightarrows M$, as well as the multiplication in G, are all restrictions of the composition $\cdot$ in the groupoid Φ, and as such, do not deserve special notation like $\vdash$.

An arrow $u : x \to y$ in a groupoid $\Phi \rightrightarrows N$ gives rise to a conjugation group-isomorphism $\Phi(x,x) \cong \Phi(y,y)$, sending $r \in \Phi(x,x)$ to $u^{-1} \cdot r \cdot u \in \Phi(y,y)$ (where x and y are $\in N$). In particular, consider the case $\Phi \rightrightarrows M + \{*\}$ of a principal groupoid with $\pi : P \to M$ the associated principal bundle; let $y = *$. Then $u \in P_x = \Phi(x,*)$ induces a group isomorphism $r \mapsto u^{-1} \cdot r \cdot y$

$$PP^{-1}(x,x) \cong P^{-1}P.$$

Therefore, all the groups $PP^{-1}(x,x)$ are isomorphic to the group $P^{-1}P$, the structural group G of the principal bundle; but not canonically so, unless G is commutative.

The classical notion of *reduction* of a principal G-bundle to a subgroup H can also conveniently be formulated in terms of principal groupoids. If $H \subseteq G$ is a subgroup, we get for each $u \in P_x$, by the conjugation isomorphism induced by u, a subgroup of $PP^{-1}(x,x)$, namely $u \cdot H \cdot u^{-1}$.

Let now $\Phi \rightrightarrows M + \{*\}$ be a principal groupoid, with associated principal bundle $P \to M$. A *reduction* of it is a subgroupoid Ψ of $\Phi \rightrightarrows M + \{*\}$, $\Psi \subseteq \Phi$ with the same set of objects, and which is transitive.

So $\Psi \rightrightarrows M + \{*\}$ is itself a principal groupoid; the associated principal bundle $Q \to M$ is a subbundle of $P \to M$; the associated group $H := \Psi(*,*)$ is a subgroup of $G = \Phi(*,*)$. The classical terminology is that $Q \to M$ is a *reduction of the principal G-bundle $P \to M$ to the subgroup $H \subseteq G$*.

A reduction Ψ of $P \to M$ to $\{e\} \subseteq G$ is tantamount to a cross-section of $P \to M$. For, all vertex groups $\Psi(y,y)$ of Ψ are trivial ($y \in M + \{*\}$), in particular, if $x \in M$, there is exactly one element $\psi(x)$ in $\Psi(x,*) = P_x$, and this ψ is a cross-section of $P \to M$. Conversely, a cross-section ψ of $P \to M$ gives a subgroupoid of PP^{-1}consisting of arrows of the form $\psi(x) \cdot \psi(y)^{-1}$, whose vertex groups are trivial; extend it in the unique way to a subgroupoid Ψ of Φ.

A subgroup $H \subseteq G$ gives rise to an equivalence relation $\equiv_H$ on each fibre P_x of $P \to M$ ($x \in M$) namely, for u, v in P_x, $u \equiv_H v$ iff $u^{-1} \cdot v \in H$, i.e. if the unique $g \in G$ with $u \cdot g = v$ is in H.

5.7 Principal connections

In the present section we shall continue to work algebraically with principal groupoids $\Phi \rightrightarrows M + \{*\}$, so we shall keep composing from left to right; P is therefore defined as the set of arrows with *domain* in M, and with *codomain* $*$. The map $\pi : P \to M$ is domain-formation. Composition in Φ will be denoted by a multiplication dot $\cdot$. Recall the full subgroupoid $PP^{-1} \rightrightarrows M$ of $\Phi \rightrightarrows (M + \{*\})$; it acts on the bundle $P \to M$ from the left. Also we have the associated group $G = \Phi(*,*) = P^{-1}P$, which acts, fibrewise, on P, on the right.

Proposition 5.7.1 *There is an equivalence between the following two kinds of data:*

(1) a connection in the groupoid $PP^{-1} \rightrightarrows M$

(2) a bundle connection on $P \to M$, whose transport mappings $P_y \to P_x$ preserve the right G-action.

Such data are called a *principal connection* on the principal bundle P. The data in the form (2) are the traditional way to present these data.

Proof. Given a connection ∇ in the groupoid $PP^{-1} \rightrightarrows M$, we produce a bundle connection on the bundle $P \to M$ as follows (we write the action of the graph $M_{(1)} \rightrightarrows M$ on P on the left, using the symbol $\dashv$): for $x \sim y$, and $a \in P_y$,

$$(x,y) \dashv a := \nabla(x,y) \cdot a;$$

in terms of the composition in the groupoid Φ, this is the composite

$$x \xrightarrow{\nabla(x,y)} y \xrightarrow{\quad a \quad} *.$$

The right equivariance for the right action by $G = \Phi(*,*)$ follows from associativity of the groupoid composition.

Conversely, given a bundle connection on $P \to M$, written in terms of a left action $\dashv$ by $M_{(1)}$ on $P \to M$, we define $\nabla(x,y) : x \to y$ to be

$$((x,y) \dashv a) \cdot a^{-1} \tag{5.7.1}$$

where a is an arbitrary element in P_y (such elements exist, since Φ is transitive). To see that this is independent of the $a \in P_y$ chosen, consider another one, say $b \in P_y$. Then again by transitivity, there exists a $g \in G$ with $b = a \cdot g$. Then

$$((x,y) \dashv b) \cdot b^{-1} = ((x,y) \dashv (a \cdot g)) \cdot (a \cdot g)^{-1}, \tag{5.7.2}$$

and since we assume that the bundle connection $\dashv$ is right G-equivariant, $(x,y) \dashv (a \cdot g) = ((x,y) \dashv a) \cdot g$, and then (5.7.2) immediately rewrites into (5.7.1).

Bundle connections in bundles $E \to M$ with some fibrewise algebraic structure may sometimes be encoded as principal connections in a principal groupoid, or equivalently, in principal bundles: this happens if all the fibres E_x are isomorphic. To be concrete, let us consider a *vector* bundle $E \to M$ such that all the fibres are linearly isomorphic, say isomorphic to R^n. Then the groupoid $GL(E \to M)$ appears as PP^{-1} of a principal groupoid $\Phi \rightrightarrows M + \{*\}$ where Φ is the groupoid of linear isomorphisms between vector spaces which are either $= E_x$ for some $x \in M$, or $= R^n$, where R^n is placed as the fibre over the isolated point $*$. Then a groupoid connection on $PP^{-1} \rightrightarrows M$ amounts to a linear bundle connection on the bundle $E \to M$.

Consider now a principal bundle $\pi : P \to M$ where not only M, but also P is a manifold. We keep doing calculations in the enveloping groupoid $\Phi \rightrightarrows (M + \{*\})$, and the group G of the principal bundle is thus $\Phi(*,*)$.

Then a principal connection ∇ on P gives rise to a G-valued 1-form ω on P, called the *connection form*: let $\pi : P \to M$ be the structural map (=domain formation for arrows with codomain $*$ and domain in M). For u and v neighbours in P, with $\pi(u) = a$, $\pi(v) = b$, put

$$\omega(u,v) := u^{-1} \cdot (\nabla(a,b) \cdot v). \tag{5.7.3}$$

Note that both u and $\nabla(a,b) \cdot v$ are in the π-fibre over a, so that the "fraction" $u^{-1} \cdot (\nabla(a,b) \cdot v)$ makes sense as an element of $P^{-1}P$.

The defining equation is equivalent to

$$u \cdot \underbrace{\omega(u,v)}_{\in P^{-1}P} = \underbrace{\nabla(\pi(u),\pi(v))}_{\in PP^{-1}} \cdot v. \tag{5.7.4}$$

Let us agree that (for u,v in P a pair of neighbours in P) $\nabla(u,v)$ denotes $\nabla(\pi(u),\pi(v))$. Then the equation (5.7.4) may be written

$$u \cdot \omega(u,v) = \nabla(u,v) \cdot v. \tag{5.7.5}$$

It is possible to represent the relationship between ∇ and the associated ω by means of a simple figure:

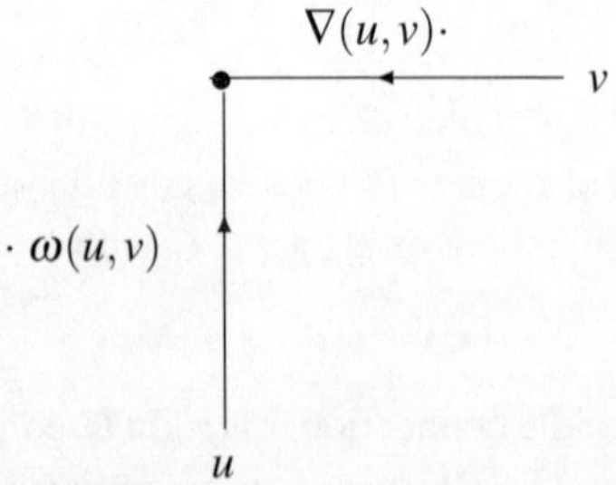

The figure reflects something geometric, namely that $\omega(u,v)$ acts inside the fibre ("vertically"), whereas the action of ∇ defines a notion of *horizontality*.

Note that if u and v are in the same fibre, say in the fibre over $a \in M$, there is a unique $g \in G$ with $u \cdot g = v$. In terms of the principal groupoid defining P, this g is $g = u^{-1} \cdot v$; if furthermore $u \sim v$, $\omega(u,v) = g$. This follows from (5.7.5), since $\nabla(a,a)$ is an identity arrow. We conclude that if u and v are neighbours in P and in the same fibre, then

$$\omega(u,v) = u^{-1} \cdot v. \tag{5.7.6}$$

We shall relate *flatness* of a principal connection to *closedness* of the corresponding 1-form. Let ∇ be a principal connection in a principal bundle $\pi : P \to M$; then for any infinitesimal 2-simplex (x,y,z) in M, we have the curvature $R_\nabla(x,y,z) \in \Phi(x,x) = PP^{-1}(x,x)$; and we have the connection form ω, which is a G-valued 1-form on P (where $G = P^{-1}P$).

Then the $R = R_\nabla$ and ω are related as follows: for any infinitesimal 2-simplex u,v,w in P, we have

$$u \cdot d\omega(u,v,w) = R(\pi(u),\pi(v),\pi(w)) \cdot u. \tag{5.7.7}$$

This is a trivial calculation, using (5.7.5) three times; we use the abbreviated notation $\nabla(u,v)$ for $\nabla(\pi(u),\pi(v))$:

$$\begin{aligned} u \cdot d\omega(u,v,w) &= u \cdot \omega(u,v) \cdot \omega(v,w) \cdot \omega(w,u) \\ &= \nabla(u,v) \cdot v \cdot \omega(v,w) \cdot \omega(w,u) \\ &= \nabla(u,v) \cdot \nabla(v,w) \cdot w \cdot \omega(w,u) \\ &= \nabla(u,v) \cdot \nabla(v,w) \cdot \nabla(w,u) \cdot u \\ &= R(u,v,w) \cdot u \end{aligned}$$

(where $R(u,v,w)$ is an abbreviation for $R(\pi(u),\pi(v),\pi(w))$).

The relationship between $d\omega$ and R in particular implies a relationship between closedness of the 1-form ω and flatness of the connection ∇. Let us call a map $\pi : P \to M$ between manifolds a *submersion* if for any $u \in P$, and any infinitesimal k-simplex X in M with first vertex $\pi(u)$, there exists an infinitesimal k-simplex X' in P with first vertex the given u, and mapping by π to X. Clearly, if $P \to M$ is locally a product $M \times F \to M$, it satisfies this condition.

Proposition 5.7.2 *If the principal connection ∇ is flat, the connection form ω is closed; the converse holds provided that π is a submersion.*

Proof. Let u, v, w be an infinitesimal 2-simplex in P, and assume that ∇ is flat, so all values of R are identity arrows in PP^{-1}. To see that $d\omega(u,v,w) = e$, it suffices to see that $u \vdash d\omega(u,v,w) = u$ (the right action by G on P being free). But this is immediate from (5.7.7). The converse implication goes by the same calculation run backwards, once it gets started: given an infinitesimal 2-simplex x, y, z in M: given u above x, we use the "submersion" assumption to pick an infinitesimal 2-simplex u, v, w, above x, y, z; then the calculation gives that $\nabla(x,y) \cdot \nabla(y,z) \cdot \nabla(z,x)$ acts trivially on u, so is an identity arrow, so ∇ is flat.

Since R measures the curvature of ∇, the relationship expressed by (5.7.7), and by the proposition in particular, motivates the name *curvature form* for the G-valued 2-form $d\omega$ on P.

Consider a cross-section k of $P \to M$ (an "abstract framing"). It gives rise to a principal connection, i.e. to a connection ∇_k in the groupoid $PP^{-1} \rightrightarrows M$; it is given by

$$\nabla_k(x,y) = k(x) \cdot k(y)^{-1}.$$

It is trivial to verify that connections coming about in this way from a framing are flat.

We shall consider the question of a converse. The following is a generalization of Proposition 3.7.3 concerning which affine connections locally come about from framings. As there, we make an assumption on closed 1-forms; now it is the assumption that closed $P^{-1}P$-valued 1-forms on M locally are exact; we shall assume that the bundle $P \to M$ locally admits cross-sections. Also, we assume that P itself is a manifold. (A slight modification of the proof will reveal that this last hypothesis is redundant.)

Proposition 5.7.3 *Under these assumptions, if ∇ is flat, then it locally comes about from a framing.*

Proof. Because of the local nature of the conclusion, we may as well assume that $P \to M$ admits not only local sections, but admits a global section $h : M \to P$. We know from flatness of ∇ and from Proposition 5.7.2 that the connection form ω on P is closed. Therefore also the form $h^*(\omega)$ on M is a closed form. As a closed $P^{-1}P$-valued 1-form on M, it is therefore locally exact; so locally, we may find $g : M \to P^{-1}P$ with

$$h^*(\omega)(x,y) = g(x)^{-1} \cdot g(y)$$

for $x \sim y$ in M, i.e. with $\omega(h(x), h(y)) = g(x)^{-1} \cdot g(y)$. Therefore, the relation (5.7.4) between ∇ and ω gives, with $u := h(x)$, $v := h(y)$ that $\nabla(x,y) \cdot h(y) = h(x) \cdot [g(x)^{-1} \cdot g(y)]$, or by rearranging,

$$\nabla(x,y) = h(x) \cdot g(x)^{-1} \cdot g(y) \cdot h(y)^{-1},$$

and this shows that the framing k given by $k(z) := h(z) \cdot g(z)^{-1}$ satisfies $k(x) \cdot k(y)^{-1} = \nabla(x,y)$.

This proves the proposition. (An alternative proof: prove the proposition for constant principal bundles, and then prove that a (local) cross-section, as the h used in the proof, provides a (local) isomorphism of the given bundle with a constant one.)

A principal connection ∇ in P gives rise to a bundle connection $\tilde{\nabla}$ on $P \to M$, i.e. a left action $\dashv$

$$\tilde{\nabla}(x,y) \dashv u := \nabla(x,y) \cdot u,$$

where the dot on the right-hand side is simply the composition in the principal groupoid defining the principal bundle. As a bundle connection in the bundle $P \to M$, $\tilde{\nabla}$ gives rise to a geometric distribution $\approx$ in P transverse to the fibres, by the recipe described in Section 2.6 – the *horizontality distribution* for ∇. And we know (Proposition 2.6.15) that if ∇ is flat, then $\approx$ is involutive (and vice versa, under a mild hypothesis). Note that the distribution $\approx$ is right invariant, in the sense that $u \approx v$ implies $u \cdot g \approx v \cdot g$; for,

$$u \approx v \text{ iff } \nabla(x,y) \cdot v = u \text{ iff } \nabla(x,y) \cdot v \cdot g = u \cdot g \text{ iff } v \cdot g \approx u \cdot g,$$

since the left action by ∇ commutes with the right action by G. There is no need to distinguish $\tilde{\nabla}$ from ∇ itself, nor to keep $\dashv$, as it is just another name for the composition in the principal groupoid.

Summarizing some of the above, we have, for a principal bundle $\pi : P \to M$ with group G,

Proposition 5.7.4 *Given a principal connection ∇ on $\pi : P \to M$ (π assumed to be a submersion), if the corresponding G-valued 1-form is closed, the distribution $\approx$ corresponding to ∇ is involutive.*

Note that for $u \sim v$ in P, above $x \sim y$ in M,

$$u \approx v \text{ iff } \nabla(x,y) \cdot v = u \text{ iff } u \cdot \omega(u,v) = u \text{ iff } \omega(u,v) = e. \tag{5.7.8}$$

We consider again a principal bundle $P \to M$ with group G, and with a principal connection ∇, with associated G-valued connection form ω. To say, as in the proposition, that the connection form is closed can also be expressed: all values of the curvature form $d\omega$ are in the subgroup $\{e\}$ of G. We generalize this into other subgroups H of G.

Besides the "∇-horizontal" distribution $\approx$ considered above ($u \approx v$ if $\omega(u,v) = e$), any Lie subgroup H of G gives rise to another coarser distribution $\sim_H$, given by

$$u \sim_H v \text{ iff } \omega(u,v) \in H$$

for $u \sim v$ in P. For $H = \{e\}$, $\sim_H$ is the horizontal distribution. We may say that the relation $u \sim_H v$ expresses "horizontality modulo H". With $P \to M$, ∇ and ω as above, we have:

Proposition 5.7.5 *Assume $d\omega$ takes all its values in the subgroup $H \subseteq G$. Then the distribution $\sim_H$ is involutive.*

Proof. Let u, v, w be an infinitesimal 2-simplex in P with two of its sides horizontal mod H, say (u,v) and (u,w), i.e. $\omega(u,v) \in H$ and $\omega(u,w) \in H$. Consider $d\omega(u,v,w)$:

$$d\omega(u,v,w) = \omega(u,v) \cdot \omega(v,w) \cdot \omega(w,u).$$

By assumption, this threefold product is in H. Since two of the factors are in H by assumption, then so is the third, $\omega(v,w) \in H$, i.e. $v \sim_H w$.

Recall from Section 2.6 that we call a distribution $\approx$ on M *totally an-holonomic* if for any two points x and y in M, there is a connected set which is an integral set for $\approx$ and which contains both x and y. If a distribution $\approx'$ is coarser than $\approx$ (so $x \approx y$ implies $x \approx' y$), integral sets for $\approx$ are also integral for $\approx'$. It follows that if $\approx$ is totally an-holonomic, then so is $\approx'$. The coarsest distribution on M is by definition the neighbour relation $\sim$. Since any subset of M is an integral subset for this distribution, it follows that M is connected iff $\sim$ is totally an-holonomic.

If M is connected, an involutive distribution is totally an-holonomic iff M itself is a leaf (assuming that we can use the Frobenius Theorem). If a manifold admits a totally an-holonomic distribution, it is connected. And if the distribution is furthermore involutive, it is trivial, meaning equal to $\sim$.

Let now $P \to M$ be a principal bundle (with P and M manifolds), with group G; let ∇ be a principal connection on P, with ω as connection form (a G-valued 1-form on P). Let $H \subset G$ be a Lie subgroup (i.e. with $\mathfrak{M}_G(e) \cap H = \mathfrak{M}_H(e)$).

Proposition 5.7.6 *Assume that the horizontality distribution $\approx$ for ∇ is totally an-holonomic. Assume also that all values of $d\omega$ are in H, and that G is connected. Then $H = G$.*

Proof. The distribution $\sim_H$ is coarser than the horizontality distribution, and so $\sim_H$ is likewise totally an-holonomic. Since $\sim_H$ is furthermore involutive, by Proposition 5.7.5, P itself is the only leaf. From this, we want to deduce that $H = G$; since G is connected, it suffices, by "linear sufficiency" (see Proposition A.6.7) to see that $\mathfrak{M}_G(e) \subseteq H$. Pick any $x \in P$, above $a \in M$, say. Let $g \in \mathfrak{M}_G(e)$; then $x \cdot g \sim x$. Since P is an integral manifold for $\sim_H$, it follows that $x \cdot g \sim_H x$, or equivalently $g \in H$.

We shall, at the end of the next section, give heuristic arguments why Proposition 5.7.6 is a version of the Ambrose–Singer Theorem, asserting that, with suitable provisos, the *curvature of a principal connection generates its local holonomy*.

5.8 Holonomy of connections

Recall that for any space M, we have the "codiscrete" groupoid $M \times M \rightrightarrows M$.

Let $\Phi = (\Phi \rightrightarrows M)$ be a groupoid. A functor $\overline{\nabla}$ from $M \times M \rightrightarrows M$ to Φ is called a *(total) trivialization* of Φ.

So for x and y in M, $\overline{\nabla}(x,y)$ is an arrow $x \to y$ in Φ; $\overline{\nabla}(x,x)$ is id_x; also $\overline{\nabla}(x,y)$ is inverse of $\overline{\nabla}(y,x)$, and

$$\overline{\nabla}(x,y) \cdot \overline{\nabla}(y,z) = \overline{\nabla}(x,z) \tag{5.8.1}$$

for all x,y,z in M.

Sometimes, we call such trivialization a *total* trivialization, because there is a more general notion, *partial trivialization* of a groupoid $\Phi \rightrightarrows M$ *along* a map $f : N \to M$; this is by definition a total trivialization $\overline{\nabla}$ of the groupoid $f^*(\Phi)$, thus for n_1 and n_2 in N, $\overline{\nabla}(n_1,n_2)$ is an arrow $f(n_1) \to f(n_2)$ in Φ, and $\overline{\nabla}(n,n) = \mathrm{id}_{f(n)}$, for all $n \in N$, and similarly for the other equations: $\overline{\nabla}(n_2,n_1) = \overline{\nabla}(n_1,n_2)^{-1}$ and $\overline{\nabla}(n_1,n_2) \cdot \overline{\nabla}(n_2,n_3) = \overline{\nabla}(n_1,n_3)$.

Trivializations pull back in an evident sense. If $\Phi \rightrightarrows M$ is a groupoid, and if $f : N \to M$ is an arbitrary map, a total trivialization $\overline{\nabla}$ of Φ gives rise to a total trivialization $f^*(\overline{\nabla})$ of $f^*(\Phi)$, i.e. to a partial trivialization of Φ along f, $f^*(\overline{\nabla})(n_1,n_2) = \overline{\nabla}(f(n_1), f(n_2))$ for $n_1, n_2 \in N$. More generally, if $\overline{\nabla}$ is a partial trivialization of $\Phi \rightrightarrows M$ along $f : N \to M$, and $g : P \to N$ is any map, we get an induced partial trivialization of Φ along $f \circ g : P \to M$, in an evident way.

Example. Recall from Example 5.1.5 the groupoid $M \times M \times G \rightrightarrows M$ given by a space M and a group G. This groupoid carries a total trivialization $\overline{\nabla}$ given by $\overline{\nabla(x,y)} := (x,y,e)$ where $e \in G$ is the neutral element. But conversely:

Proposition 5.8.1 *Given a groupoid $\Phi \rightrightarrows M$ and a total trivialization $\overline{\nabla}$ of it, then for each $z \in M$, there is a canonical isomorphism between the groupoids Φ and $M \times M \times G \rightrightarrows M$, where G is the group $\Phi(z,z)$.*

Proof/Construction. Let $\phi : m_1 \to m_2$ be an arrow in $\Phi \rightrightarrows M$. Then $\overline{\nabla}(z,m_1) \cdot \phi \cdot \overline{\nabla}(m_2,z) \in \Phi(z,z) = G$, and so $(m_1, m_2, \overline{\nabla}(z,m_1) \cdot \phi \cdot \overline{\nabla}(m_2,z))$ is an arrow $m_1 \to m_2$ in $M \times M \times G \rightrightarrows M$. Conversely, to an arrow (m_1,m_2,g) in $M \times M \times G \rightrightarrows M$, we associate the arrow $\overline{\nabla}(m_1,z) \cdot g \cdot \overline{\nabla}(z,m_2)$ in $\Phi \rightrightarrows M$.

Exercise 5.8.2 Given a groupoid $\Phi \rightrightarrows M$, construct a bijective correspondence between the set of trivializations along maps $D \to M$, and deplacements in Φ.

A trivialization $\overline{\nabla}$ of a groupoid $\Phi = (\Phi \rightrightarrows M)$ where M is a manifold gives rise to a connection ∇ in Φ by restricting $\overline{\nabla} : M \times M \to \Phi$ to the subset $M_{(1)} \subseteq M \times M$, i.e. $\nabla(x,y) = \overline{\nabla}(x,y)$ for $x \sim y$ in M. Then (5.8.1) for $\overline{\nabla}$ implies that the connection ∇, obtained by restriction in this way, is flat in the sense of (5.2.4). A *complete integral* for a connection ∇ on a groupoid $\Phi \rightrightarrows M$ is a trivialization $\overline{\nabla}$ of Φ which extends the given ∇, $\overline{\nabla}(x,y) = \nabla(x,y)$ whenever $(x,y) \in M_{(1)}$. Clearly a necessary condition for ∇ to admit a complete integral is that it is flat. (Also, the reflexivity law and symmetry law assumed for connections in groupoids, $\nabla(x,x) = \mathrm{id}_x$ and $\nabla(y,x) = (\nabla(x,y))^{-1}$ follow from existence of complete integrals.)

Complete integrals in this sense are rare. More common are "*partial integrals along maps*": a *partial integral* of the connection ∇ in Φ *along* $f : N \to M$ is a complete integral of $f^*(\nabla)$.

If $g : N' \to N$ is a map, and $\Psi \rightrightarrows N$ a groupoid, then if $H : N \times N \to \Psi$ is a trivialization of Ψ, then $H \circ (g \times g)$ is a trivialization of $g^*(\Psi)$. And if H is a complete integral of a connection ∇ on Ψ, then $H \circ (g \times g)$ is a complete integral of $g^*(\nabla)$ (where $g^*(\nabla)(u,v) := \nabla(g(u),g(v))$). We express these properties by saying that complete integrals, and trivializations, *pull back*.

Holonomy

Let M be a manifold. Many groupoids with connection $(\Phi \rightrightarrows M, \nabla)$ have the property that unique partial integrals exist along any map (path) $R \to M$; in this case, we say that the pair $(\Phi \rightrightarrows M, \nabla)$ *admits path integration.*

If $(\Phi \rightrightarrows M, \nabla)$ admits path integration, we thus have for each path $\gamma : R \to M$ a unique complete integral for the induced connection $\gamma^*(\nabla)$ on R; we denote it $\int_\gamma \nabla$. So its value on $(s,t) \in R \times R$ is an arrow in Φ

$$\gamma(s) \xrightarrow{(\int_\gamma \nabla)(s,t)} \gamma(t). \tag{5.8.2}$$

Note that the fact that a trivialization is a functor implies a subdivision law for these "integrals"; in fact, it is a version of (5.8.1):

$$(\int_\gamma \nabla)(a,b) \cdot (\int_\gamma \nabla)(b,c) = (\int_\gamma \nabla)(a,c) \tag{5.8.3}$$

(left-to-right composition notation in Φ used).

Note that if $(\Phi \rightrightarrows M, \nabla)$ admits path integration, ∇ is "flat along any path", i.e. for any path γ in M, $\gamma^*(\nabla)$ is flat (=curvature free), i.e. satisfies (5.2.4).

For $s \sim t$ in R, we have that

$$(\int_\gamma \nabla)(s,t) = \nabla(\gamma(s), \gamma(t)); \tag{5.8.4}$$

this equation expresses the requirement that the complete integral of $\gamma^*(\nabla)$ agrees with $\gamma^*(\nabla)$ on pairs of neighbour points.

Remark 5.8.3 There is a condition related to (5.8.4), but stronger, namely that this equation holds not just under the assumption that $s \sim t$, but under the weaker assumption that $\gamma(s) \sim \gamma(t)$. However, this would not be realistic for paths with self-intersection, say.

From the assumed uniqueness of complete integrals, invariance under reparametrization follows. Let $\rho : R \to R$ be any map ("reparametrization"), then for any path $\gamma : R \to M$,

$$(\int_{\gamma \circ \rho} \nabla)(s,t) = (\int_\gamma \nabla)(\rho(s), \rho(t)). \tag{5.8.5}$$

For, as Φ-valued functions of $s,t \in R$ both sides are partial trivializations of Φ along $\gamma \circ \rho$, and for $s \sim t$, both return the value $\nabla(\gamma(\rho(s)), \gamma(\rho(t)))$.

Path connections

Recall that a connection in a groupoid $\Phi \rightrightarrows M$ in the present exposition is construed as a morphism of reflexive symmetric graphs $M_{(1)} \to \Phi$. Besides the graph $M_{(1)} \rightrightarrows M$ – which is the main actor presently – there is another, likewise reflexive symmetric, graph, namely the graph $P(M) \rightrightarrows M$ of paths $R \to M$, as described in Example 5.2.3. We can then consider morphisms of graphs $P(M) \to \Phi$; however, the subdivision law (5.8.3) motivates us to put a similar condition on graph maps $P(M) \to \Phi$, and this leads to the following definition (essentially due to Virsik, 1971). We shall use left-to-right notation for composition in Φ. Recall the notation $\gamma \mid [a,b]$ from the Appendix (Section 9).

Definition 5.8.4 *A* path connection *in a groupoid $\Phi \rightrightarrows M$ is a morphism Ω : $P(M) \to \Phi$ of reflexive symmetric graphs over M, satisfying the* subdivision law*:*

$$\Omega(\gamma \mid [a,c]) = \Omega(\gamma \mid [a,b]) \cdot \Omega(\gamma \mid [b,c]) \tag{5.8.6}$$

for all $a,b,c \in R$ and all $\gamma : R \to M$.

Let $\Phi \rightrightarrows M$, ∇ be a groupoid with a connection that admits path integration, so that $\left(\int_\gamma \nabla\right)(s,t) \in \Phi$ is defined for all paths $\gamma : R \to M$ and all $s,t \in R$. Under these assumptions, we may put

Definition 5.8.5 *The* holonomy *of* ∇ along γ *is* $(\int_\gamma \nabla)(0,1)$. *The* holonomy *of* ∇, *denoted* hol_∇, *is the map* $P(M) \to \Phi$ *sending* γ *to* $(\int_\gamma \nabla)(0,1)$.

Proposition 5.8.6 *The map* $\text{hol}_\nabla : P(M) \to \Phi$ *is a path connection.*

Proof. We have to see that hol_∇ is a morphism of reflexive symmetric graphs. The fact that hol_∇ preserves the "book-keeping" (domain- and codomain-formation) follows from (5.8.2) with $s = 0$ and $t = 1$. To see that the reflexivity structure is preserved means to prove $\text{hol}_\nabla(j(m)) = i(m)$; here $j(m)$ denotes the constant path at m, so $j(m)(t) = m$ for all $t \in R$; and $i(m)$ denotes the identity arrow $m \to m$ in Φ. Let $\rho : R \to R$ be the map with constant value $0 \in R$. Then $j(m) \circ \rho = j(m)$, and so by reparametrization (5.8.5),

$$\left(\int_{j(m)\circ\rho} \nabla\right)(0,1) = \left(\int_{j(m)} \nabla\right)(\rho(0),\rho(1)) = \left(\int_{j(m)} \nabla\right)(0,0)$$

which is an identity arrow since $(\int_{j(m)} \nabla)$ is a functor. It then has to be the identity at m, since book-keeping is preserved. This proves $\text{hol}_\nabla(j(m)) = i(m)$. The proof that the symmetry structure is preserved is similar: now we consider the map $\rho : R \to R$ "reflection in $\frac{1}{2}$", and use that $(\int_\gamma \nabla)(0,1)$ is the inverse arrow in Φ of $(\int_\gamma \nabla)(1,0)$, again because $\int_\gamma \nabla$ is a functor.

The subdivision law is a consequence of the law (5.8.3).

Remark 5.8.7 One may define the notion of *piecewise path* in a space M; it is a finite sequence of endpoint-matching paths $R \to M$, see Section A.6 in the Appendix. The construction of hol_γ can be extended from paths to piecewise paths, in an evident way.

It is reasonable to say that we have obtained the path connection hol_∇ by *integrating* the connection ∇. There is a "differentiation" process going the other way, from path connections to connections. (In fact, in a special case, these processes amount to the usual integration and diffentation of functions $R \to R$, see Example 5.8.8 below.) We describe the differentiation process.

Consider a groupoid $\Phi \rightrightarrows M$, and a morphism of symmetric reflexive graphs over M, $\sigma : P(M) \to \Phi$. We don't assume the subdivision law for σ, as we do for path connections. Since σ is a morphism of graphs over M, $P(M) \to \Phi$, we may precompose it by the graph morphism $[-,-] : M_{(1)} \to P(M)$ over M considered in Example 5.2.4. Both these graph morphisms preserve reflexive symmetric structure. We thus obtain a morphism $M_{(1)} \to \Phi$, of reflexive symmetric graphs $M_{(1)} \to \Phi$ over M, in other words, we obtain a connection in the groupoid $\Phi \rightrightarrows M$, which we denote σ'; we say that we have *differentiated* $\sigma : P(M) \to \Phi$ into a connection σ'.

This process from graph morphisms to connections is natural: since any map $f : N \to M$ between manifolds preserves affine combinations of neighbour points, it follows that we have a commutative square of graph morphisms,

$$\begin{array}{ccc} N_{(1)} & \longrightarrow & M_{(1)} \\ {\scriptstyle [-,-]}\downarrow & & \downarrow{\scriptstyle [-,-]} \\ P(N) & \longrightarrow & P(M) \end{array}$$

From this follows the naturality of the "differentiation process", in the sense that

$$(f^*(\sigma))' = f^*(\sigma').$$

Here f^* denotes the process of pulling back graph morphisms along a map $f : N \to M$, arising from the fact that f induces a graph map over f from $P(N)$

to $P(M)$, and also induces a graph map over f from $N_{(1)}$ to $M_{(1)}$. Both these induced graph maps are maps preserving reflexive and symmetric structure.

Example 5.8.8 (Elementary Calculus) Let $M \subseteq R$ be an open subset with the property that anti-derivatives exist unique up to unique constant, for functions $f : M \to R$. For any $f : M \to R$, there is a connection ∇ in the groupoid $M \times M \times (R,+)$, namely

$$\nabla(x,y) = (x,y,f(x)\cdot(y-x)).$$

There is a unique complete integral for ∇, namely $(a,b) \mapsto (a,b,\int_a^b f(u)\,du)$ for $a,b \in M$. Therefore for any $\gamma : R \to M$,

$$(\int_\gamma \nabla)(a,b) = (\gamma(a),\gamma(b),\int_{\gamma(a)}^{\gamma(b)} f(u)\,du);$$

hence

$$\mathrm{hol}_\nabla(\gamma) = (\gamma(0),\gamma(1),\int_{\gamma(0)}^{\gamma(1)} f(u)\,du),$$

for $\gamma \in P(M)$. To calculate hol'_∇, we have for $x \sim y \in M$

$$\begin{aligned}\mathrm{hol}'_\nabla(x,y) &= \mathrm{hol}_\nabla([x,y])\\ &= ([x,y](0),[x,y](1),\int_{[x,y](0)}^{[x,y](1)} f(u)\,du) = (x,y,\int_x^y f(u)\,du),\end{aligned}$$

which by the Fundamental Theorem of Calculus is $(x,y,f(x)\cdot(y-x))$, which is $\nabla(x,y)$. Thus, $(\mathrm{hol}_\nabla)' = \nabla$.

We shall investigate to what extent a similar result holds for a general $(\Phi \rightrightarrows M,\nabla)$ (assuming that (Φ,∇) admits integration, so that holonomy is defined). We need a weak flatness assumption on ∇; namely that $\nabla(x,y)\cdot\nabla(y,z) = \nabla(x,z)$ whenever the infinitesimal 2-simplex x,y,z is "1-dimensional" in the sense that there exist $a \sim b$ in M so that x,y, and z all belong to the image of $[a,b] : R \to M$. This is a reasonable condition; it follows, essentially from Proposition 3.1.14, that the condition holds for any locally constant groupoid whose vertex groups are manifolds (even "manifolds modelled on KL vector spaces").

Theorem 5.8.9 (Fundamental Theorem of Calculus Part 1) *Let $(\Phi \rightrightarrows M,\nabla)$ admit integration, and assume the above weak flatness condition for ∇. Then*

$$(\mathrm{hol}_\nabla)' = \nabla.$$

Proof. Let $a \sim b$ in M. We have the path $[a,b] : R \to M$, and by definition we have

$$(\mathrm{hol}_\nabla)'(a,b) = \mathrm{hol}_\nabla([a,b]) = \Gamma(0,1), \tag{5.8.7}$$

where $\Gamma : R \times R \to \Phi$ is the trivialization of $[a,b]^*(\nabla)$. We are going to describe Γ explicitly. We claim that

$$\Gamma(s,t) = \nabla([a,b](s),[a,b](t))$$

will do the job, for $s,t \in R$. (Note that even though s and t may not be neighbours in R, $[a,b](s)$ and $[a,b](t)$ are neighbours in M, so that it makes sense to apply ∇.) First, the Γ thus described is a functor $(R \times R) \to \Phi$, since ∇ is flat on infinitesimal 2-simplices in the image of $[a,b]$, by the weak flatness assumption. And clearly for $s \sim t$, it returns the value of $[a,b]^*(\nabla)$ on s,t. These two properties characterize the trivialization of $[a,b]^*(\nabla)$, so this proves the claim. Substituting $s = 0$ and $t = 1$, we see

$$\Gamma(0,1) = \nabla([a,b](0),[a,b](1)) = \nabla(a,b),$$

and combining this with (5.8.7) proves the desired equality of $(\mathrm{hol}_\nabla)'(a,b)$ and $\nabla(a,b)$, and thus the theorem.

We have, more trivially, the other half of the Fundamental Theorem of Calculus; we consider a groupoid $\Phi \rightrightarrows M$ with a path connection σ; we assume that the connection σ' in Φ admits path integration, so that it makes sense to talk about $\mathrm{hol}_{\sigma'}$.

Theorem 5.8.10 (Fundamental Theorem of Calculus Part 2) *Let $\sigma : P(M) \to \Phi$ be a path connection in Φ. Then*

$$\mathrm{hol}_{\sigma'} = \sigma.$$

Proof. Let $\gamma : R \to M$ be a path. Consider the map $h : R \times R \to \Phi$ given by

$$h(s,t) := \sigma(\gamma \circ [s,t])$$

for $s,t \in R$. Since for $s,t,u \in R$, $\gamma \circ [s,u]$ subdivides into $\gamma \circ [s,t]$ and $\gamma \circ [t,u]$, it follows from the subdivision rule assumed for σ that $h(s,t) \cdot h(t,u) = h(s,u)$, so h is a functor $R \times R \to \Phi$ above γ, or equivalently, a functor above R from $R \times R \to \gamma^*(\Phi)$, i.e. a complete trivialization of $\gamma^*(\Phi)$. Also, h extends $\gamma^*(\sigma')$; for, if $s \sim t$ in R,

$$\gamma^*(\sigma')(s,t) = \sigma'(\gamma(s),\gamma(t)) = \sigma([\gamma(s),\gamma(t)]) = \sigma(\gamma \circ [s,t]),$$

the last equality because γ preserves affine combinations of neighbour points. Thus, h is a trivialization of $\gamma^*(\Phi)$, extending $\gamma^*(\sigma')$; thus it is the unique such, and is the trivialization defining the holonomy $\mathrm{hol}_{\sigma'}(\gamma)$. So we have the first equality sign in

$$\mathrm{hol}_{\sigma'}(\gamma) = h(0,1) = \sigma(\gamma \circ [0,1]) = \sigma(\gamma),$$

the last since $[0,1]$ is the identity map of R. This proves the theorem.

"Curvature generates the local holonomy"

Given a groupoid $\Phi \rightrightarrows N$ with a connection ∇ which admits path integration, every closed path γ at $x \in N$ gives rise to an element $\mathrm{hol}_\nabla(\gamma)$ in the group $\Phi(x,x)$, and the subgroup generated by these elements (or perhaps better, by elements $\mathrm{hol}_\nabla(\gamma)$ for γ a closed *piecewise* path), may be called the *holonomy group of* ∇ *at* x.

For the case of a principal groupoid $\Phi \rightrightarrows M + \{*\}$, we can encode the holonomy group of ∇ more uniformly in terms of the associated principal bundle $P \to M$ with group $G = \Phi(*,*)$; namely, to each $u \in P_x$, we may consider (left-to-right composition, $u : x \to *$)

$$\mathrm{Hol}_\nabla(\gamma, u) := u^{-1} \cdot \mathrm{hol}_\nabla(\gamma) \cdot u \in G.$$

This Hol_∇ may be viewed as the path-connection version of the (likewise G-valued) connection form ω of ∇, and the subgroup generated by the $\mathrm{Hol}_\nabla(\gamma,u)$s likewise deserves the name of the *holonomy group* of ∇.

The Ambrose–Singer Theorem (Ambrose and Singer, 1953) gives conditions under which the *curvature of* ∇ *generates its holonomy group*. Here, "curvature" is understood in terms of the curvature *form* $d\omega$, where ω is the connection form of ∇ (a G-valued 1-form on P), and "holonomy" is understood in terms of the G-valued Hol_∇.

The theorem only deals with *local* holonomy, meaning holonomy along *contractible* paths. In the formulation we shall give, this corresponds to the assumption that G is a *connected* group. Secondly, we need the assumption that the connection ∇ is totally an-holonomic, in the sense of Proposition 5.7.6; this *assumption* replaces a major *construction* in the proof of the full theorem, namely the construction of totally an-holonomic subbundles of the given P. To say that ∇ is totally an-holonomic is to say that G itself is the holonomy group. Thus, the theorem we present below (and which is just a verbal reformulation of Proposition 5.7.6) is only a "baby version" of the classical theorem.

Theorem 5.8.11 *Let ∇ be a principal connection on the principal bundle $P \to M$ with group G, a connected Lie group. Assume that the horizontality distribution for ∇ is totally an-holonomic. If a Lie subgroup H of G contains all values of the curvature form $d\omega$, then $H = G$, i.e. the curvature generates the holonomy.*

6

Lie theory; non-abelian covariant derivative

We consider in this chapter the relationship, on the infinitesimal level, between multiplicative and additive algebraic structure. This is relevant for the theory of group-valued differential forms, and for the relationship between their exterior derivatives, when formed multiplicatively, and when formed additively. The setting involves a space G equipped with a group structure. In Sections 6.7, 6.8 and 6.9, we assume that G is a manifold (so G is a Lie group); in the other sections, we have more general assumptions that will be explained.

6.1 Associative algebras

We begin with some observations in the case where the group G admits some *enveloping* associative algebra; this means an associative unitary algebra A, and a multiplication preserving injection $G \to A$ (so we consider G as a subgroup of the multiplicative monoid of A). The algebra A should be thought of as an auxiliary thing, not necessarily intrinsic to G, and in particular, it does not necessarily have any universal property. Of course, under some foundational assumptions on the category $\mathscr{E}$, there exists a universal such A, the group algebra $R(G)$ of G, but we need to assume that A is a KL algebra, in the sense that the underlying vector space of A is KL, and it is not clear that $R(G)$ has this property, except when G is finite.

We present some examples of enveloping algebras at the end of this chapter.

We shall also consider group *bundles*, and enveloping KL algebra *bundles*. Such occur whenever a vector bundle is given; see Example 6.10.1 below.

So consider an associative algebra $A = (A, +, \cdot)$ whose underlying vector space is a KL vector space. The multiplicative unit of A is denoted 1, or sometimes e.

Recall that, even when a KL vector space A may not be finite dimensional, it is possible to define a (first-order) neighbour relation $\sim$ on it, by the "covariant determination" considered in Section 2.1; in particular $a \sim 0$ in A iff there exists a linear $F : V \to A$ from a finite-dimensional vector space such that $F(y) = a$ for some $y \sim 0$ in V.

Proposition 6.1.1 *Let $a \sim 0$ in A. Then $a \cdot a = 0$. More generally, for any $u \in A$, $a \cdot u \cdot a = 0$.*

Proof. There exists, as we observed, some linear map $F : V \to A$ (V a finite-dimensional vector space) and some $y \in D(V)$ witnessing $a \sim 0$ (so $F(y) = a$). Since the multiplication $\cdot$ on A is bilinear, we therefore have a bilinear map

$$V \times V \xrightarrow{F \times F} A \times A \xrightarrow{\quad \cdot \quad} A, \tag{6.1.1}$$

and $a \cdot a = F(y) \cdot F(y)$. But since $y \in D(V)$, the bilinear map (6.1.1) kills (y, y), hence $a \cdot a = 0$. For the more general assertion in the proposition, one replaces the multiplication map $A \times A \to A$ in (6.1.1) by the, likewise bilinear, map $A \times A \to A$ given by $(b, c) \mapsto b \cdot u \cdot c$.

Proposition 6.1.2 *Let $a \sim 0$ in A. Then for any $b \in A$, $a \cdot b - b \cdot a \sim 0$.*

Proof. It is clear that linear maps between vector spaces preserve the property of being ~ 0. Now the map $x \mapsto x \cdot b - b \cdot x$ is clearly linear.

Proposition 6.1.3 *Let $a \sim 0$. Then $1 + a$ has a two-sided multiplicative inverse, namely $1 - a$.*

Proof. This is an immediate consequence of Proposition 6.1.1; for

$$(1 + a) \cdot (1 - a) = 1 + a - a - a \cdot a = 1,$$

since $a \cdot a = 0$ by Proposition 6.1.1. Similarly, $(1 - a) \cdot (1 + a) = 1$.

We now consider a subgroup G of the group $U(A)$ of multiplicative units of A.

From Proposition 6.1.3 we have that if $a \sim 0$, then $1 + a$ is invertible, $(1 + a)^{-1} = 1 - a$. If further $1 + a$ belongs to $G \subseteq A$, then so does $1 - a$, since G by assumption is stable under multiplicative inversion in A. If we have that both $a \sim 0$ and $b \sim 0$, with $1 + a$ and $1 + b$ in G, we may therefore form their group-theoretic commutator in G,

$$\{1 + a, 1 + b\} = (1 + a) \cdot (1 + b) \cdot (1 + a)^{-1} \cdot (1 + b)^{-1};$$

using $(1 + a)^{-1} = 1 - a$ and similarly for b, we rewrite this as the product

$$\{1 + a, 1 + b\} = (1 + a) \cdot (1 + b) \cdot (1 - a) \cdot (1 - b).$$

Multiplying out by the distributive law, we get 16 terms, seven of them contain either a factor $\pm a$ twice, or a factor $\pm b$ twice, and so vanish, by Proposition 6.1.1 (second clause). Of the remaining nine terms, six cancel out by simple additive calculus, and this leaves three terms, $1 + a \cdot b - b \cdot a$. So we have proved

Proposition 6.1.4 *If a and b are neighbours of 0 in A, and $1+a \in G$, $1+b \in G$, then*

$$\{1+a, 1+b\} = 1 + [a,b] \in G,$$

where $[a,b]$ denotes the usual "algebraic" commutator $a \cdot b - b \cdot a$.

From Proposition 6.1.2 follows that $[a,b] \sim 0$ if a or b is ~ 0. Therefore we get as a consequence of this proposition that if g and h are ~ 1, then so is their group-theoretic commutator $\{g,h\}$.

We define the map $l : G \to A$ by putting $l(g) = g - 1$. For $g \sim 1$, one should think of $l(g)$ as the *logarithm* of g. (If G is all of $U(A)$, then $l : \mathfrak{M}(1) \to \mathfrak{M}(0) = \mathfrak{M}(e)$ has an inverse exp, given by $\exp(a) = 1 + a$, which one may similarly think of as an exponential.)

Using l, Proposition 6.1.4 may be formulated:

if $g, h \in G$ and are ~ 1 in A (in particular, if they are ~ 1 in G), then

$$l(\{g,h\}) = [l(g), l(h)]. \tag{6.1.2}$$

The map l does not convert products into sums in general, but at least it has the property

Proposition 6.1.5 *Assume $g_1, \ldots, g_n \in G$ have $l(g_i) \cdot l(g_j) = 0$ for all i, j. Then*

$$l(g_1 \cdot \ldots \cdot g_n) = l(g_1) + \ldots + l(g_n).$$

Proof. Write $g_i = 1 + l(g_i)$. Then the product occurring on the left is $\prod(1 + l(g_i))$, which when multiplied out yields $1 + \sum l(g_i)$, plus terms which contain factors $l(g_i) \cdot l(g_j)$ and therefore vanish. Applying l gives the result.

Let M be a manifold, and let $\omega : M_{(1)} \to G \subseteq A$ have $\omega(x,x) = 1$ for all $x \in M$. Then $l\omega : M_{(1)} \to A$ has $l\omega(x,x) = 0$ for all x, and therefore is an

A-valued 1-form. Hence by (2.2.4), $l\omega(y,x) = -l\omega(x,y)$. So

$$\omega(y,x) = 1 + l\omega(y,x) = 1 - l\omega(x,y),$$

which is the multiplicative inverse of $1 + l\omega(x,y) = \omega(x,y)$, by Proposition 6.1.3 (which we may use, since $l\omega(x,y) \sim 0$, by Proposition 3.1.4). Summarizing,

$$\textit{if } \omega(x,x) = 1 \textit{ for all } x, \textit{ then } \omega(y,x) = \omega(x,y)^{-1} \quad \text{for all } y \sim x. \tag{6.1.3}$$

6.2 Differential forms with values in groups

The notion of (simplicial) differential k-form with values in a group G was studied for $k = 1$ in Section 3.7. The generalization to $k \geq 2$ is straightforward, but we cannot do much with it unless the group G admits some enveloping associative KL algebra A, as in the previous section. So we assume this throughout the present section. The unit e of G thus equals the multiplicative unit 1 of A.

Definition 6.2.1 *Let M be a manifold, and let G be a group which admits some enveloping algebra. Then a (simplicial) differential k-form with values in G is a law ω which to each infinitesimal k-simplex $(x_0, x_1, \ldots, x_k)$ in M, associates an element $\omega(x_0, x_1, \ldots, x_k) \in G$, subject to the condition that $\omega(x_0, x_1, \ldots, x_k) = 1$ if two of the x_is are equal.*

We use the notation $l(g) = g - 1$, as in the previous section. It follows that $l\omega$ defined by

$$l\omega(x_0, x_1, \ldots, x_k) := \omega(x_0, x_1, \ldots, x_k) - 1$$

is a differential k-form with values in A, and in particular by Theorem 3.1.5, it is alternating. Also, the values of $l\omega$ are ~ 0 by Proposition 3.1.4. It follows that all the values of ω itself are ~ 1; and also, it follows from Proposition 6.1.3 that ω is alternating in the multiplicative sense, $\omega(\sigma\underline{x}) = (\omega(\underline{x}))^{\pm 1}$, where $\underline{x}$ is an infinitesimal k-simplex, $\sigma \in \mathfrak{S}_{k+1}$, and where the sign in the exponent is $+$ if σ is even, and $-$ if odd.

The definition here is compatible with the one given for $k = 0, 1$, and 2 in Section 3.7. There, we also defined $d.\omega$ in case ω is a k-form with $k = 0$ or $k = 1$, but noted that these "multiplicative exterior derivatives" ramified into a "right" and a "left" version.

It turns out that for $k \geq 2$, there is no such ramification. Let ω be a k-form with values in G. We define the value of $d.\omega$ on an infinitesimal $k+1$-simplex by just rewriting the standard simplicial formula (3.2.1) in multiplicative notation; thus

$$(d.\omega)(x_0,x_1,\dots,x_{k+1})$$
$$:=\omega(x_1,\dots,x_{k+1})\cdot\omega(x_0,x_2,\dots,x_{k+1})^{-1}\cdot\dots\cdot\omega(x_0,x_1,\dots,x_k)^{(-1)^{k+1}}; \tag{6.2.1}$$

the point is that we don't have to worry about the order of the $k+2$ factors in the product; we shall prove (Proposition 6.2.3 below) that they commute. (The same $k+2$-fold product is also found in the formula (6.3.1) below, if you ignore the symbol "$\nabla(x_0,x_1) \dashv$".)

The factors that enter into $d.\omega$ are

$$\omega(x_0,x_1,\dots,\widehat{x_i},\dots,x_{k+1})$$

or their inverses. First, let us prove

Proposition 6.2.2 *For $G \subseteq A$, as above, and with $i \geq 1$, $j \geq 1$,*

$$l(\omega(x_0,x_1,\dots,\widehat{x_i},\dots,x_{k+1}))\cdot l(\omega(x_0,x_1,\dots,\widehat{x_j},\dots,x_{k+1})) = 0,$$

provided $k \geq 2$.

Proof. In a standard coordinatized situation $M \subseteq V$, we may write $l\omega(x_0,\dots,x_k)$ as

$$\Omega(x_0;x_1-x_0,\dots,x_k-x_0)$$

with $\Omega : M \times V^k \to A$ being k-linear in the arguments after the semicolon. Among the indices $h = 1,\dots,k+1$, there is at least one h which is neither $= i$ nor $= j$ (using that $k \geq 2$). Then in the product

$$\Omega(x_0;x_1-x_0,\dots,\widehat{i},\dots)\cdot\Omega(x_0;x_1-x_0,\dots,\widehat{j},\dots),$$

$x_h - x_0$ appears in each of the two factors in a linear position, thus the product vanishes since $x_h \sim x_0$.

Proposition 6.2.3 *Let $k \geq 2$. Then the $k+1$ factors that enter into the definition (6.2.1) of $d.\omega$ commute.*

Proof. Consider the ith and jth factor. Without loss of generality we may assume that i and j are ≥ 1. We have then to prove that the group-theoretic commutator of $\omega(x_0,\dots,\hat{i},\dots,x_{k+1})$ and $\omega(x_0,\dots,\hat{j},\dots,x_{k+1})$ is 1. By (6.1.2), it suffices to see that the algebraic commutator of the corresponding l-values is 0. This follows immediately from Proposition 6.2.2: both the two terms $a\cdot b$ and $-b\cdot a$ that make up the algebraic commutator $[a,b]$ in question, are in fact 0.

Consider a differential k-form on M with values in $G \subseteq A$, as above, and with $k \geq 2$. Then ω has a coboundary in the multiplicative sense, as defined in (6.2.1), and which we, as there, denote $d.\omega$; and $l\omega$ has a coboundary in the standard additive sense (3.2.1), which we now denote $d_+ l\omega$, with the decoration "+", for contrast.

Proposition 6.2.4 *Let ω be a k-form on M with values in G, $k \geq 2$. Then $l(d.\omega) = d_+(l\omega)$.*

Proof. This follows by combining Lemma 6.1.5 with Proposition 6.2.2.

Thus, for $k \geq 2$, the comparison between $d.$ and d_+ is easy. The case $k = 1$ is more interesting; here $d.$ is defined by the formula (3.7.1). We shall prove

Proposition 6.2.5 *Let ω be a 1-form on a manifold M, with values in $G \subseteq A$, and let $l\omega(x,y) = \omega(x,y) - 1$, and similarly for 2-forms. Then for any infinitesimal 2-simplex x,y,z in M,*

$$l(d.\omega)(x,y,z) = d_+ l\omega(x,y,z) + l\omega(x,y) \cdot l\omega(y,z). \tag{6.2.2}$$

Note again that the $d.$ on the left-hand side utilizes the multiplication $\cdot$ of G, the d_+ on the right-hand side utilizes the $+$ of A. Note also that since $l\omega(x,y) \cdot l\omega(y,z)$ depends linearly on $y - x$, we may, by the Taylor principle, replace the y in the last factor by x; we get a formula equivalent to (6.2.2):

$$l(d.\omega)(x,y,z) = d_+ l\omega(x,y,z) + l\omega(x,y) \cdot l\omega(x,z). \tag{6.2.3}$$

Proof. Consider a G-valued 1-form ω on a manifold M and consider an infinitesimal 2-simplex x,y,z in M. Then

$$\begin{aligned} d.\omega(x,y,z) &= \omega(x,y) \cdot \omega(y,z) \cdot \omega(z,x) \\ &= (1 + l\omega(x,y)) \cdot (1 + l\omega(y,z)) \cdot (1 + l\omega(z,x)). \end{aligned}$$

This we may multiply out, using the distributive law in the algebra A; we get

$$\begin{aligned} &1 + l\omega(x,y) + l\omega(y,z) + l\omega(z,x) \\ &+ l\omega(x,y) \cdot l\omega(y,z) + l\omega(x,y) \cdot l\omega(z,x) + l\omega(y,z) \cdot l\omega(z,x) \\ &+ l\omega(x,y) \cdot l\omega(y,z) \cdot l\omega(z,x). \end{aligned} \tag{6.2.4}$$

Lemma 6.2.6 *The three terms in the middle line are equal except for sign. The term in the third line is 0.*

Proof. We recognize each of the three terms in the middle line as values of the cup product 2-form $l\omega \cup. l\omega$, applied to permutation instances of x,y,z – for the middle term, this may require an argument:

$$l\omega(x,y)\cdot l\omega(z,x) = (-l\omega(y,x))\cdot(-l\omega(x,z)) = (l\omega \cup . \, l\omega)(y,x,z)$$

by cancelling the two minus signs. Now (y,x,z) is an odd permutation, the two other terms come from even permutation instances; the fact that $l\omega \cup . \, l\omega$ is alternating (Theorem 3.1.5) then gives the first assertion of the lemma. For the second assertion, we consider the A-valued 3-form $l\omega \cup . \, l\omega \cup . \, l\omega$ (for associative $A \times A \xrightarrow{\cdot} A$, like here, the corresponding $\cup .$ is associative); we recognize the last line in (6.2.4) as this 3-form applied to the 3-simplex (x,y,z,x), but since this simplex has a repeated entry, the form vanishes on it. This proves the lemma.

So in the expression (6.2.4) for $d.\omega$, only the first line and one term from the second line survive, so we have

$$\begin{aligned}(d.\omega)(x,y,z) &= 1 \, + \, l\omega(x,y) + l\omega(y,z) + l\omega(z,x)\\ &\qquad + l\omega(x,y)\cdot l\omega(y,z)\\ &= 1 + (d_{+}l\omega)(x,y,z) + l\omega(x,y)\cdot l\omega(y,z).\end{aligned}$$

Subtracting 1 on each side gives the desired equation. Thus Proposition 6.2.5 is proved.

Here is an alternative expression for $l(d.\omega)$:

Proposition 6.2.7 *Let ω be as in Proposition 6.2.5. Then for any infinitesimal 2-simplex (x,y,z) in M, we have*

$$\big(\omega(x,y)\cdot\omega(y,z)\cdot\omega(z,x)\big) - 1 = \omega(x,y)\cdot\omega(y,z) - \omega(x,z). \qquad (6.2.5)$$

Proof. Write $\omega(u,v) = 1 + l\omega(x,y)$, as above. The proof of Lemma 6.2.6 gives that the left-hand side of (6.2.5) equals

$$l\omega(x,y) + l\omega(y,z) + l\omega(z,x) + l\omega(x,y)\cdot l\omega(y,z). \qquad (6.2.6)$$

Let us calculate the right-hand side of (6.2.5). Rewriting $\omega(x,y)$ as $1 + l\omega(x,y)$, and similarly for x,z and y,z, gives

$$\big((1 + l\omega(x,y))\cdot(1 + l\omega(y,z))\big) - \big(1 + l\omega(x,z)\big);$$

replacing $l\omega(x,z)$ by $-l\omega(z,x)$, and multiplying out, we get six terms, two of which are 1 and -1, which cancel; the remaining four are those of (6.2.6), and this proves the proposition.

The vector space A carries two bilinear structures, partly the given multiplication $\cdot$, partly the algebraic commutator $[-,-] : A \times A \to A$. Each of these bilinear structures gives rise to a wedge product on differential forms with values in A, and we denote these $\wedge$ and $[-,-]$, respectively. Note that the last term on the right-hand side of (6.2.2) expresses $(l\omega \wedge l\omega)(x,y,z)$.

Then the content of Proposition 6.2.5 may also be stated:

Proposition 6.2.8 *Let ω be a 1-form with values in $G \subseteq A$. Then we have an equality of $(A,+)$-valued 2-forms*

$$l(d.\omega) = d_+ l\omega + \tfrac{1}{2}[l\omega, l\omega]. \tag{6.2.7}$$

This follows by combining the equation (6.2.2) with Corollary 3.5.4.

For later use, we record an assertion much analogous to Proposition 6.2.3. We consider a k-form ρ on M with values in a group G, with G acting (from the left, say) on a KL vector space W, and we consider also a k-form β on M with values in W.

Proposition 6.2.9 *Let $k \geq 2$, and let $(x_0, x_1, \ldots, x_{k+1})$ be an infinitesimal $k+1$-simplex in M. Then for any $i, j = 0, 1, \ldots, k+1$*

$$\rho(x_0, x_1, \ldots, \hat{i}, \ldots, x_{k+1}) \dashv \beta(x_0, x_1, \ldots, \hat{j}, \ldots, x_{k+1}) = \beta(x_0, x_1, \ldots, \hat{j}, \ldots, x_{k+1}). \tag{6.2.8}$$

Proof. It suffices to do this in a standard coordinatized situation $M \subseteq V$; without loss of generality, we may assume that i and j are ≥ 1. Since $k \geq 2$, we may pick an index $h = 1, \ldots, k+1$ which is neither i nor j. Then $x_h - x_0$ appears linearly in the β-factor, and so by the Taylor principle, we may replace x_h in the ρ-factor in (6.2.8) by x_0. Since ρ is a G-valued form, $\rho(x_0, x_1, \ldots, x_0, \ldots) = 1 \in G$, and after this replacement, the two sides in (6.2.8) clearly are equal.

In a similar vein, one may prove that if $g \in G$, and if ν is a G-valued 1-form and θ is a W-valued 1-form, then for $x_0 \sim x_1$,

$$(\nu(x_0,x_1) \cdot g \cdot \nu(x_0,x_1)^{-1}) \dashv \theta(x_0,x_1) = g \dashv \theta(x_0,x_1). \tag{6.2.9}$$

We leave the proof as an easy exercise.

6.3 Differential forms with values in a group bundle

We discussed in Section 3.8 the notion of covariant derivative for (combinatorial) differential forms on a manifold M with values in a vector bundle $E \to M$, in the presence of a linear bundle connection ∇ in $E \to M$. Also, we discussed in Section 6.2 a theory of (simplicial) differential forms with values in a group G, assuming G admits some enveloping associative KL algebra. We now combine these generalizations.

We replace the algebra A and the group G by *bundles* over M, equipped with connections; thus $A \to M$ is a bundle of KL algebras equipped with a bundle connection ∇, whose transport laws preserve the algebra structure (such arise naturally when a vector bundle $E \to M$, with a linear connection, is given, see Example 6.10.5 below). So the fibres A_x are KL algebras; $G \subseteq A$ is a subbundle, with G_x a subgroup of the multiplicative monoid of A_x. We assume that $G \subseteq A$ is stable under ∇, in the evident sense. Finally, we assume that $A \to M$ is locally constant as an algebra bundle $A \to M$, i.e. we assume that it locally is of the form $M \times A_0 \to M$ with A_0 a KL algebra.

By a (simplicial) *differential k-form on M with values in $G \to M$*, we understand, in analogy with Section 3.8, a law ω which to each infinitesimal k-simplex $(x_0, x_1, \ldots, x_k)$ in M associates an element $\omega(x_0, x_1, \ldots, x_k) \in G_{x_0}$, with the property that the value is 1 (= the identity element in the group G_{x_0}) if $x_i = x_0$ for some $i \neq 0$. We let $\Omega^k(G \to M)$ denote the set of these.

We shall define a "covariant derivative" map

$$d_{\cdot}^{\nabla} : \Omega^k(G \to M) \to \Omega^{k+1}(G \to M).$$

Let $\omega \in \Omega^k(G \to M)$, and let $(x_0, \ldots, x_{k+1})$ be an infinitesimal $k+1$-simplex in M. For $k \geq 2$, we define the value of $d_{\cdot}^{\nabla}(\omega)$ on an infinitesimal $k+1$-simplex $(x_0, x_1, \ldots, x_{k+1})$ by the same expression as for the covariant derivative in vector bundles (cf. (3.8.1)), except that we replace "plus" by "times"; thus

$$d_{\cdot}^{\nabla}\omega(x_0, x_1, \ldots, x_{k+1}) := \left[\nabla(x_0, x_1) \dashv \omega(x_1, \ldots, x_{k+1})\right] \cdot \prod_{i=1}^{k+1} \omega(x_0, \ldots, \widehat{x_i}, \ldots, x_{k+1})^{(-1)^i}; \tag{6.3.1}$$

just as for group-valued forms, it turns out that for $k \geq 2$, the order of the factors in (6.3.1) is irrelevant. Only for $k = 1$ and $k = 0$, the order matters, and here, we take the order so as to generalize the coboundary formulas (3.7.1) and (3.7.2) (i.e. the "right-handed versions"). Explicitly, for ω a 1-form,

$$d_{\cdot}^{\nabla}\omega(x, y, z) := \omega(x, y) \cdot \left[\nabla(x, y) \dashv \omega(y, z)\right] \cdot \omega(x, z)^{-1}. \tag{6.3.2}$$

For g in $\Omega^0(G)$, so g is a $G \to M$-valued 0-form, i.e. a cross-section of $G \to M$, we define

$$d_{\cdot}^{\nabla} g(x, y) := g(x)^{-1} \cdot \left[\nabla(x, y) \dashv g(y)\right]. \tag{6.3.3}$$

If ω is a (simplicial) k-form with values in the bundle G, we get a k-form $l\omega$ with values in $A \to M$,

$$l\omega(x_0, x_1, \ldots, x_k) = \omega(x_0, x_1, \ldots, x_k) - 1_{x_0}.$$

We want to compare the covariant derivatives of ω and $l\omega$. By a small variation of the argument leading to Corollary 6.2.4 we shall prove

Proposition 6.3.1 *Let ω be a $G \to M$-valued simplicial k-form on M. Then if $k = 1$, we have*

$$l(d^{\nabla}\omega)(x,y,z) = d_{+}^{\nabla} l\omega(x,y,z) + l\omega(x,y) \cdot l\omega(x,z); \tag{6.3.4}$$

for $k \geq 2$, we have

$$l(d^{\nabla}\omega) = d_{+}^{\nabla}(l\omega). \tag{6.3.5}$$

Proof. We consider first (6.3.4). The proof is then a matter of modifying the proof leading to Proposition 6.2.5; this is most easily done in a coordinatized situation where M is an open subset of a finite-dimensional vector space V, and where $A = M \times A_0$. Then

$$\omega(x,y) = 1 + \theta(x; y - x)$$

(more precisely, $\omega(x,y) = (x, 1 + \theta(x; y - x)) \in M \times A_0$), with $\theta : M \times V \to A_0$ linear in the second variable. Less pedantically, $\theta = l\omega$. Also, we claim that there exists a map $\Gamma : M \times V \times V \to A_0$, bilinear in the last two arguments, such that for $x \sim y$ and $y \sim z$ in M,

$$\nabla(x,y) \dashv (1 + \theta(y; z - y)) = 1 + \theta(y; z - y) + \Gamma(x; y - x, z - y).$$

(We identify $\nabla(x,y) : A_y \to A_x$ notationally with a map $A_0 \to A_0$, for simplicity.) The difference $\nabla(x,y) \dashv (1 + \theta(y; z - y)) - (1 + \theta(y; z - y))$ is 0 if $x = y$, because $\nabla(x,x)$ is the identity map, and is 0 if $y = z$, because $\nabla(x,y)$ preserves $1 \in A_0$, being an algebra connection. Then for an infinitesimal 2-simplex x, y, z in M,

$$d^{\nabla}\omega(x,y,z) = (1 + \theta(x; y - x)) \cdot (1 + \theta(y; z - y) + \Gamma(x; y - x, z - y)) \cdot (1 + \theta(x; x - z));$$

these we multiply out by distributivity, much as in (6.2.4) (with θ-expressions instead of $l\omega$s), except that we now have some correction terms involving Γ; thus, instead of the first line in (6.2.4) (the 0- and 1-order terms), we get

$$1 + \theta(x; y - x) + \theta(y; z - y) + \theta(x; x - z) + \Gamma(x; y - x, z - y)$$

which is $1+d_{+}^{\nabla} l\omega$. Corresponding to the second line in (6.2.4) (the second-order terms), we get some terms which only involve θs, and some which also involve some Γ; those that only involve θ give

$$\theta(x;y-x)\cdot\theta(y;z-y)+\theta(x;y-x)\cdot\theta(x;x-z)+\theta(y;z-y)\cdot\theta(x;x-z); \tag{6.3.6}$$

the terms which involve Γ can all be seen to vanish for degree reasons, for instance in $\Gamma(x;y-x,z-y)\cdot\theta(x;x-z)$, we may replace the last occurrence of z by y, because $z-y$ occurs linearly in the Γ-factor; and then we have an expression where $y-x$ appears bilinearly. Likewise, the "third-order" terms (corresponding to the third line in (6.2.4)) vanish. So we are left with (6.3.6), and as in the proof of Lemma 6.2.6, the last two terms cancel each other, so only $\theta(x;y-x)\cdot\theta(y;z-y)$ remains. It equals, by the Taylor principle, $\theta(x;y-x)\cdot\theta(x;z-x)$. Translating back into ω and $l\omega$, the result is now clear.

The case $k\geq 2$ may be carried out along the same lines in a coordinatized situation; it is actually easier than the $k=1$ case, because the factors involving $l\omega$ that enter into $d_{\cdot}^{\nabla}$ can be seen to have pairwise product 0, just as in Proposition 6.2.2; we leave the details to the reader.

We shall next analyse $d^{\nabla}\circ d^{\nabla}$, leading to the classical formula $d^{\nabla}\circ d^{\nabla}(\theta)=lR\wedge\theta$ for any $E\to M$-valued k-form. We shall do it only for the case $k=1$. Here, as above, $E\to M$ is assumed to be a locally constant vector bundle whose fibres are KL vector spaces; ∇ is a linear connection on $E\to M$, and $R=R_{\nabla}$ is the curvature of ∇ viewed as a 2-form with values in (the gauge group bundle of the groupoid) $GL(E\to M)$. Then we also have a 2-form lR with values in the vector bundle $End(E\to M)$, namely $lR(x,y,z):=R(x,y,z)-1$, just as in Section 6.2; here, 1 denotes the identity map of E_x. The wedge product in question is with respect to the bilinear evaluation map $\dashv\colon End(E)\times_M E\to E$.

Proposition 6.3.2 *Let θ be an $E\to M$-valued k-form, and let ∇ be a linear bundle-connection in $E\to M$. Then*

$$d^{\nabla}(d^{\nabla}(\theta))=lR_{\nabla}\wedge\theta.$$

Proof. We shall prove this for the case $k=1$ only. Let (x,y,z,u) be an infinitesimal 3-simplex in M. Then each instance of $d^{\nabla}(d^{\nabla}\theta)$ is a sum of four terms of the form $d^{\nabla}(\theta)(a,b,c)$, each of the terms possibly decorated by some action instances of ∇; and each of these four terms $d^{\nabla}(\theta)(a,b,c)$ in turn is a sum of three terms, of the form $\theta(v,w)$, possibly decorated by some action instances of ∇. Altogether, we have 12 terms. If we could ignore the ∇ decorations, these 12 terms would cancel out pairwise, as in the usual formula for $d \circ d = 0$ for simplicial cochains. In the present case, six of the 12 terms have no ∇ decoration, and they cancel out pairwise. The expression $\nabla(x,y) \dashv \theta(y,z)$ occurs twice except for sign; also the expression $\nabla(x,y) \dashv \theta(y,u)$ occurs twice except for sign. So these four terms likewise cancel, and only two terms are left; they are the ones exhibited on the right-hand side in the following formula, which we have thus derived:

$$d^{\nabla}(d^{\nabla}(\theta))(x,y,z,u) = \nabla(x,y) \dashv \nabla(y,z) \dashv \theta(z,u) - \nabla(x,z) \dashv \theta(z,u). \quad (6.3.7)$$

To prove that this equals $(lR \wedge \theta)(x,y,z,u)$, we may assume that the vector bundle is a constant bundle $M \times W \to M$, and the linear connection ∇ may be identified with a $GL(W)$-valued 1-form ω on M,

$$\nabla(x,y) \dashv (y,w) = (x, \omega(x,y)(w)) = (x, w + l\omega(x,y)(w))$$

with $l\omega$ a 1-form with values in the vector space $End(W)$. With this notation, we may rewrite the right-hand side in (6.3.7) as

$$(\omega(x,y) \circ \omega(y,z) - \omega(x,z))(\theta(z,u)).$$

Using Proposition 6.2.7, this may in turn be written

$$[\omega(x,y) \circ \omega(y,z) \circ \omega(z,x) - 1](\theta(u,z)).$$

The square bracket here is $lR(x,y,z)$, and the expression we have now is $(lR \wedge \theta)(x,y,z,u)$. This proves the proposition.

6.4 Bianchi identity in terms of covariant derivative

If we have a connection ∇ in a groupoid $\Phi \rightrightarrows M$, we get a bundle connection in the group bundle $gauge(\Phi)$, since Φ acts on the bundle $gauge(\Phi)$ by conjugation; this bundle connection is denoted $ad\nabla$. (We are here talking about the *left* action.) In particular, if ρ is a k-form with values in the group bundle $gauge(\Phi)$, and ∇ is a connection in Φ, we may define a $k+1$-form $d^{ad\nabla}\rho$ with values in $gauge(\Phi)$.

We now have the following reformulation of the combinatorial Bianchi identity (Theorem 5.2.11):

Theorem 6.4.1 (Combinatorial Bianchi Identity, 2) *Let ∇ be a connection in a Lie groupoid $\Phi \rightrightarrows M$, and let R be its curvature, $R \in \Omega^2(gauge(\Phi))$. Then $d^{ad\nabla}(R)$ is the "zero" 3-form, i.e. takes only the neutral group elements in the fibres as values.*

Proof. Let x, y, z, u form an infinitesimal 3-simplex. We have by definition of $d_{\cdot}^{ad\nabla}$ that

$$(d_{\cdot}^{ad\nabla}(R))(xyzu) = ad(\nabla(xy))R(yzu) \cdot R(xzu)^{-1} \cdot R(xyu) \cdot R(xyz)^{-1}$$

(omitting commas between the input entries, for ease of reading). Now the two middle terms may be interchanged, by arguments such as those of Proposition 6.2.3. We then get the expression in the combinatorial Bianchi identity in Theorem 5.2.11, and by this theorem, it has value id_x.

We shall derive the classical Bianchi identity from the combinatorial one. We consider a locally constant vector bundle $E \to M$, equipped with a linear connection ∇, or equivalently, with a connection in the groupoid $GL(E \to M)$. Then its curvature R is a simplicial 2-form with values in the group bundle $gauge(GL(E \to M))$, which is a group bundle with a group connection $ad\nabla$ (cf. also Example 6.10.5 below). According to the combinatorial Bianchi identity, in the form of Theorem 6.4.1, $d_{\cdot}^{ad\nabla}R$ is the "zero" form; as there, we have decorated the covariant derivation symbol $d^{ad\nabla}$ with the symbol $\cdot$ (symbol for the multiplication in $gauge(GL(E \to M))$), the algebraic structure which is used for calculating the covariant derivative. Now the group bundle $gauge(GL(E \to M))$ has an enveloping algebra bundle $A = End(E \to M)$, with $A_x = End(E_x)$, and we have the covariant derivation $d_{+}^{ad\nabla}$ with respect to addition in this bundle.

The fact that $d_{\cdot}^{ad\nabla}R$ is the "zero" form implies that $l(d_{\cdot}^{ad\nabla}R) = 0$ as an $End(E \to M)$-valued 2-form. Applying (6.3.5) (with $k = 2$, and with ∇ replaced by $ad\nabla$) then yields

Theorem 6.4.2 (Classical Bianchi Identity) *For ∇ a linear connection in a locally constant vector bundle $E \to M$, the $End(E \to M)$-valued curvature lR of ∇ satisfies*

$$d_{+}^{ad\nabla}(l(R)) = 0.$$

Here, $l(R)$ is a (simplicial) 2-form with values in the vector bundle $End(E \to M)$. The correspondences of Section 4.7 between simplicial and classical vector-space-valued forms and their coboundaries extend to vector-bundle-valued forms; and $l(R)$ corresponds to the classical curvature of the linear connection ∇ in $E \to M$. The equation $d_{+}^{ad\nabla}(l(R)) = 0$ then expresses that the classical curvature of ∇ has vanishing covariant derivative w.r.t. $ad\nabla$.

6.5 Semidirect products; covariant derivative as curvature

All forms in the present section are simplicial differential forms on a manifold M; the value groups or value group bundles vary, and will be specified.

We consider a groupoid $\Phi \rightrightarrows M$ (composition from left to right), and a vector bundle $E \to M$ on which it acts, from the left, say, by linear maps, i.e. for $g : x \to y$ in Φ, $g \dashv$ is a linear isomorphism $E_y \to E_x$. Then there is a new groupoid (the semidirect product) $E \ltimes \Phi \rightrightarrows M$, where an arrow $x \to y$ is a pair (u, g) with $u \in E_x$ and $g : x \to y$ in Φ. Composition is given by

$$(u,g)\cdot(v,h) := (u + (g \dashv v), g\cdot h).$$

Here is a display of the book-keeping involved:

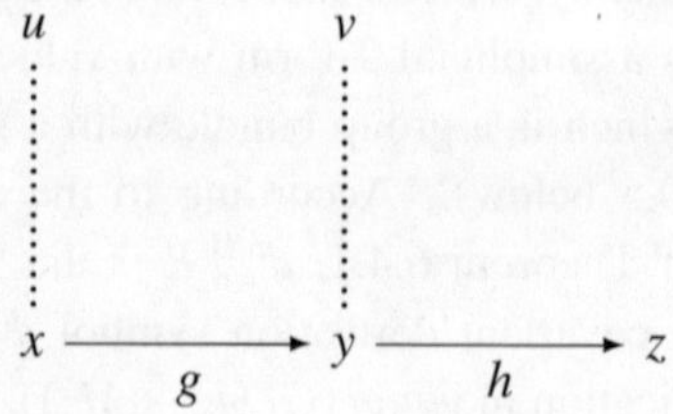

There is a morphism of groupoids (i.e. a functor) $\square : E \ltimes \Phi \to \Phi$, "forgetting the E-part".

Exercise 6.5.1 Show that the groupoid $E \ltimes GL(E \to M)$ acts on the left on the bundle $E \to M$,

$$(u,g) \dashv v := u + (g \dashv v),$$

and that by this action it may be identified with the groupoid of invertible *affine* maps between the fibres of $E \to M$.

A connection $\overline{\nabla}$ in the groupoid $E \ltimes \Phi$ amounts to the following data: for $x \sim y$, an element $\theta(x,y) \in E_x$, and an arrow $\nabla(x,y) : x \to y$ in Φ. The condition that $\overline{\nabla}(x,x)$ is the identity arrow at x implies that $\theta(x,x) = 0$, and also that ∇ is a connection in the groupoid $\Phi \rightrightarrows M$. So $\overline{\nabla}$ may be identified with a pair (θ, ∇), where θ is a 1-form with values in the vector bundle $E \to M$, and ∇ is a connection in Φ.

We consider the curvature $R_{\overline{\nabla}}$ of the connection $\overline{\nabla} = (\theta, \nabla)$. So for (x,y,z) an infinitesimal 2-simplex in M, we consider

$$(\theta(x,y), \nabla(x,y)) \cdot (\theta(y,z), \nabla(y,z)) \cdot (\theta(z,x), \nabla(z,x)) \in (E \ltimes \Phi)(x,x).$$

Since $\square$ is a functor, and $\nabla = \square \circ \overline{\nabla}$, it is clear that the second component is $R_\nabla(x,y,z) \in \Phi(x,x)$. The first component is in E_x. What is it?

Proposition 6.5.2 *Assume that $E \to M$ is locally of the form $M \times W \to M$ with W a KL vector space. Let $\overline{\nabla} = (\theta, \nabla)$ be a connection in $E \ltimes \Phi \rightrightarrows M$. Then the first component of $R_{\overline{\nabla}}(x,y,z)$ is $(d^{\nabla}\theta)(x,y,z) \in E_x$. In other words*

$$R_{(\theta,\nabla)} = (d^{\nabla}\theta, R_{\nabla}).$$

Proof. Calculating $\overline{\nabla}(x,y) \cdot \overline{\nabla}(y,z) \cdot \overline{\nabla}(z,x)$ by the recipe for composition in $E \ltimes \Phi$ gives, for its first component,

$$\theta(x,y) + \nabla(x,y) \dashv \theta(y,z) + (\nabla(x,y) \cdot \nabla(y,z)) \dashv \theta(z,x).$$

This we must compare with $d^{\nabla}\theta(x,y,z)$, which by definition is

$$(\nabla(x,y) \dashv \theta(y,z)) - \theta(x,z) + \theta(x,y).$$

Two of the three terms here match two of the terms in the previous expression; to match the remaining terms we need that

$$\nabla(x,y) \cdot \nabla(y,z) \dashv \theta(z,x) = -\theta(x,z).$$

This we carry out in a "coordinatized" situation, i.e. where $E = M \times W$, with M an open subset of a finite-dimensional vector space and W a KL vector space; in this case θ may be written $\theta(x,y) = (x, \Theta(x; y-x))$ with $\Theta : M \times V \to W$, linear in the second argument; and $\nabla(x,y)(y,w) = (x, \omega(x,y)(w))$ where ω is a $GL(W)$-valued 1-form (in analogy with Proposition 3.7.5); we can then apply the Taylor principle in the calculation of $(\nabla(x,y) \cdot \nabla(y,z)) \dashv \theta(z,x)$; we get (identifying the fibres of $E \to M$ with W)

$$(\nabla(x,y) \cdot \nabla(y,z)) \dashv \theta(z,x) = \omega(x,y) \cdot \omega(y,z) \dashv \Theta(z; x-z)$$

and replace z by x in the middle factor; then we end up with $\Theta(z; x-z)$ $(= \Theta(x; x-z))$, and the result is then immediate.

In particular, we see that the torsion of an affine connection λ is part of the curvature of a connection in a certain semidirect product, namely the connection determined by λ together with the solder form. (Combine Proposition 6.5.2 and Theorem 4.8.1.)

Affine Bianchi identity

It is convenient to have the notion of covariant derivative of bundle-valued forms described explicitly for the case of *constant* group or vector bundles. We shall do it for the case of constant group bundles $M \times H \to M$ (with H a group). (The vector bundle case is then a further special case, which the reader may want to make explicit.) We do not assume that the connection ∇ in the constant bundle $M \times H \to M$ is the trivial connection; rather, it is encoded by an $Aut(H)$-valued 1-form ν on M, so that for $x \sim y$ in M,

$$\nabla(x,y)(y,h) = (x, \nu(x,y)(h)).$$

Differential k-forms $\overline{\omega}$ with values in the bundle $M \times H \to M$ may be identified with H-valued k-forms,

$$\overline{\omega}(x_0,x_1,\ldots,x_k) = (x_0, \omega(x_0,x_1,\ldots,x_k)).$$

The formula for coboundary $d.$ is changed into $d^{\nu}_{.}$; explicitly, for $k \geq 2$, the $k+2$ factors in the formula (6.2.1) for $d.\omega$ are unchanged, except that the first factor is modified by ν, i.e. the factor $\omega(x_1,\ldots,x_{k+1}) \in H$ is replaced by $\nu(x_0,x_1)(\omega(x_1,\ldots,x_{k+1}))$. This modification is, for the special case at hand, the modification by $\nabla(x_0,x_1)$ in the first factor of (6.3.1). For $k = 1$, the formula for d^{ν} can similarly be read out from (6.3.2),

$$d^{\nu}_{.}\,\omega(x,y,z) := \omega(x,y) \cdot \nu(x,y)(\omega(y,z)) \cdot \omega(x,z)^{-1}.$$

An H-valued 1-form σ on M gives rise to an $Aut(H)$-valued 1-form $ad\sigma$,

$$(ad\sigma)(x,y)(h) := \sigma(x,y) \cdot h \cdot \sigma(x,y)^{-1}.$$

Recall from general group theory that if a group G acts by group homomorphisms on an (additively written) group W, from the left, say, there is a semidirect product group $W \ltimes G$; its underlying set is $W \times G$, and the group multiplication is given by

$$(w_1,g_1) \cdot (w_2,g_2) := (w_1 + (g_1 \dashv w_2), g_1 \cdot g_2).$$

Projection to G is a group homomorphism; the neutral element is $(0,1)$. Let us also record the formula for conjugation in $W \ltimes G$:

$$(t,n) \cdot (b,r) \cdot (t,n)^{-1} = (t + (n \dashv b) - (n \cdot r \cdot n^{-1}) \dashv t, n \cdot r \cdot n^{-1}). \qquad (6.5.1)$$

If $H = W \ltimes G$, an H-valued k-form on M may be identified with a pair (β,ρ) of k-forms, where β takes values in W and ρ takes values in G. An H-valued 1-form (θ,ν) gives rise to an $Aut(H)$-valued 1-form $ad(\theta,\nu)$, as described above.

We consider the special case where W is a KL vector space, and where $G = GL(W)$, the group of linear automorphisms of W. This group is a subgroup of the algebra $End(W)$ of linear endomorphisms of W, and the underlying vector space of this algebra is a KL vector space. We consider $H = W \ltimes GL(W)$.

Let (θ,ν) be an H-valued 1-form on M, and let (β,ρ) be an H-valued k-form on M, with $k \geq 2$. We want to describe the H-valued $k+1$-form

$d^{ad(\theta,v)}(\beta,\rho)$. Recall that if ρ is a $GL(W)$-valued form, we get an $End(W)$-valued form $l\rho$ by subtracting $1 \in GL(W)$. We shall prove

Proposition 6.5.3 *We have*

$$d^{ad(\theta,v)}(\beta,\rho) = (d^{v}\beta \pm (l\rho \wedge \theta), d^{ad\,v}\rho)$$

with the sign being minus if k is even, plus if k is odd.

Here, $\wedge$ refers to the bilinear map $End(W) \times W \to W$ sending (r,w) to $r(w)$.

Proof. The assertion about the second component follows immediately because the projection $\square : W \ltimes GL(W) \to GL(W)$ is a group homomorphism. We evaluate $d^{ad(\theta,v)}(\beta,\rho)$ on an infinitesimal $k+1$-simplex $(x_0,x_1,\ldots,x_{k+1})$; the value is by definition a product in $W \ltimes GL(W)$ of $k+2$ factors. It follows from Proposition 6.2.9 that the product of the last $k+1$ of them may be calculated in the direct product $W \times GL(W)$, rather than in the semidirect $W \ltimes GL(W)$. In particular, the W-component of these $k+1$ factors is

$$\sum_{1}^{k+1} \pm\beta(x_0,x_1,\ldots,\hat{i},\ldots,x_{k+1}). \tag{6.5.2}$$

On the other hand, the first factor is special, since it involves conjugation by $(\theta(x_0,x_1), v(x_0,x_1))$; explicitly, this factor is

$$(\theta(x_0,x_1), v(x_0,x_1)) \cdot (\beta(x_1,\ldots,x_{k+1}), \rho(x_1,\ldots,x_{k+1})) \cdot (\theta(x_0,x_1), v(x_0,x_1))^{-1}.$$

We use the general formula (6.5.1) for conjugation in $W \ltimes G$, with $t = \theta(x_0,x_1)$, $n = v(x_0,x_1)$, $b = \beta(x_1,\ldots,x_{k+1})$, and $r = \rho(x_1,\ldots,x_{k+1})$. The W component of the first factor is by (6.5.1) equal to

$$t + n(b) - (n \cdot r \cdot n^{-1})(t), \tag{6.5.3}$$

but from (6.2.9) follows that for $n = v(x_0,x_1)$ and $t = \theta(x_0,x_1)$, the conjugation by n has no effect. So the the sum (6.5.3) may be rewritten

$$\theta(x_0,x_1) + v(x_0,x_1)(\beta(x_1,\ldots,x_{k+1})) - \rho(x_1,\ldots,x_{k+1})(\theta(x_0,x_1)).$$

The middle term here goes together with the terms of (6.5.2) to yield $d^{v}\beta(x_0,\ldots,x_{k+1})$; the remaining two terms yield $-l\rho(x_1,\ldots,x_{k+1})(\theta(x_0,x_1))$. However,

$$\begin{aligned} -l\rho(x_1,\ldots,x_{k+1})(\theta(x_0,x_1)) &= l\rho(x_1,\ldots,x_{k+1})(\theta(x_1,x_0)) \\ &= (l\rho \wedge \theta)(x_1,\ldots,x_{k+1},x_0); \end{aligned}$$

which in turn is $\pm(l\rho \wedge \theta)(x_0, x_1, \ldots, x_{k+1})$ where the sign is the sign of the cyclic permutation of $k+2$ letters, so is minus if k is even, plus if k is odd. This proves the proposition.

Consider a vector bundle $E \to M$, and the groupoid $GL(E \to M)$. Consider a connection $\overline{\nabla} = (\theta, \nabla)$ in $E \ltimes GL(E \to M)$, as above; by Proposition 6.5.2, its curvature is $(d^{\nabla}\theta, R_{\nabla})$. By the Bianchi identity for the connection $\overline{\nabla} = (\theta, \nabla)$,

$$d^{ad\overline{\nabla}}(d^{\nabla}\theta, R_{\nabla}) \equiv (0,1).$$

On the other hand, by Proposition 6.5.3, the W-component of this 2-form is $d^{\nabla}(d^{\nabla}\theta) - lR_{\nabla} \wedge \theta$. This provides an alternative proof of $d^{\nabla}(d^{\nabla}\theta) = lR_{\nabla} \wedge \theta$ for $E \to M$-valued 1-forms θ.

The following "affine Bianchi identity" is a special case.

Corollary 6.5.4 *The torsion* $\tau := d^{\nabla}\theta$ *satisfies*

$$d^{\nabla}\tau = lR_{\nabla} \wedge \theta.$$

This is just the special case obtained from Proposition 6.3.2 by taking $E \to M$ to be $TM \to M$, and taking θ to be the solder form of M. This equation "d of the *torsion* equals *curvature* wedge *solder*" is sometimes called the *affine Bianchi identity*, cf. Bishop and Crittenden (1964, §6.2).

6.6 The Lie algebra of G

We continue to assume that the group G can be embedded as a subgroup of the multiplicative monoid of a KL algebra A, so that we can utilize the calculations of Section 6.1. Then $e \in G$ equals the multiplicative unit $1 \in A$. The algebra A is to be thought of as an auxiliary thing, not intrinsically tied to G in the same way as $T_e(G)$ is; we shall assume that G is a manifold, hence micro-linear, so that in particular $T_e(G)$ carries a Lie algebra structure, as described by (4.9.3). Ultimately, the Lie bracket on $T_e(G)$ was constructed in terms of the group-theoretic commutator of elements of G. In the present section, we shall provisionally denote this bracket operation on $T_e(G)$ by double square brackets $[[-,-]]$, for contrast with the algebraic commutator $[-,-]$ on A; so now (4.9.3) reads

$$[[\xi, \eta]](d_1 \cdot d_2) = \{\xi(d_1), \eta(d_2)\}, \tag{6.6.1}$$

where curly brackets denote group-theoretic commutator.

Let us collect some of the constructions involved, in the following diagram:

$$\begin{array}{ccccc} & & T_e(G) & \xrightarrow{L = \text{ principal part}} & A \\ & & \uparrow{\scriptstyle \log_e} & & \uparrow{\scriptstyle \text{inclusion}} \\ G & \supseteq & \mathfrak{M}(e) & \xrightarrow[l = \text{ subtract } e]{} & \mathfrak{M}(0) \end{array} \qquad (6.6.2)$$

It is easy to see that the diagram commutes. For, let $g \sim e$ in G. Then the principal part of the tangent vector $\log_e(g)$, i.e. of $d \mapsto (1-d)\cdot e + d \cdot g = e + d\cdot(g-e)$, is $g-e$, i.e. it is $l(g)$.

The map $L : T_e(G) \to A$ is clearly injective. By construction, it satisfies

$$d \cdot L(\tau) = \tau(d) - e \qquad (6.6.3)$$

for $\tau \in T_eG$ and $d \in D$.

We now have three binary operations on A, on G, and on $T_e(G)$, respectively:

(1) the algebraic commutator on A,

$$[a,b] := a\cdot b - b\cdot a;$$

(2) the group commutator on G,

$$\{x,y\} := x\cdot y\cdot x^{-1}\cdot y^{-1};$$

(3) the Lie bracket $[[-,-]]$ on $T_e(G) = \mathfrak{g}$ as given in Section 4.9, and characterized by (6.6.1).

Recall that $\mathfrak{M}(e)$ is stable under the group-theoretic commutator formation, see Proposition 6.1.4 and the remarks immediately following it.

Theorem 6.6.1 *The maps in the diagram (6.6.2) preserve these operations:*

$$l(\{g,h\}) = [l(g), l(h)],$$

$$L([[\xi,\eta]]) = [L(\xi), L(\eta)],$$

$$\log_e(\{g,h\}) = [[\log_e(g), \log_e h]],$$

where g and h are in $\mathfrak{M}(e)$, ξ and η in $T_x(G)$.

Proof. The first of these equations was proved in Proposition 6.1.4, in the form of (6.1.2). The third follows purely formally from the first two, using injectivity of L. So it remains to prove the second equation.

Let ξ and η belong to $T_e(G)$. To prove $L([[\xi,\eta]]) = [L(\xi),L(\eta)]$, it suffices to prove for all $(d_1,d_2) \in D \times D$ that

$$d_1 \cdot d_2 \cdot L([[\xi,\eta]]) = d_1 \cdot d_2 \cdot [L(\xi),L(\eta)]$$

for all d_1 and d_2 in D (this is just a matter of cancelling the two universally quantified d_is, one at a time). We calculate

$$d_1 \cdot d_2 \cdot L([[\xi,\eta]]) = [[\xi,\eta]](d_1 \cdot d_2) - e$$

(by (6.6.3))

$$= \{\xi(d_1),\eta(d_2)\} - e$$

(by (6.6.1))

$$= [\xi(d_1) - e, \eta(d_2) - e]$$

(by Proposition 6.1.4)

$$= [d_1 \cdot L(\xi), d_2 \cdot L(\eta)]$$

(using (6.6.3) on each "factor"); but this equals the right-hand side of the desired equation, by the bilinearity of the algebraic commutator $[-,-]$. This proves the theorem.

Because of the theorem, we may henceforth denote the Lie bracket $[[-,-]]$ in $T_e(G)$ by the same symbol $[-,-]$ as the algebraic commutator in A.

6.7 Group-valued vs. Lie-algebra-valued forms

We consider a Lie group G; the unit is denoted e. We have in Section 3.7 considered (simplicial) 1- and 2-forms on a manifold M with values in G. We shall compare these to (simplicial as well as classical) differential forms with values in $T_e(G) = \mathfrak{g}$, the Lie algebra of G. To a G-valued simplicial 1-form ω, we get a simplicial $T_e(G)$-valued 1-form $\tilde{\omega}$ by applying $\log_e : \mathfrak{M}(e) \to T_e(G)$. (Note that if $x \sim y$ in M, then $\omega(x,y) \sim e$ in G, so that we may apply $\log_e$ to it.) Thus $\tilde{\omega}$ is defined by

$$\tilde{\omega}(x,y) := \log_e \omega(x,y).$$

Similarly if θ is a simplicial G-valued 2-form, we get a $T_e(G)$-valued simplicial 2-form $\tilde{\theta}$,

$$\tilde{\theta}(x,y,z) := \log_e \theta(x,y,z).$$

Now since $T_e(G)$ is a KL vector space, there is by Theorem 4.7.1 a bijective correspondence between simplicial $T_e(G)$-valued forms μ on M, and classical $T_e(G)$-valued forms $\overline{\mu}$ on M. Recall that the passage from $\overline{\mu}$ to μ was given explicitly in formula (4.7.1). Applying this to the case where μ is $\tilde{\omega}$ or $\tilde{\theta}$, and substituting the defining equations for $\tilde{\omega}$ and $\tilde{\theta}$ in terms of $\log_e$, we thus get that for G-valued simplicial 1- and 2-forms ω and θ on M, there exist unique classical $T_e(G)$-valued forms $\overline{\omega}$ and $\overline{\theta}$ satisfying

$$\overline{\omega}(\log_x(y)) = \log_e(\omega(x,y)) \tag{6.7.1}$$

and similarly

$$\overline{\theta}(\log_x(y), \log_x(z)) = \log_e(\theta(x,y,z)). \tag{6.7.2}$$

The theorem to be proved in this section concerns a simplicial G-valued 1-form ω and its coboundary $\theta = d.\omega$ which is a G-valued simplicial 2-form. The decoration "." on the coboundary operator d is to remind us that it is the multiplication in G which is used to produce $d.\omega$ out of ω, by the recipe in Section 3.7.

Let ω be a G-valued simplicial 1-form on a manifold M. So we have also the simplicial G-valued 2-form $d.\omega$. Let $\overline{\omega}$ and $\overline{d.\omega}$ be the corresponding classical $T_e(G)$-valued forms given by (6.7.1) and (6.7.2), respectively (with $\theta = d.\omega$). We shall prove

Theorem 6.7.1 *In the above situation*

$$\overline{d.\omega} = \tfrac{1}{2}\{\overline{d}\overline{\omega} + \tfrac{1}{2}[\overline{\omega}, \overline{\omega}]\}.$$

Here, $\overline{d}$ denotes the exterior derivative for classical differential forms (as in Theorem 4.7.2), and the square bracket denotes the wedge product of classical differential forms with respect to the bilinear Lie bracket on $T_e(G)$. The factor $\frac{1}{2}$ outside the curly bracket is essentially a matter of convention: it comes from the factors $\frac{1}{2}$ which occur in Theorem 4.7.2 (for $k = 1$) and in Theorem 4.7.3 (for $k = l = 1$), comparing simplicial coboundary with exterior derivative, and cup product with wedge product, respectively.

Proof. Let $\overline{\omega}$ and $\overline{d.\omega}$ be the classical $T_e(G)$-valued 1- and 2-forms om M corresponding to the G-valued forms ω and $d.\omega$, respectively. By the way the correspondence between G-valued forms and classical $T_e(G)$-valued forms was set up, $\overline{\omega}$ is also the classical 1-form corresponding to the simplicial $T_e(G)$-valued 1-form $\log_e \omega$ by the correspondence of Theorem 4.7.1, and likewise $\overline{d.\omega}$ corresponds to the $T_e(G)$-valued simplicial 2-form $\log_e(d.\omega)$.

First, we claim that we have the following equality of $T_e(G)$-valued simplicial forms:

$$\log_e d.\omega = d_+ \log_e \omega + \tfrac{1}{2}[\log_e \omega, \log_e \omega], \tag{6.7.3}$$

the square bracket here being the cup product w.r.t. to the Lie bracket on $T_e(G)$. Since L is a Lie algebra homomorphism by Theorem 6.6.1, and is injective, it suffices to see that

$$L(\log_e d.\omega) = L(d_+ \log_e \omega) + \tfrac{1}{2}[L(\log_e \omega), L(\log_e \omega)],$$

with the square bracket now denoting cup product w.r.t. the algebraic commutator $[-,-]$ on A. By the commutativity of the diagram (6.6.2), and using $L \circ d_+ = d_+ \circ L$, this in turn is equivalent to $l(d.\omega) = d_+ l\omega + \frac{1}{2}[l\omega, l\omega]$, which is true by Proposition 6.2.5 (in the form of equation (6.2.7)); thus (6.7.3) is proved.

Now, if $\overline{\omega}$ is the classical $T_e(G)$-valued 1-form corresponding to the simplicial $T_e(G)$-valued 1-form $l\omega$, it follows from Theorem 4.7.2 that $\frac{1}{2}\overline{d\omega}$ is the classical 2-form corresponding to $d_+ l\omega$, and from Theorem 4.7.3 it follows that $\frac{1}{2}[\overline{\omega}, \overline{\omega}]$ (= wedge product w.r.t. $[-,-]$) is the classical 2-form corresponding to $[l\omega, l\omega]$. Since the correspondence between classical and simplicial $T_e(G)$-valued differential forms is clearly linear, the formula of Theorem 6.7.1 follows.

Recall that for a Lie group G, we have a canonical simplicial G-valued 1-form ω on G, namely what we in Section 3.7 called the *Maurer–Cartan* form, given by

$$\omega(x,y) = x^{-1} \cdot y,$$

which trivially is closed, $d.\omega = 0$, where 0 denotes the 2-form with constant value $e \in G$. Since $\overline{0} = 0$ by the process $\theta \mapsto \overline{\theta}$ of (6.7.2), we get immediately from the formula in Theorem 6.7.1:

Corollary 6.7.2 (Maurer–Cartan equation) *For ω the Maurer–Cartan form on a Lie group G,*

$$\overline{d\omega} + \tfrac{1}{2}[\overline{\omega}, \overline{\omega}] = 0.$$

(The T_eG-valued 1-form $\overline{\omega}$ is the classical left-invariant Maurer–Cartan form.)

6.8 Infinitesimal structure of $\mathfrak{M}_1(e) \subseteq G$

A main aspect of Lie theory is the theory describing how infinitesimal algebraic structure around e contains information about the whole group G. In the classical treatment, and partly also in the previous sections, the infinitesimal structure is encapsuled in $T_e(G)$, where the group multiplication on G cunningly induces a Lie algebra structure on $T_e(G)$.

In the present section, we shall present that aspect of Lie theory which deals with $\mathfrak{M}(e) = \mathfrak{M}_1(e)$, the (first-order) neighbourhood around $e \in G$. One may see this as a paraphrasing of the treatment of "formal groups" in Serre (1965, LG.4).

We note that we cannot expect the multiplication map $G \times G \to G$ to restrict to a map $\mathfrak{M}(e) \times \mathfrak{M}(e) \to \mathfrak{M}(e)$. Consider e.g. the most basic of all Lie groups, $(R, +)$. Here $\mathfrak{M}(e)$ is $D \subseteq R$, and we know that D is not stable under addition.

So in the general case, $x \sim e$ and $y \sim e$ does not imply $x \cdot y \sim e$. It turns out, however, that if not only $x \sim e$ and $y \sim e$, but also $x \sim y$, then we *can* conclude $x \cdot y \sim e$.

For notation, we let $\{x,y\}$ denote the group-theoretic commutator of x and y,

$$\{x,y\} = x \cdot y \cdot x^{-1} \cdot y^{-1},$$

for any $x, y \in G$. Also, recall that if $x \sim y$ in G, there is a map $[x,y] : R \to G$ given by affine combinations, $t \mapsto (1-t)x + ty$. (We omit the multiplication dot $(1-t) \cdot x + t \cdot y$ previously used in connection with such affine combinations, because this dot is now reserved for the multiplication in G. Also beware that the square brackets used in most of the present section have nothing to do with Lie bracket or algebraic commutator.)

Theorem 6.8.1 *(1) If $x \sim e$ in G, then $x^{-1} \sim e$, in fact, it is the affine combination*

$$x^{-1} = 2e - x$$

(= the mirror image of x in e), and as such, does not depend on the multiplication $\cdot$.

(2) If $x \sim e$ in G, then the map $[e,x] : R \to G$ is a group homomorphism $(R, +) \to (G, \cdot)$. All the points in the image of this map are mutual neighbours.

(3) If $x \sim e$ and $y \sim e$, then $x \cdot y \sim e$ iff $x \sim y$.

(4) If $x \sim e$, $y \sim e$, and $x \sim y$, the group commutator $\{x,y\}$ is an affine combination of the mutual neighbour points $x \cdot y$, x, y, e:

$$\{x,y\} = 2(x \cdot y) - 2x - 2y + 3e, \tag{6.8.1}$$

and in particular $\{x,y\} \sim e$*; and* $x \cdot y$ *is an affine combination of the mutual neighbour points* $\{x,y\}, x, y, e$*:*

$$x \cdot y = x + y + \frac{1}{2}\{x,y\} - \frac{3}{2}e. \tag{6.8.2}$$

In particular, if x *and* y *commute,* $x \cdot y = x + y - e$*, and as such,* $x \cdot y$ *does not depend on the multiplication* $\cdot$.

Proof. Note that assertion (1) is a special case of (2), by considering $-1 \in R$. To prove assertions (2)–(5), we need to coordinatize the situation.

If U is an open subset of G containing e, then for $x \sim e$, we have $x \cdot y \sim y$. For, right multiplication by y is a map $G \to G$, hence $x \sim e$ implies $x \cdot y \sim e \cdot y = y$. So if both x and y are $\sim e$, then $x \cdot y \sim y \sim e$, so $x \cdot y$ is a second-order neighbour of e, hence $x \cdot y \in U$. Similarly a product of k factors $\sim e$ gives a kth-order neighbour of e, hence such product is in U as well.

If $U \subset G$ is an open subset containing e, it follows that the multiplication restricts to a map

$$\mathfrak{M}(e) \times \mathfrak{M}(e) \to \mathfrak{M}_2(e) \subseteq U. \tag{6.8.3}$$

In particular, let us pick a coordinate neighbourhood U around e which identifies it with an open subset of a finite-dimensional vector space V, and with e identified with $0 \in V$. So $\mathfrak{M}(e)$ gets identified with $D(V)$, and from (6.8.3), we get a map

$$m : D(V) \times D(V) \to U \subseteq V. \tag{6.8.4}$$

Since $x \cdot e = x$, it follows that $m(x,0) = x$, and similarly $m(0,y) = y$, for x and y in $D(V)$. From KL then follows that there is a unique bilinear $B : V \times V \to V$ such that for all $x, y \in D(V)$, we have $m(x,y) = x + y + B(x,y)$, or, returning to $x \cdot y$ notation,

$$x \cdot y = x + y + B(x,y) \tag{6.8.5}$$

for all $x, y \in D(V) = \mathfrak{M}(e)$.

Since $x \sim e$, we have the map $[e,x] : R \to G$ given by affine combinations of neighbour points. All its values are neighbour points of e, and hence the map factors through the open subset U. Thus it gets identified with the map $[0,x] : R \to V$ (recalling that $e = 0$ under this identification). This is the map $t \mapsto tx$. The assertion (2) then amounts to $tx \cdot sx = (t+s)x$. We calculate the left-hand side, using (6.8.5); we get

$$tx \cdot sx = tx + sx + B(tx, sx).$$

But the last term vanishes since it depends in a bilinear way on $x \in D(V)$. This proves (2).

To prove (3) amounts by (6.8.5) to proving that for $x \in D(V)$ and $y \in D(V)$,

$$x+y+B(x,y) \in D(V) \text{ iff } x-y \in D(V).$$

Since x and y are in $D(V)$, $x-y \in D(V)$ iff $(x,y) \in \widetilde{D}(2,V)$. The result now follows from Proposition 1.2.18.

To prove (4), we first have to calculate the group commutator $\{x,y\}$ in terms of B. The calculations involved in this are almost explicitly to be found in e.g. Serre (1965, LG.4 §7); now they just come in a different conceptual garment. We first calculate $x \cdot y \cdot x^{-1}$ in coordinate terms:

Lemma 6.8.2 *For $x \in D(V)$ and $y \in D(V)$, we have*

$$x \cdot y \cdot x^{-1} = y + B(x,y) - B(y,x). \tag{6.8.6}$$

Proof. It suffices to prove

$$x \cdot y = (y + B(x,y) - B(y,x)) \cdot x.$$

Note that, for x fixed, $y+B(x,y)-B(y,x)$ depends linearly on y, so is in $D(V)$ since $y \in D(V)$; so we can calculate both sides here using (6.8.5). The right-hand side gives

$$y+B(x,y)-B(y,x)+x+B(y+B(x,y)-B(y,x),x),$$

but the "nested" appearances of B expressions vanish, since $x \in D(V)$, so we are left with $x+y+B(x,y)$, and this is the left-hand side of the desired equation, by (6.8.5) again.

We note that the expression in (6.8.6) is in $D(V)$; for, it depends linearly on y, and $y \in D(V)$. Also, it is $\sim -y$; for, subtracting $-y$ yields $2y+B(x,y)-B(y,x)$ which is in $D(V)$ since it depends linearly on $y \in D(V)$. So we may calculate $(x \cdot y \cdot x^{-1}) \cdot y^{-1}$ by (6.8.5) again; using (6.8.6), this yields

$$[y+B(x,y)-B(y,x)]-y+B([y+B(x,y)-B(y,x)],-y);$$

here the two isolated ys kill each other, and the last term is 0 since it depends bilinearly on $y \in D(V)$; so we are left with $B(x,y)-B(y,x)$. We record the result in the following:

Lemma 6.8.3 *For $x \in D(V)$ and $y \in D(V)$, we have*

$$\{x,y\} = B(x,y) - B(y,x). \tag{6.8.7}$$

Note that these two lemmas do not depend on $x \sim y$. However, when $x \sim y$, we have $(x,y) \in \widetilde{D}(2,V)$, and for such (x,y), the bilinear $B(x,y)$ behaves as if it were alternating (see Section 1.3), hence $-B(y,x) = B(x,y)$. We have then the following expression for $\{x,y\}$:

$$\{x,y\} = 2B(x,y). \tag{6.8.8}$$

Since the right-hand side here depends linearly on x and on y, it follows that $\{x,y\}$ is $\sim x$, $\sim y$, and also $\sim 0 = e$. Substituting $B(x,y) = \frac{1}{2}\{x,y\}$ in (6.8.5), we get, still assuming $x \sim y$,

$$x \cdot y = x + y + \frac{1}{2}\{x,y\}; \tag{6.8.9}$$

equivalently, since $e = 0$

$$x \cdot y = x + y + \frac{1}{2}\{x,y\} - \frac{3}{2}e. \tag{6.8.10}$$

The right-hand side here is an affine combination of mutual neighbour points, and as such is preserved by the identification of the open neighbourhood of e in G with an open neighbourhood of 0 in V, and this proves (6.8.2). The equation (6.8.1) comes about by equational rewriting. The assertion about commuting elements x, y now follows because $\{x,y\} = e$ if x and y commute.

Example 6.8.4 Consider the set

$$G := \{(x_1, x_2) \in R^2 \mid x_2 \text{ is invertible}\}.$$

It is an open subset of R^2, so in particular, it is a manifold. It carries a group structure given by

$$(x_1, x_2) \cdot (y_1, y_2) := (x_1 + x_2 y_1, x_2 y_2).$$

(This is a semidirect product; we can also see it as the group of affine isomorphisms $R \to R$; with the notation of the Appendix, $(x_1, x_2) \in G$ defines the affine map $||x_1 \mid x_2|| : R \to R$ given by $t \mapsto x_1 + t x_2$ which is invertible since x_2 is invertible.) The unit e is $(0,1)$, and the inverse of (x_1, x_2) is $(-x_2^{-1} x_1, x_2^{-1})$. So this is a Lie group.

To see the above calculations in coordinates, we describe an open set $U \subseteq R^2$, and a bijection $G \to U$ (taking $e \in G$ to $0 \in R^2$) by taking $U = \{(x_1, x_2) \mid x_2 + 1 \text{ is invertible}\}$, and the bijection $G \to U$ is simply "subtracting $(0,1)$". The group structure on G gets transported via this bijection to a group structure on U, which is easily seen to be

$$(x_1, x_2) \cdot (y_1, y_2) = (x_1 + y_1 + x_2 y_1, x_2 + y_2 + x_2 y_2),$$

with $(0,0)$ as multiplicative unit. The right-hand side here may be rewritten

$$(x_1, x_2) + (y_1, y_2) + (x_2 y_1, x_2 y_2),$$

so the bilinear $B : V \times V \to V$, which was the basis for our calculation (here with $V = R^2$), is given by the third term in this expression. (The example is atypical in the sense that the formula for the multiplication in terms of $+$ and B applies to all elements of the group, not just to neighbours of the multiplicative unit $(0,0)$.)

Let x denote (x_1, x_2), and similarly for y. Then if x and y are not only neighbours of $e = (0,0)$, but also mutual neighbours (i.e. $(x,y) \in \widetilde{D}(2,2)$), we have $B(x,y) = (x_2 y_1, 0)$.

For such x, y, the commutator $\{x,y\}$ can be calculated to be $(-d, 0)$, where d is the determinant of the 2×2 matrix (x,y). The expression (6.8.8) gives $2x_2 y_1$, but for $(x,y) \in \widetilde{D}(2,2)$, this equals minus the determinant.

Jacobi identity from Hall identity

Let us record a reformulation of Lemma 6.8.3; since by (6.8.5) $B(x,y) = x \cdot y - x - y$ and $B(y,x) = y \cdot x - y - x$ for $x \sim 0 = e$ and $y \sim 0 = e$, we have for such x, y that $B(x,y) - B(y,x) = x \cdot y - y \cdot x$; therefore, Lemma 6.8.3 implies

$$\{x,y\} = x \cdot y - y \cdot x. \tag{6.8.11}$$

The right-hand side here may be denoted $[x,y]$, since it looks like the commutator construction in associative algebras ("algebraic commutator"). But note that there is no associative algebra around, and there is no a priori reason why a Jacobi identity should hold. For the rest of this section $[x,y]$ denotes this "algebraic commutator" construction. For x and y neighbours of $0 = e$, we have by (6.8.11)

$$\{x,y\} = [x,y]. \tag{6.8.12}$$

We may similarly, for $x, y, z \in D(V)$, calculate $\{\{x,y\}, z^y\}$. (Here, z^y denotes $y \cdot z \cdot y^{-1}$.) Note that both arguments in the outer $\{-,-\}$ are in $D(V)$, so we may use the formula (6.8.12) on it; we also use the formula for the $\{-,-\}$ inside the first argument; for the second argument, we use Lemma 6.8.2. We get

$$[\{x,y\}, z^y] = [[x,y], z^y] = [[x,y], z + B(y,z) - B(z,y)];$$

the terms involving B vanish by expansion of the outer (bilinear) $[-,-]$, because of repeated occurrence of y. We are left with $[[x,y],z]$; let us record this also:

Proposition 6.8.5 *For x, y and $z \in D(V)$, we have*

$$\{\{x,y\}, z^y\} = [[x,y], z].$$

This can be used to prove the Jacobi identity for the bilinear $[-,-]: V \times V \to V$ (essentially following the exposition in Serre, 1965, LG.4). One has in any group G the beautiful 42-letter identity of Ph. Hall, which with our conventions reads that the cyclic product of $\{\{x,y\},z^y\}$ is e. Now each of the three factors in this cyclic product is in $D(V)$ whenever $x,y,z \in D(V)$, and equals the cyclic product of $[[x,y],z]$, by Proposition 6.8.5. We expand this product of three factors using (6.8.5); starting e.g. with

$$[[x,y],z] \cdot [[y,z],x] = [[x,y],z] + [[y,z],x] + B([[x,y],z],[[y,z],x]);$$

in the B term here, x occurs twice in linear position, so that the B term vanishes; we get

$$[[x,y],z] \cdot [[y,z],x] = [[x,y],z] + [[y,z],x],$$

and similarly

$$[[x,y],z] \cdot [[y,z],x] \cdot [[z,x],y] = [[x,y],z] + [[y,z],x] + [[z,x],y].$$

Thus the cyclic product of $\{\{x,y\},z^y\}$ is $[[x,y],z] + [[y,z],x] + [[z,x],y]$, but on the other hand, the cyclic product is 0, by the Hall identity. Thus $[[x,y],z] + [[y,z],x] + [[z,x],y] = 0$. Since this holds for x,y,z in $D(V)$, and B is bilinear, it holds for all $x,y,z \in V$, by KL.

Exercise 6.8.6 The cyclic product referred to in Ph. Hall's identity is in full

$$\{\{x,y\},z^y\} \cdot \{\{y,z\},x^z\} \cdot \{\{z,x\},y^x\}.$$

Each of the three $\cdot$ factors is a 14-fold product. Thus altogether, there are 42 factors. The first $\cdot$ factor, for instance, resolves into the following 14-fold product:

$$\begin{aligned}&\{x \cdot y \cdot x^{-1} \cdot y^{-1}, y \cdot z \cdot y^{-1}\}\\ &\quad = (x \cdot y \cdot x^{-1} \cdot y^{-1}) \cdot (y \cdot z \cdot y^{-1}) \cdot (y \cdot x \cdot y^{-1} \cdot x^{-1}) \cdot (y \cdot z^{-1} \cdot y^{-1});\end{aligned}$$

this product of 14 factors reduces to a product of 10 factors:

$$x \cdot y \cdot x^{-1} \cdot z \cdot x \cdot y^{-1} \cdot x^{-1} \cdot y \cdot z^{-1} \cdot y^{-1}.$$

Writing these 10 factors next to their two cyclically permuted versions gives a product of 30 factors, and these factors cancel two by two in an elegant pattern, leaving e.

6.9 Left-invariant distributions

Let G be a Lie group. Since left multiplication by $x \in G$ is a bijection $G \to G$, we have that

$$x \sim y \text{ iff } e \sim x^{-1} \cdot y \sim e \text{ iff } x^{-1} \cdot y \in \mathfrak{M}(e).$$

For $S \subset \mathfrak{M}(e)$, we get a refinement $\approx$ of $\sim$ by putting

$$x \approx y \text{ iff } x^{-1} \cdot y \in S.$$

If $e \in S$, the relation $\approx$ is reflexive. If further S is stable under multiplicative inversion (this is equivalent to S being stable under reflection in e, by Theorem 6.8.1 item (3)), we get that $\approx$ is symmetric. Under these assumptions, $\approx$ is thus a pre-distribution on G. It is evidently left invariant in the sense that $x \approx y$ implies $z \cdot x \approx z \cdot y$.

Conversely, given a left-invariant pre-distribution $\approx$ on G, the set $S \subseteq \mathfrak{M}(e)$ given as $\{z \in \mathfrak{M}(e) \mid z \approx e\}$ contains e and is stable under multiplicative inversion. It is clear that this defines a bijective correspondence between left-invariant pre-distributions on G, and subsets $S \subseteq \mathfrak{M}(e)$ containing e and stable under multiplicative inversion.

If S is furthermore a *linear* subset of $\mathfrak{M}(e)$, the pre-distribution $\approx$ is a distribution, and vice versa.

Proposition 6.9.1 *Let $S \subseteq \mathfrak{M}(e)$ be a subset containing e and stable under multiplicative inversion, and let $\approx$ be the corresponding pre-distribution. Then $\approx$ is involutive iff S is stable under multiplication of mutual neighbours.*

Proof. Assume $\approx$ involutive. Let $x \in S$, $y \in S$ and $x \sim y$. Then also $x^{-1} \in S$. Now x^{-1} is an affine combination of x and e (cf. Theorem 6.8.1), and since e, x, y are mutual neighbours, $x^{-1} \sim y$ as well. So for e, x^{-1}, y, we have $x^{-1} \approx e$, $y \approx e$ and $x^{-1} \sim y$. By the assumed involutivity of $\approx$, we conclude $x^{-1} \approx y$, which means that $x \cdot y \in S$.

Conversely, assume S has the stability property stated, and let $x \approx y$, $x \approx z$ and $y \sim z$. The two first statements translate into $x^{-1} \cdot y \in S$, $x^{-1} \cdot z \in S$ and the third one implies that $x^{-1} \cdot y \sim x^{-1} \cdot z$. The same type of "inversion" argument as in the first part of the proof yields that $(x^{-1} \cdot y)^{-1} \sim x^{-1} \cdot z$, and then we can use the stability assumption on these two elements to conclude $y^{-1} \cdot z \in S$, that is, $y \approx z$.

In the rest of this section, we assume the validity of the Frobenius Theorem.

Theorem 6.9.2 *Let $S \subseteq \mathfrak{M}(e)$ be a linear subset stable under multiplication of mutual neighbours, i.e. $x \in S, y \in S$ and $x \sim y$ implies $x \cdot y \in S$. Then there is a unique maximal connected subgroup H of G with $H \cap \mathfrak{M}(e) = S$.*

Proof. By the proposition above, the left-invariant distribution $\approx$ defined by S is involutive, hence by the Frobenius Theorem, G gets partitioned into leaves; let H be the leaf through e. It is an easy consequence of left invariance of $\approx$ that the partition into leaves is likewise left invariant, i.e. if K is a leaf and $z \in G$, $z \cdot K$ is likewise a leaf. This in particular implies that the leaf H is stable under multiplication. Since the partition is stable under left multiplication, $x^{-1} \cdot H$ is a leaf, but if $x \in H$, this leaf contains $x^{-1} \cdot x = e$, so equals H, so $x \in H$ implies that $x^{-1} \in H$ as well, so H is stable under multiplicative inversion. It follows that H is a subgroup of G, and it is connected, since any leaf by definition is so. The fact that $H \cap \mathfrak{M}(e) = S$ is now a special case of (2.6.4).

Finally, if H' is a subgroup with $H' \cap \mathfrak{M}(e) = S$, it is easy to see that H' is an integral subset for $\approx$, and so if H' is connected, we have $H' \subseteq H$ by the maximality property of H as a leaf. This proves the uniqueness assertion of the theorem.

A variant of this theorem involves the group-theoretic commutator rather than the product itself:

Theorem 6.9.3 *Let $S \subseteq \mathfrak{M}(e)$ be a linear subset such that $x \in S$, $y \in S$ and $x \sim y$ implies $\{x, y\} \in S$. Then there is a unique maximal connected subgroup H of G with $H \cap \mathfrak{M}(e) = S$.*

Proof. Since a linear subset is clearly stable under affine combinations of mutual neighbours, it follows from Theorem 6.8.1 (item (4)) that S is stable under multiplication of mutual neighbours, so the previous theorem applies and gives the conclusion.

Corollary 6.9.4 *For any Lie group G, there is a unique maximal connected subgroup H with $\mathfrak{M}(e) \subseteq H$.*

Proof. We know from Theorem 6.8.1 (3) that $\mathfrak{M}(e)$ is closed under multiplication of mutual neighbours.

It is also the case that a subgroup H of G with $\mathfrak{M}(e) \subseteq H$ is an *open* subgroup of G, but this depends on "submersions have open image". For, the inclusion $H \hookrightarrow G$ is easily seen to be a submersion.

6.10 Examples of enveloping algebras and enveloping algebra bundles

If G is a Lie group, one has the Lie algebra $T_e(G)$, and hence one has an associative algebra A, namely the universal enveloping algebra of $T_e(G)$; this, however, is *not* known to be an example of an enveloping algebra of G in the sense we have been using it; for one thing, there may be no inclusion map $G \to A$, and secondly, it is not clear whether the underlying vector space of A is KL.

The examples we give now are of a different kind, and in some sense more elementary.

Example 6.10.1 Let $A = gl(n,R)$, the algebra of $n \times n$ matrices with entries from R; let $G = GL(n,R) \subseteq gl(n,R)$ be the group of invertible matrices. This is a particularly simple example, since here one can prove that the Lie algebra $\mathfrak{g}$ of G is $gl(n,R) = A$ (with algebraic commutator $xy - yx$ as Lie bracket).

Example 6.10.2 We take $A = gl(n,R)$ like in Example 6.10.1, but with G the group $SL(n,R)$ of matrices of determinant 1. Its Lie algebra consists of matrices of trace 0 ; they do not form a subalgebra of $gl(n,R)$, "algebra" meaning "associative algebra".

Both these examples have A finite dimensional (hence KL), and with G a manifold.

Example 6.10.3 (cf. Demazure and Gabriel, 1970, §II.4.5; see also Kock, 1981/2006, §I.12 and in particular Kock, 1983b). Let G be a group (a Lie group, say). Let A be the vector space "of distributions† on G with compact support"; it can in the present synthetic context be construed as the vector space $Lin_R(R^G,R)$ of linear maps $R^G \to R$. It is a KL vector space (but not finite dimensional in general). Multiplication is convolution of distributions (using the multiplication of G). Every $g \in G$ gives rise to a punctual distribution, namely the Dirac distribution δ_g at g. (To make sure that this is an example, one needs to prove that the map $g \mapsto \delta_g$ is injective, which is probably not generally possible, on the meagre axiomatic basis we are using here.)

If G is a finite group, $Lin_R(R^G,R)$ is the standard group algebra $R[G]$ (the vector space with the elements of G as basis). In general, the distributions of compact support on a group form a kind of group algebra for it.

Example 6.10.4 Let M be a manifold. We have the vector space R^M of functions on M. Let A be the vector space of R-linear maps $R^M \to R^M$; it becomes

† Here, we are talking about distributions in the sense of Schwartz – they are not related to the geometric distributions $\approx$ studied in Section 2.6.

an associative unitary algebra by taking composition of functions as multiplication. The group G of invertible maps $M \to M$ (= the group of diffeomorphisms) is a subgroup of the multiplicative monoid of this algebra: to $f : M \to M$, associate the linear map $R^M \to R^M$ "precompose by f". (To make sure that this is an example, one needs some injectivity, as in the previous example.)

Example 6.10.5 Consider a locally trivial vector bundle $E \to M$ whose fibres E_x are finite-dimensional vector spaces. Then we have a locally trivial algebra bundle $End(E) \to M$ with fibre the algebra $End(E_x, E_x)$ of linear endomorphisms of E_x; the multiplication is composition of endomorphisms. There is a subbundle $G \to M$ which is a group bundle, namely $G_x = GL(E_x)$, the general linear group of linear automorphisms of E_x. (Equivalently, G is the gauge group bundle of the groupoid $GL(E) \rightrightarrows M$.)

If $E \to M$ is equipped with a linear bundle connection ∇, then $End(E) \to M$ acquires an algebra connection $ad\nabla$: for $\phi \in End(E_y)$ and $x \sim y$, put

$$(ad\nabla) \dashv \phi := \nabla(x,y) \circ \phi \circ \nabla(y,x) \in End(E_x)$$

(conjugation by $\nabla(x,y) : E_y \to E_x$). It restricts to a group connection in $G \to M$ (in fact, viewing ∇ as a connection in the groupoid $GL(E) \rightrightarrows M$, this is the connection $ad\nabla$ in the gauge group bundle of $GL(E) \rightrightarrows M$).

7

Jets and differential operators

In this chapter, we use the same notation and notational shortcuts as in Section 2.7. In particular, if $\pi : E \to M$ is a bundle over manifold M, we let $J^k(E)$ denote the bundle over M whose fibre $J^k(E)_x$ over $x \in M$ is the set of k-jet sections $j : \mathfrak{M}_k(x) \to E$ (so $\pi(j(y)) = y$ for all $y \in \mathfrak{M}_k(x)$). Also, the notation $J^k(\pi)$ will be used.

Most of the content of the chapter is paraphrased from Palais *et al.* (1965).

7.1 Linear differential operators and their symbols

Let $\pi : E \to M$ and $\pi' : E' \to M$ be bundles over a manifold M, and let $x \in M$. A *differential operator of order* $\leq k$ *at* x from E to E' is a map

$$(J^k(E))_x \xrightarrow{\quad d \quad} E'_x.$$

The main interest of this notion is when E and E' are vector bundles over M. In most of this chapter, bundles are assumed to be vector bundles, locally trivial, and with KL vector spaces as fibres. Then $J^k(E)$ is a vector bundle over M (it can be proved likewise to be locally trivial with KL fibres), and one may ask that the d above is a linear map. We pose

Definition 7.1.1 *A linear differential operator of order* $\leq k$ *from* E *to* E', *at* $x \in M$, *is a linear* $d : (J^k(E))_x \to E'_x$.

The vector space of linear differential operators of order $\leq k$ at x is denoted $\mathrm{Diff}^k_x(E, E')$; these vector spaces form, as x ranges, a vector bundle over M, denoted $\mathrm{Diff}_k(E, E')$.

Since there is a (linear) "restriction" map $J^k_x(E) \to J^l_x(E)$ for $l \leq k$, any differential operator of order $\leq l$ gives, by composition with this restriction map, rise to a differential operator of order $\leq k$.

Annular jets

An element $j \in J_x^1(E)$, i.e. a section 1-jet $j : \mathfrak{M}_1(x) \to E$, is called an *E-valued (combinatorial) cotangent at x* if $j(x) = 0 \in E_x$. If $E \to M$ is a product bundle $M \times R \to M$, such j is of the form $j(y) = (y, \omega(y))$, where $\omega : \mathfrak{M}_1(x) \to R$ takes x to $0 \in R$; such ω is what we elsewhere have called a (combinatorial) cotangent at x, cf. e.g. Definition 2.7.3. This should explain the choice of terminology "E-valued cotangent".

The notion of E-valued cotangent is the special case $k = 1$ of the following notion. Let k be a non-negative integer.

Definition 7.1.2 *An annular k-jet section at x of the vector bundle $E \to M$ is a k-jet section $j : \mathfrak{M}_k(x) \to E$ such that $j(y) = 0 \in E_y$ for all $y \in \mathfrak{M}_{k-1}(x)$.*

The reason for the name is that we may visualize $\mathfrak{M}_k(x)$ as a disk around x, containing the slightly smaller disk $\mathfrak{M}_{k-1}(x)$; so an annular jet $j : \mathfrak{M}_k(x) \to E$ only takes non-trivial values in the "annulus" between the two disks.

Let $A_x^k(E) \subseteq J_x^k(E)$ be the subset consisting of annular k-jet sections; it is clearly a linear subspace. By construction, we have a short exact sequence of vector spaces (exactness in the right-hand end depends on the possibility of extending $k-1$-jet sections to k-jet sections; such "extension principle" is discussed below).

$$0 \longrightarrow A_x^k(E) \longrightarrow J_x^k(E) \longrightarrow J_x^{k-1}(E) \longrightarrow 0. \qquad (7.1.1)$$

We pose (for $E \to M$ and $E' \to M$ vector bundles over M, and k a non-negative integer):

Definition 7.1.3 *A k-*symbol *from E to E′ at $x \in M$ is a linear map $A_x^k(E) \to E'_x$.*

The vector space of k-symbols at x is denoted $Sb_x^k(E, E')$; these vector spaces form, as x ranges, a vector bundle over M, denoted $Sb^k(E, E')$, the *k-symbol bundle* from E to E'.

Since $A_x^k(E) \subseteq J_x^k(E)$, a differential operator d from E to E' of order $\leq k$ at x restricts to a k-symbol at x, denoted $sb^k(d)$, the *k-symbol of d* at x.

It is clear that if d comes about by restriction from an operator of order $l < k$, then $sb^k(d) = 0$. For, an annular k-jet j at x vanishes on $\mathfrak{M}_l(x)$, and hence $d(j) = 0$.

A k-symbol s at x is not a differential operator, in the sense of Definition 7.1.1, since there is no canonical way to provide a value of s at a k-jet section j, unless j is annular. (On the other hand, if the short exact sequence (7.1.1) is provided with a linear splitting, we can extend symbols to genuine differential operators.)

For V,W vector spaces, let us denote by $[V,W]$ the vector space of linear maps $V \to W$. By applying the contravariant functor $[-,E'_x]$ to the short exact sequence (7.1.1), we get a sequence of vector spaces (whose exactness in the right-hand end depends on further assumptions; the exactness in the left-hand end depends on the extension principle below)

$$0 \longrightarrow [J_x^{k-1}(E),E'_x] \longrightarrow [J_x^k(E),E'_x] \longrightarrow [A_x^k(E),E'_x] \longrightarrow 0 \qquad (7.1.2)$$

which by definition of $\mathrm{Diff}(E,E')$ and $Sb^k(E,E')$ is the same as

$$0 \longrightarrow \mathrm{Diff}_x^{k-1}(E,E') \longrightarrow \mathrm{Diff}_x^k(E,E') \xrightarrow{sb^k} Sb_x^k(E,E') \longrightarrow 0. \qquad (7.1.3)$$

The most important example is when both $E \to M$ and $E' \to M$ are just $(M \times R) \to M$. Then a differential operator of order $\leq k$ at $x \in M$, from E to E' (called simply "a differential operator of order $\leq k$", but without "from" and "to") is tantamount to a law d which to a k-jet $\mathfrak{M}_k(x) \to R$ associates an element in R, in a linear way. In this case, one often omits E and E' from the notation. Thus J_x^k is the set of R-valued k-jets at $x \in M$; in particular A_x^1 is the set of 1-jets from $x \in M$ to $0 \in R$, i.e. it is the set of combinatorial cotangents at x – which by (4.5.1) may be identified with $T_x^*M = [T_xM,R]$. Note that we let M be understood from the context, in order not to overload notation. Strictly, $J_x^k = J_x^k(M \times R \to M)$, and similarly for A_x^k.

Example, with $M = R^n$ and $k = 1$: $f \mapsto \partial f/\partial x_i \mid 0$ is a differential operator at $0 \in R^n$ of order ≤ 1. This kind of operator is what gives name to the *differential* operators. This example is a special case of the following.

Example 7.1.4 Let τ be a tangent vector at $x \in M$ (where M is a manifold); we get a differential operator D_τ of order ≤ 1 at x,

$$D_\tau : J^1(M \times R)_x \to (M \times R)_x \cong R$$

as described in Section 4.4 (writing $M \times R$ for the constant bundle $M \times R \to M$); it is characterized by

$$d \cdot D_\tau(j) = j(\tau(d)) - j(x),$$

where $d \in D$ and where $j : \mathfrak{M}(x) \to R$ is a 1-jet. Let us calculate the 1-symbol $sb_1(D_\tau) \in Sb_1(M \times R, M \times R)$,

$$T_x^*M \cong A_x^1 \xrightarrow{sb_1(D_\tau)} R.$$

Let $j \in A^1_x$, so $j : \mathfrak{M}(x) \to R$ has $j(x) = 0$ (so j is a combinatorial cotangent at x). Restricting D_τ to such j gives, after multiplication by any $d \in D$,

$$d \cdot sb_1(D_\tau)(j) = d \cdot D_\tau(j) = j(\tau(d))$$

since $j(x) = 0$. On the other hand, the classical cotangent $\overline{j} : T_x M \to R$, corresponding to the cotangent j (cf. (4.5.1)) is given by

$$d \cdot \overline{j}(\tau) = j(\tau(d)).$$

Comparing, and cancelling the universally quantified d, we thus have

$$sb_1(D_\tau)(j) = \overline{j}(\tau);$$

so under the identification of combinatorial and classical cotangents, the symbol of D_τ is just: "evaluation at τ".

Alternative presentation of $Sb^k(E, E')$

Recall that we assume that the vector bundles $E \to M$ under consideration are locally trivial, and with KL vector spaces as fibres; so, locally there exist fibrewise linear $E \cong M \times W$, with W a KL vector space. Then we have an "extension principle"

for $l < k$, every l-jet section $j : \mathfrak{M}_l(x) \to E$ extends (but not canonically) to a k-jet section $\tilde{j} : \mathfrak{M}_k(x) \to E$.

This holds, under the general assumptions made: for, pick locally around x an isomorphism $E \cong M \times W$. Then $j : \mathfrak{M}_l(x) \to E$ is of the form $j(y) = (y, \overline{j}(y))$ for some $\overline{j} : \mathfrak{M}_l(x) \to W$. Since M is a manifold, and the question is local, we may assume $M = V$ with V a finite-dimensional vector space, and with $x = 0 \in V$. Then $\mathfrak{M}_l(x) = D_l(V)$. The map $\overline{j} : D_l(V) \to W$ extends by KL to a polynomial map $V \to W$ (of degree $\leq l$). This polynomial map restricts to a map $D_k(V)$, and, under the various identifications, provides the desired extension $\tilde{j} : \mathfrak{M}_k(x) \to E$. Of course, $\tilde{j}$ depends on the trivializations chosen.

We remark that to give an R-valued cotangent ω_x at every point x of a manifold M is tantamount to giving a (combinatorial) R-valued 1-form ω on M,

$$\omega(x, y) = \omega_x(y).$$

Similarly for 1-forms with values in a vector space. However, given a vector *bundle* $E \to M$, the data of an E-valued cotangent ω_x at every point $x \in M$ is *not* the same as the data of an $E \to M$-valued 1-form ω on M in the sense of Section 3.8. For, for an E-valued cotangent ω_x,

$$\omega_x(y) \in E_y$$

whereas for an E-valued 1-form ω,

$$\omega(x,y) \in E_x.$$

Pictorially, the E-valued cotangents are "horizontal", the E-valued 1-forms "vertical"; but for suitable vector bundles, there is a "verticalization" procedure:

Lemma 7.1.5 *[Verticalization of annular section jets] There is a canonical bijective correspondence between annular k-jet sections $j : \mathfrak{M}_k(x) \to E$, and annular k-jets $\hat{j} : \mathfrak{M}_k(x) \to E_x$.*

Proof. A trivialization of $E \to M$ over $\mathfrak{M}_k(x)$ amounts to a $\mathfrak{M}_k(x)$-parametrized family of linear isomorphisms $g(y) : E_y \to E_x$, with $g(x)$ the identity map of E_x. Such a trivialization provides a passage $j \mapsto \hat{j}$ from k-jet sections $\mathfrak{M}_k(x) \to E$ to k-jets $\mathfrak{M}_k(x) \to E_x$, by the formula (writing $g(y)(e)$ as $g(y;e)$)

$$\hat{j}(y) = g(y; j(y)).$$

This clearly provides a bijection between k-jet sections, on the one hand, and k-jets with values in the fixed E_x, on the other; and this passage does not depend on j being annular. It does in general depend on the choice of trivialization g. But if j is annular, we prove by "degree calculus" that it does not depend on the choice of g. Two choices g_1 and g_2 differ by a $\mathfrak{M}_k(x)$-parametrized family of linear endomorphisms $h(y) : E_x \to E_x$ ($y \in \mathfrak{M}_k(x)$), with $h(x) = 0$, as follows:

$$g_2(y;e) = g_1(y;e) + h(y; g_1(y;e)).$$

In particular

$$g_2(y; j(y)) - g_1(y; j(y)) = h(y; g_1(y; j(y))).$$

But since $h(x; -) = 0$ and $j(y) = 0$ for $y \in \mathfrak{M}_{k-1}(x)$, it follows by degree calculus that $h(y; g_1(y; j(y)))$ is $= 0$ on $\mathfrak{M}_k(x)$ (apply Proposition 2.7.7 to the bilinear evaluation map $[E_x, E_x] \times E_x \to E_x$, and to the two maps $h : \mathfrak{M}_k(x) \to [E_x, E_x]$ and $\mathfrak{M}_k(x) \to E_x$ given by $y \mapsto h(y; -)$ and $y \mapsto g_1(y; j(y))$).

This verticalization lemma implies a linear isomorphism $A^k(E)_x \cong A^k(E_x)$, the latter being the vector space of annular jets into the *constant* vector space E_x, and for this $A^k(E_x)$, we have a more concrete algebraic presentation, namely $A^k(E_x) \cong A^k_x \otimes E_x$ (where A^k_x = annular jets $\mathfrak{M}_k(x) \to R$). This isomorphism is given by the map

$$A^k_x \otimes E_x \longrightarrow A^k(E_x)$$

which sends $\psi \otimes e$ to the annular E_x-valued jet $y \mapsto \psi(y) \cdot e$ (where $y \in \mathfrak{M}_k(x)$), where $\psi : \mathfrak{M}_k(x) \to R$ is an annular jet, and $e \in E_x$.

It follows that we have an isomorphism of vector spaces

$$Sb^k_x(E,E') \cong [A^k_x \otimes E_x, E'_x] \tag{7.1.4}$$

which to a k-symbol s at x associates the linear map characterized by $\psi \otimes e \mapsto \psi \cdot \tilde{e}$, where $\tilde{e}$ is an (arbitrarily chosen) k-jet section through $e \in E_x$, and where $\psi \in A^k_x$. This is well defined, again by degree calculus, and a local trivialization of $E \to M$. By the adjointness which characterizes $\otimes$, the isomorphism can be re-interpreted as an isomorphism

$$Sb^k_x(E,E') \cong [A^k_x, [E_x, E'_x]]. \tag{7.1.5}$$

The vector space $[A^k_x, [E_x, E'_x]]$ admits, by pure algebra, some other presentations, e.g. as the vector space of k-homogeneous maps $T^*_x M \to [E_x, E'_x]$; this is essentially the concrete presentation given in Palais *et al.* (1965, p. 54). We shall not pursue k-symbols further, except for the case $k = 1$.

For $k = 1$, we have as a special case of (7.1.4) the isomorphism

$$Sb^1_x(E,E') \cong [T^*_x M \otimes E_x, E'_x] \tag{7.1.6}$$

given by sending a 1-symbol s at $x \in M$ into the map $T^*_x M \otimes E_x \to E'_x$, characterized by

$$\psi \otimes e \mapsto s(\psi \cdot \tilde{e}).$$

In this formula, we consider $T^*_x M$ as the vector space of combinatorial cotangents ψ at x. In terms of classical cotangents, the description looks like this (recalling (4.5.2) for the correspondence $\psi \leftrightarrow \overline{\psi}$):

$$\overline{\psi} \otimes e \mapsto s([y \mapsto \overline{\psi}(\log_x y) \cdot \tilde{e}(y)]).$$

We have a map $\beta : T_x M \to Sb^1_x(E,E)$, namely the map which under the identification in (7.1.6) is described as follows: to $\tau \in T_x M$, β associates the linear map $T^*_x M \otimes E_x \to E'_x$ characterized by $\overline{\psi} \otimes e \mapsto \overline{\psi}(\tau) \cdot e$ for $\overline{\psi} \in T^*_x M$, $e \in E_x$.

Under suitable assumptions on E_x and E'_x (e.g. if $E_x = E'_x$ is finite dimensional of dimension ≥ 1), this map β is actually monic; for, the right-hand side of (7.1.6) is in turn isomorphic to $T^{**}_x M \otimes [E_x, E'_x]$. If now the linear map $i : R \to [E_x, E_x]$ which takes $1 \in R$ to the identity map of E_x is monic, we have monic linear maps

$$T_x M \xrightarrow{\ \eta\ } T^{**}_x M \quad \cong \ T^{**}_x M \otimes R \xrightarrow{\ \mathrm{id} \otimes i\ } T^{**}_x M \otimes [E_x, E_x]$$

(with η the canonical map to the double dual, the last map is $T^{**}_x M \otimes i$) and the composite here is, modulo the identifications, equal to β, which thus is monic.

7.2 Linear deplacements as differential operators

We continue to consider a locally trivial vector bundle $E \to M$ whose fibres are KL vector spaces.

We shall analyse the Lie algebroid of the groupoid $GL(E \to M)$, i.e. the bundle $\mathscr{A}(GL(E \to M))$ of deplacements in this groupoid.

Let ξ be a deplacement in the groupoid $GL(E \to M) \rightrightarrows M$; we call such a *linear* deplacement; it is given by a tangent vector $\overline{\xi} : D \to M$ at x (the *anchor* of ξ), and for each $d \in D$, there is a linear isomorphism $\xi(d) : E_x \to E_{\overline{\xi}(d)}$, with $\xi(0)$ the identity map of E_x. The deplacement ξ defines a first-order linear differential operator D_ξ at x from E to itself, i.e. a map $D_\xi : J^1(E)_x \to E_x$; it is given by the formula

$$d \cdot D_\xi(f) = \left[\xi(d)^{-1}(f(\overline{\xi}(d)))\right] - f(x) \tag{7.2.1}$$

for $f : \mathfrak{M}(x) \to E$ a 1-jet section at $x \in M$. In particular, if f is an annular 1-jet section at x,

$$d \cdot D_\xi(f) = \xi(d)^{-1}(f(\overline{\xi}(d))). \tag{7.2.2}$$

We shall calculate the 1-symbol of this differential operator. In the following proposition, e denotes an element of E_x, $\overline{\psi}$ denotes an element of $T^*_x M$ (a classical cotangent), so $\overline{\psi} : T_x M \to R$ is a linear map. The combinatorial cotangent corresponding to $\overline{\psi}$ is denoted ψ.

Proposition 7.2.1 *The 1-symbol $sb^1(D_\xi)$ corresponds under the isomorphism (7.1.6) to the map given by $\overline{\psi} \otimes e \mapsto \overline{\psi}(\overline{\xi}) \cdot e$.*

Proof. By the description (7.1.6), $sb^1(D_\xi)$ corresponds to the map

$$\overline{\psi} \otimes e \mapsto D_\xi(\psi \cdot \tilde{e}).$$

Now for $d \in D$,

$$\begin{aligned} d \cdot sb^1(D_\xi)(\psi \cdot \tilde{e}) &= d \cdot D_\xi(\psi \cdot \tilde{e}) \\ &= \xi(d)^{-1}\left(\psi(\overline{\xi}(d)) \cdot \tilde{e}(\overline{\xi}(d))\right) \\ &= \psi(\overline{\xi}(d)) \cdot \xi(d)^{-1}\left(\tilde{e}(\overline{\xi}(d))\right) \end{aligned}$$

since $\xi(d)^{-1}$ is linear; since now $\psi(\overline{\xi}(d))$ is 0 when $d = 0$, it follows from the Taylor principle that we may replace the two ds in the second factor by 0s, so that the equation may be continued

$$\begin{aligned} &= \psi(\overline{\xi}(d)) \cdot e \\ &= d \cdot \overline{\psi}(\overline{\xi}) \cdot e \end{aligned}$$

by the formula for the correspondence $\psi \leftrightarrow \overline{\psi}$. Cancelling the universally quantified d gives the result.

Proposition 7.2.1 may be seen as part of a more comprehensive assertion, best formulated in diagrammatic terms. Consider the following diagram (whose bottom row is the "symbol exact sequence" (7.1.3) for $k = 1$)

$$\begin{array}{ccccccccc} 0 & \longrightarrow & [E_x,E_x] & \longrightarrow & \mathscr{A}(GL(E \to M))_x & \xrightarrow{\alpha} & T_xM & \longrightarrow & 0 \\ & & \Big\downarrow \cong & & \Big\downarrow & & \Big\downarrow \beta & & \\ 0 & \longrightarrow & \mathrm{Diff}^0_x(E,E) & \longrightarrow & \mathrm{Diff}^1_x(E,E) & \longrightarrow & Sb^1_x(E,E) & \longrightarrow & 0 \end{array} \qquad (7.2.3)$$

The central vertical map is the "linear deplacements as differential operators"; α is the anchor map. The right-hand square commutes by Proposition 7.2.1. The map $[E_x,E_x] \to \mathscr{A}(GL(E \to M))_x$ sends a linear $a : E_x \to E_x$ into the deplacement ξ with $\xi(d) = \mathrm{id} - d \cdot a$, which is a deplacement along the zero tangent at x (= the map "$d \mapsto x$ for all d"). Then the left-hand square commutes. The exactness of the top row is easy; for the exactness in the right-hand ends, one must use the assumption that the bundle $E \to M$ is locally trivial.

It follows from standard diagram-chasing arguments in abelian categories that the right-hand square is a pull back. We therefore have the following proposition (cf. Mackenzie, 1987, §III.2, where $\mathscr{A}(GL(E \to M))$ is denoted $CDO(E)$):

Proposition 7.2.2 *The bundle $\mathscr{A}(GL(E \to M))$ sits in a pull-back diagram of vector bundles over M,*

$$\begin{array}{ccc} \mathscr{A}(GL(E \to M)) & \xrightarrow{\alpha} & TM \\ \Big\downarrow & & \Big\downarrow \beta \\ \mathrm{Diff}^1(E,E) & \xrightarrow[sb^1]{} & Sb^1(E,E) \end{array}$$

(The map β was described at the end of Section 7.1; it is not monic without further assumptions on E; for instance, $E \to M$ might be the zero vector bundle, $E_x = 0$ for all x. But under a not too strong regularity assumption, as discussed there, it is monic, and since the right-hand square is a pull back, it then follows that the central vertical map is monic.)

7.3 Bundle-theoretic differential operators

We may "globalize" the pointwise notions of Sections 7.1 and 7.2: if $E \to M$ and $E' \to M$ are bundles over a manifold M, a map $J^k E \to E'$ (commuting with the structural maps to M) is called a *(global) differential operator* of order $\leq k$ from E to E', since for each $x \in M$ it restricts to a differential operator at x, as considered previously. (There is a notion of "differential operator along a map", of which both the global and the pointwise notion are special cases; see Remark 7.3.2 below.) We also call such $J^k E \to E'$ a *bundle-theoretic differential operator*, to contrast this notion with the notion of *sheaf*-theoretic differential operator to be considered in Section 7.4 below.

Bundle-theoretic differential operators may be composed. The main ingredient in this composition is the inclusion of holonomous jets into non-holonomous ones, cf. Section 2.7,

$$J^{k+l}(E) \subseteq J^l(J^k(E)),$$

for E a bundle over a fixed manifold M. This inclusion gives almost immediately a way to compose differential operators on bundles over M: given $d : J^k(E) \to E'$ and $\delta : J^l(E') \to E''$ we can produce

$$J^{k+l}(E) \longrightarrow J^l(J^k(E)) \xrightarrow{J^l(d)} J^l(E') \xrightarrow{\quad \delta \quad} E''$$

thus providing a differential operator from E to E'' of order $\leq k+l$. (The second map here, $J^l(d)$, is just applying the functorality of the jet bundle construction J^l, cf. Section 2.7.)

Remark 7.3.1 Some readers may find the following generality enlightening in the construction of the bundle $J^k(E) \to M$ of k-jets of sections of a bundle $E \to M$. Let there be given a pair of parallel maps $N \rightrightarrows M$, where M is a manifold; call the two maps α and β. (For the application to k-jet bundles, take N to be $M_{(k)}$, the kth neighbourhood of the diagonal, and take α and β to be the projections.) If now $E \to M$ is a bundle over M, one may form a new bundle over M, namely $\beta_*(\alpha^*(E))$. (Recall that β_* is the right adjoint of the pull-back functor β^*.) The fibre of $\beta_*\alpha^*E$ over $m \in M$ consists of laws which associate to each $g \in N$ with $\beta(g) = m$ an element in E above $\alpha(g)$.

Given a morphism ϕ of parallel pairs with common codomain M, from (α', β') to (α, β), there is a "restriction map"

$$\beta'_*\alpha'^*E \to \beta_*\alpha^*E;$$

it comes about as follows: we have by assumption that $\alpha' = \alpha \circ \phi$ and $\beta' = \beta \circ \phi$, so we may rewrite $\beta'_* \alpha'^* E$ as the right-hand side of the following map

$$\beta_* \alpha^* E \longrightarrow \beta_* \phi_* \phi^* \alpha^* E$$

where the map itself comes about from the unit for the adjunction $\phi^* \dashv \phi_*$, by whiskering on the left with β_* and on the right by α^*.

This construction explains, in general terms, the "restriction" map from $J^k(E) \to J^l(E)$ (where $k \geq l$), since then the inclusion of $M_{(l)} \subseteq M_{(k)}$ is such a morphism ϕ of parallel pairs.

Remark 7.3.2 If $f : M' \to M$ is a map between manifolds, and $E \to M$ and $E' \to M'$ are bundles, a differential operator (of order $\leq k$) from $E \to M$ to $E' \to M'$ *along* f is a map of bundles over M', $f^* J^k E \to E'$, or, equivalently, a map of bundles over M, $J^k E \to f_* E'$.

The following construction applies to linear differential operators between locally constant vector bundles with KL vector spaces as fibres. These assumptions imply that an l-jet section at $x \in M$ extends (not canonically) to a k-jet section ($k \geq l$), and they also imply that degree calculus is available.

Let $\xi : E \to M$ and $\eta : E' \to M$ be two such vector bundles.

Let $\psi \in C^\infty(M) = R^M$, and let d be a linear differential operator from E to E' of order $\leq k$. We construct a new linear differential operator, likewise from E to E', denoted $[d, \psi]$, of order $\leq k - 1$, by the following recipe. Given $s \in J^{k-1}(E)_x$, extend in some way s to a k-jet section s'. Then since the function $\psi - \psi(x)$ vanishes at x, it follows by degree calculus that the k-jet $(\psi - \psi(x)) \cdot s'$ does not depend on the choice of the extension s' of s, and so we may define

$$[d, \psi](s) := d\big((\psi - \psi(x)) \cdot s'\big). \tag{7.3.1}$$

The construction may be "localized": we just need that d is a differential operator *at* x, $d : J^k(E)_x \to E'$, as explained in Section 7.1.

7.4 Sheaf-theoretic differential operators

If $\xi : E \to M$ is a vector bundle, we denote by $C^\infty(\xi)$ the space of sections $S : M \to E$ of ξ. (It exists as a space, by virtue of cartesian closedness of $\mathscr{E}$.) It carries a structure of module over R. But it is also a module over the ring $C^\infty(M) = R^M$ of functions $\psi : M \to R$:

$$(\psi \cdot f)(x) := \psi(x) \cdot f(x),$$

using the multiplication in E_x by scalars in R.

Let $\xi : E \to M$ and $\eta : E' \to M$ be two vector bundles over the manifold M.

Definition 7.4.1 *A* sheaf-theoretic differential operator *of order $\leq k$ from ξ to η is an R-linear map $D : C^\infty(\xi) \to C^\infty(\eta)$ with the property that for any $S \in C^\infty(\xi)$, $D(S)(x)$ only depends on the k-jet of S at x.*

To a linear (bundle-theoretic) differential operator d of order $\leq k$ from ξ to η, one immediately associates a sheaf-theoretic one, namely the $\widetilde{d}$ given by

$$(\widetilde{d}(S))(x) := d(j_x^k(S)).$$

It is clear that if D is a sheaf-theoretic differential operator of order $\leq k$, and $\psi : M \to R$ is a function, we get two new operators $\psi \cdot D$ and $D \cdot \psi$, given by, respectively

$$\big((\psi \cdot D)(S)\big)(x) := \psi(x) \cdot D(S)(x) \text{ and } (D \cdot \psi)(S) := D(\psi \cdot S); \qquad (7.4.1)$$

they are likewise of order $\leq k$. So given any R-linear $D : C^\infty(\xi) \to C^\infty(\eta)$, and any $\psi \in C^\infty(M)$, we may form the R-linear $D \cdot \psi - \psi \cdot D : C^\infty(\xi) \to C^\infty(\eta)$, which we denote $[D, \psi]$. Then $[D, \psi] = 0$ for all $\psi \in C^\infty(M)$ iff D is $C^\infty(M)$-linear.

Proposition 7.4.2 *Given a bundle-theoretic differential operator d of order $\leq k$ and a function $\psi : M \to R$, then*

$$\widetilde{[d, \psi]} = \widetilde{d} \cdot \psi - \psi \cdot \widetilde{d} (= [\widetilde{d}, \psi]).$$

Proof. Consider $S \in C^\infty(\xi)$. Consider for fixed $x \in M$ the value $\widetilde{[d, \psi]}(S)(x) \in E'$. With s denoting the $k-1$-jet of S at x, we have

$$\widetilde{[d, \psi]}(S)(x) = d\big((\psi - \psi(x)) \cdot s\big);$$

here, s also denotes the k-jet of S at x, which will serve as a possible s' in the recipe for $[d, \psi](s)$. Since $\psi \cdot s$ clearly is the k-jet of $\psi \cdot S$ at x, the right-hand side here is the desired $\big(\widetilde{d} \cdot \psi - \psi \cdot \widetilde{d}\big)(S)(x)$. This proves the proposition.

Soft vector bundles

We have described a process which to a bundle-theoretic differential operator d associates a sheaf-theoretic one, $\widetilde{d}$. We give sufficient conditions for this process to be a bijection.

For this purpose, we consider *soft* vector bundles $\xi : E \to M$; by this, we understand a locally trivial vector bundle whose fibres are KL vector spaces, and which have the following softness properties(for any k):

(1) any k-jet of a section of ξ extends to a global section $M \to E$ (equivalently, for each $x \in M$, the restriction map $C^\infty(\xi) \to J^k(E)_x$ is surjective)

(2) any global section $f : M \to E$ which vanishes on $\mathfrak{M}_k(x)$ may be written as a sum of functions of the form

$$(\Pi_{i=0}^k \phi_i) \cdot h,$$

where each $\phi_i : M \to R$ vanishes at x, and where $h : M \to E$ is a global section (a "Hadamard remainder").

The softness condition (1) is quite restrictive, in the form given; it obtains in the models based on smooth functions on paracompact manifolds, by virtue of the existence of partitions of unity, and integrals; but it does not obtain in the more algebraic models where R^M may be R, even for good manifolds M, like projective spaces.

A treatment of sheaf-theoretic differential operators in an algebraic context is therefore a little more complicated, and more genuinely sheaf theoretic. Such treatment may be found in Demazure and Gabriel (1970, §II.4.5 and §II.4.6).

From (2) follows:

(3) if $f \in C^\infty(\xi)$ is a section vanishing on $\mathfrak{M}_k(x)$, then f may be written as a sum of functions of the form $\psi \cdot h$ where $h \in C^\infty(\xi)$ vanishes on $\mathfrak{M}_{k-1}(x)$ and $\psi : M \to R$ vanishes at x.

Proposition 7.4.3 *Under the softness assumptions given, the process which to a bundle-theoretic differential operator d associates a sheaf-theoretic one, $\widetilde{d}$, is a bijection.*

Proof. Given a sheaf-theoretic $D : C^\infty(\xi) \to C^\infty(\eta)$, of order $\leq k$, say, we construct a vector bundle map $d : J^kE \to E'$ by constructing for each $x \in M$ a linear map $d_x : (J^kE)_x \to E'_x$. Let $j \in (J^kE)_x$; by softness of $E \to M$, j extends to a global section $\tilde{j}$, and we put $d_x(j) := D(\tilde{j})(x)$. The order assumption on D implies that this value does not depend on the choice of the extension $\tilde{j}$. The verification that this process does provide an inverse for the $d \mapsto \widetilde{d}$ is essentially trivial.

We proceed to analyse to what extent sheaf-theoretic differential operators $C^\infty(\xi) \to C^\infty(\eta)$ are $C^\infty(M)$-linear

Let $\xi : E \to M$ and $\eta : E' \to M$ be two vector bundles (soft, etc.). Consider an R-linear map

$$C^\infty(\xi) \xrightarrow{D} C^\infty(\eta). \tag{7.4.2}$$

An example of such D is given whenever we have a fibrewise linear map $g : E \to F$ over M; then we can define

$$D(f) := g \circ f, \tag{7.4.3}$$

and this D is clearly not only R-linear, but $C^\infty(M)$-linear: $g \circ (\phi \cdot f) = \phi \cdot (g \circ f)$ since each $g_x : E_x \to F_x$ is linear. (This D is of course $= \tilde{g}$, when we view $g : E \to E'$ as a bundle-theoretic differential operator of order ≤ 0.)

Conversely,

Proposition 7.4.4 *Let $D : C^\infty(\xi) \to C^\infty(\eta)$ be $C^\infty(M)$-linear. Then it is of the form (7.4.3) for a unique fibrewise linear g.*

Proof. Let $e \in E_x$. Choose an $\tilde{e} \in C^\infty(\xi)$ with $\tilde{e}(x) = e$, and put $g(e) := D(\tilde{e})(x) \in F_x$. To show that this does not depend on the choice of $\tilde{e}$: another choice is of the form $\tilde{e} + f$, where $f \in C^\infty(\xi)$ has $f(x) = 0$. Such f may be written $\psi \cdot h$ with $h \in C^\infty(\xi)$, and with $\psi \in C^\infty(M)$ and $\psi(x) = 0$. Then

$$\begin{aligned} D(\tilde{e} + f) &= D(\tilde{e}) + D(f) = D(\tilde{e}) + D(\psi \cdot h) \\ &= D(\tilde{e}) + \psi \cdot D(h), \end{aligned}$$

the last by the assumed $C^\infty(M)$-linearity of D. Now the last term here vanishes on x since ψ does, and so we arrive again at $D(\tilde{e})(x)$. This proves that g is well defined. It is easy to see that g is fibrewise linear. To see $D(f)(x) = g(f(x))$, we pick the $\tilde{e}$ in the recipe for $g(f(x))$ to be the given f.

It follows from Proposition 7.4.4 that we have bijective correspondences between

- fibrewise linear maps $E \to E'$
- bundle-theoretic linear differential operators from ξ to η of order ≤ 0
- sheaf-theoretic differential operators of order from ξ to $\eta \leq 0$
- $C^\infty(M)$-linear maps $C^\infty(\xi) \to C^\infty(\eta)$.

Proposition 7.4.5 *Let $D : C^\infty(\xi) \to C^\infty(\eta)$ be an R-linear map. Then D is a (sheaf theoretic) differential operator of order $\leq k$ if and only if for any $\psi \in C^\infty(M)$, $[D, \psi] : C^\infty(\xi) \to C^\infty(\eta)$ is a (sheaf theoretic) differential operator of order $\leq k - 1$.*

Proof. For the implication $\Rightarrow$, we use the corresponding fact for bundle theoretic differential operators, noted at the end of Section 7.3 and the fact that the correspondence preserves the bracket formation (Proposition 7.4.2). For the implication $\Leftarrow$: by linearity, it suffices to prove that if $f \in C^\infty(\xi)$ vanishes on $\mathfrak{M}_k(x)$, then $D(f)(x) = 0$. Now by (3) above, f may be written as a sum of functions $\psi \cdot h$ with $\psi \in C^\infty(M)$ vanishing at x and $h \in C^\infty(\xi)$ vanishing on $\mathfrak{M}_{k-1}(x)$. It suffices to prove that D vanishes on such a product $\psi \cdot h$. Since h vanishes on $\mathfrak{M}_{k-1}(x)$, we have by assumption that $([D,\psi](h))(x) = 0$, i.e we have that the function $D(\psi \cdot h) - \psi \cdot D(h)$ vanishes at x. The second term does so automatically, since $\psi(x) = 0$; hence so does the first term $D(\psi \cdot h)$, and this proves the proposition.

Proposition 7.4.6 *Let $\xi : E \to M$ and $\eta : E' \to M$ be soft vector bundles over M, and let $D : C^\infty(\xi) \to C^\infty(\eta)$ be an R-linear map. Then D is a (sheaf-theoretic) differential operator of order $\leq k$ iff for all $\psi_0, \ldots, \psi_k \in C^\infty(M)$,*

$$[[\ldots[D,\psi_0],\ldots],\psi_k] = 0.$$

Proof. This is by induction in k. The induction step is contained in Proposition 7.4.5, and the $k = 0$ case was dealt with in Proposition 7.4.4.

8

Metric notions

One may say that if there is no metric, there is no geo*metry*.† In the present synthetic context, the notion of Riemannian (and pseudo-Riemannian) metric comes, for manifolds, in a combinatorial manifestation, besides in the classical manifestation in terms of the tangent bundle. We shall utilize both manifestations, and their interplay. The combinatorial notion deals essentially with points which are *second*-order neighbours, $x \sim_2 y$.

8.1 Pseudo-Riemannian metrics

Both the combinatorial and the classical notion are subordinate to the notion of *quadratic differential form*, which likewise comes in a combinatorial and in a classical manifestation. A *Riemannian metric* on M will be a quadratic differential form with a certain positive-definiteness property.

We begin with the combinatorial notions (cf. Kock, 1998).

Definition 8.1.1 *A* (combinatorial) quadratic differential form *on a manifold M is a map $g : M_{(2)} \to R$ vanishing on $M_{(1)} \subseteq M_{(2)}$.*

Note the analogy with the notion of differential R-valued 1-form on M, which is (cf. Definition 2.2.3) a map $M_{(1)} \to R$ vanishing on $M_{(0)} \subseteq M_{(1)}$. (Recall that $M_{(0)}$ is the diagonal in $M \times M$.)

The canonical example is $M = R^n$, with

$$g(\underline{x}, \underline{y}) = \sum_{i=1}^{n} (y_i - x_i)^2,$$

whose meaning is the *square distance* between $\underline{x}$ and $\underline{y}$. (The distance itself cannot well be formulated for neighbour points (of any order), it seems.)

† I don't agree completely, since a metric depends on a choice of unit of length; geometry does not.

Quadratic differential forms are always symmetric:

Proposition 8.1.2 *Let $g : M_{(2)} \to R$ be a quadratic differential form. Then g is symmetric, $g(x,y) = g(y,x)$, for any $x \sim_2 y$. Furthermore, g extends uniquely to a symmetric $M_{(3)} \to R$.*

Proof. The assertions are local, so we may assume that M is an open subset of a finite-dimensional vector space V. Then $M_{(2)}$ may be identified with $M \times D_2(V)$ via $(x,y) \mapsto (x, y-x)$, and g gets identified with a map $M \times D_2(V) \to R$ vanishing on $M \times D_1(V)$. It follows from Taylor expansion, Theorem 1.4.2, that g may be written in the form

$$g(x,y) = G(x; y-x, y-x) \tag{8.1.1}$$

with $G(x;-,-) : V \times V \to R$ bilinear and symmetric. (This G is the *metric tensor* of the metric g, relative to the "coordinate system" V.) In terms of G, $g(y,x)$ is then $g(y,x) = G(y; x-y, x-y) = G(y; y-x, y-x)$. Taylor expanding from x in the direction of $y-x$, we thus get

$$\begin{aligned} g(y,x) &= G(x; y-x, y-x) + dG(x; y-x, y-x, y-x) \\ &= g(x,y) + dG(x; y-x, y-x, y-x). \end{aligned}$$

The second term vanishes due to trilinear occurrence of $y-x \in D_2(V)$. (*Warning:* Don't attempt to shortcut this proof by quoting the "Taylor principle" 1.4.2; it only applies when $y-x \in D_1(V)$.)

For the proof of the second assertion: a possible extension of g to $\overline{g} : M_{(3)} \to R$ has, as a map $M \times D_3(V) \to R$, the form (for $y \sim_3 x$)

$$\overline{g}(x,y) = G(x; y-x, y-x) + T(x; y-x, y-x, y-x) \tag{8.1.2}$$

with $T(x;-,-,-) : V^3 \to R$ trilinear and symmetric. In terms of G and T, we calculate $\overline{g}(y,x)$, for $y \sim_3 x$:

$$\begin{aligned} \overline{g}(y,x) &= G(y; x-y, x-y) + T(y; x-y, x-y, x-y) \\ &= G(x; x-y, x-y) + dG(x; y-x, x-y, x-y) \\ &\quad + T(x; x-y, x-y, x-y) + dT(x; y-x, x-y, x-y, x-y) \end{aligned}$$

by Taylor expansion of G and T in their non-linear variable from x in the direction $y-x$;

$$= G(x; y-x, y-x) + dG(x; y-x, y-x, y-x) - T(x; y-x, y-x, y-x) \tag{8.1.3}$$

using bi- and trilinearity for the sign changes that occur, and noting that the dT term vanishes since it contains $y - x \in D_3(V)$ in a quadrilinear position. Comparing (8.1.2) and (8.1.3), we see that

$$T(x; y-x, y-x, y-x) = \tfrac{1}{2} dG(x; y-x, y-x, y-x),$$

and from this, uniqueness of $\overline{g}$ follows. For existence, we just have to check that (for $y \sim_3 x$)

$$\overline{g}(x,y) := G(x; y-x, y-x) + \tfrac{1}{2} dG(x; y-x, y-x, y-x)$$

is indeed symmetric in x, y, and this is essentially the same calculation. This proves the proposition.

To a quadratic differential form g on M, we can to each $x \in M$ associate a bilinear form $C_x : T_x M \times T_x M \to R$ in a coordinate-free way as follows. For τ_1 and τ_2 tangent vectors at x, we consider the function c of $(d_1, d_2) \in D \times D$ given by the expression

$$c(d_1, d_2) := -\tfrac{1}{2} g(\tau_1(d_1), \tau_2(d_2)).$$

Note that $\tau_i(d_i) \sim_1 x$; hence $\tau_1(d_1) \sim_2 \tau_2(d_2)$, so that the expression makes sense. If $d_1 = 0$, we get $g(x, \tau_2(d_2))$, which is 0 since $x \sim_1 \tau_2(d_2)$. Similarly if $d_2 = 0$. As a function of (d_1, d_2), c is therefore (by KL) of the form

$$c(d_1, d_2) = d_1 \cdot d_2 \cdot C_x$$

for a unique $C_x \in R$. This C_x depends on τ_1 and $\tau_2 \in T_x M$, and so defines a map $C_x : T_x(M) \times T_x(M) \to R$. Thus $C_x : T_x M \times T_x M \to R$ is characterized by

$$d_1 \cdot d_2 \cdot C_x(\tau_1, \tau_2) = -\tfrac{1}{2} g(\tau_1(d_1), \tau_2(d_2)).$$

We claim that C_x is symmetric bilinear. It suffices to prove this in a coordinatized situation, i.e. where M is an open subset of a finite-dimensional vector space V. Then $T_x M$ may canonically be identified with V via principal-part formation, and then it suffices to see that C_x equals the $G(x; -, -)$ considered in the proof of Proposition 8.1.2. This is the content of the following proposition. We write $\langle \tau_1, \tau_2 \rangle$ for $C_x(\tau_1, \tau_2)$.

Proposition 8.1.3 *With the identifications mentioned, $\langle \tau_1, \tau_2 \rangle = G(x; a_1, a_2)$, where a_i is the principal part of the tangent vector τ_i at $x \in M$ ($i = 1, 2$).*

Proof. We calculate $d_1 d_2 \langle \tau_1, \tau_2 \rangle$ by the recipe in terms of g; we have

$$\begin{aligned} d_1 d_2 \langle \tau_1, \tau_2 \rangle &= -\tfrac{1}{2} g(\tau_1(d_1), \tau_2(d_2)) \\ &= -\tfrac{1}{2} g(x + d_1 \cdot a_1, x + d_2 \cdot a_2) \\ &= -\tfrac{1}{2} G(x + d_1 \cdot a_1; d_2 \cdot a_2 - d_1 \cdot a_1, d_2 \cdot a_2 - d_1 \cdot a_1) \end{aligned}$$

and using bilinearity and symmetry of $G(x+d_1 \cdot a_1; -, -)$, this calculates further:

$$= G(x+d_1 \cdot a_1; d_1 \cdot a_1, d_2 \cdot a_2)$$
$$= d_1 d_2 \cdot G(x+d_1 \cdot a_1; a_1, a_2)$$
$$= d_1 d_2 \cdot G(x; a_1, a_2)$$

by the Taylor principle; finally, by cancelling d_1 and d_2 (which appear universally quantified), we obtain

$$\langle \tau_1, \tau_2 \rangle = G(x; a_1, a_2),$$

proving the proposition.

For V a vector space, V^* denotes the vector space of linear maps $V \to R$. A bilinear form $B : V \times V \to R$ gives rise to a linear map $\hat{B} : V \to V^*$, namely $\hat{B}(u)(v) := B(u, v)$. From standard linear algebra, we have the notion of a *non-degenerate* bilinear form $B : V \times V \to R$: this means that $\hat{B} : V \to V^*$ is an isomorphism.

This is equivalent to saying that for any choice of bases for V and V^*, the matrix of $\hat{B}$ is an invertible matrix (which in turn is equivalent to saying that this matrix has invertible determinant).

The "standard bilinear form" or "dot product" on R^n is the symmetric $(\underline{x}, \underline{y}) \mapsto \sum x_i y_i$. It is non-degenerate.

We now say that a quadratic differential form g on a manifold M is a *pseudo-Riemannian metric* (or that it makes M into a *pseudo-Riemannian manifold*) if for every $x \in M$ and for some, or equivalently for every, local coordinate system around x by a finite-dimensional vector space V, the map $G(x; -, -) : V \times V \to R$ (as in the proof of Proposition 8.1.2) is non-degenerate.

An equivalent, but coordinate-free, description of the notion, in terms of the tangent bundle TM, is that the bilinear forms $T_xM \times T_xM \to R$ induced by g as described, are non-degenerate.

So if g is non-degenerate, we have for each x a linear isomorphism $T_xM \to (T_xM)^*$. This means in particular that g induces an isomorphism of vector bundles over M between the tangent and cotangent bundle,

$$T(M) \underset{\check{g}}{\overset{\hat{g}}{\rightleftarrows}} T^*(M). \tag{8.1.4}$$

A pseudo-Riemannian manifold (M, g) is called a *Riemannian manifold* if the bilinear forms $T_xM \times T_xM \to R$ induced by g are *positive definite*, in the sense explained in the Appendix. "Positive definite" makes sense provided we assume that the ring R has some kind of order; explicitly, we assume that R is a Pythagorean ring, in the sense of the Appendix. This implies that a notion of positivity can be derived from the algebraic structure of R.

A main theorem in differential geometry is the existence and uniqueness of a symmetric affine connection λ, the *Levi–Civita* connection, compatible with a given pseudo-Riemannian metric. We formulate this theorem in synthetic terms below, but we don't prove it here, since the "synthetic" proofs known are as elaborate as the classical ones. (For a proof in the synthetic context, see Kock, 1998.)

If g is a quadratic differential form on a manifold M, and λ an affine connection on M, we say that λ is *compatible with* g if for $x \sim_1 y$, the λ transport

$$\mathfrak{M}_1(x) \xrightarrow{\lambda(x,y,-)} \mathfrak{M}_1(y)$$

preserves g, i.e. if for any z_1 and z_2 which are $\sim_1 x$ (so that in particular $z_1 \sim_2 z_2$), we have

$$g(z_1, z_2) = g(\lambda(x,y,z_1), \lambda(x,y,z_2)).$$

Theorem 8.1.4 (Levi–Civita) *If g is a pseudo-Riemannian metric on a manifold M, there exists a unique symmetric affine connection λ on M which is compatible with g.*

The Levi–Civita connection for g, as given by the theorem, gives rise to a range of geometric concepts which can be formulated for arbitrary symmetric affine connections, and so the following section could equally well have been placed already in Chapter 2.

8.2 Geometry of symmetric affine connections

In the present section, we consider an n-dimensional manifold, equipped with a symmetric (= torsion-free) affine connection λ. We show how λ allows us to extend some of the notions and constructions, referring to the first-order neighbour relation $\sim_1$, to the second-order neighbour relation $\sim_2$. First of all, λ itself may be seen as extending the canonical "parallelogram-formation" $(x,y,z) \mapsto y - x + z$ (for x,y,z mutual 1-neighbours) to the case where y and z are 1-neighbours of x, but not necessarily mutual 1-neighbours; this is what Proposition 2.3.4 tells us.

Extending the* log-exp *bijection

Recall the log-exp bijection: if $x \in M$, we have a bijective map $\exp_x$: $D_1(T_xM) \to \mathfrak{M}_1(x)$, with inverse $\log_x$. We shall use the symmetric affine connection λ assumed presently on M to extend this "first-order" exp to a "second-order" exp, which is to be a bijection

$$\exp_x^{(2)} : D_2(T_xM) \to \mathfrak{M}_2(x) \subseteq M.$$

It is an easy consequence of Proposition 1.3.3 and Taylor expansion, Theorem 1.4.2, that for a finite-dimensional vector space V and a manifold M, a symmetric map $f : D_1(V) \times D_1(V) \to M$ factors uniquely across the addition map $D_1(V) \times D_1(V) \to D_2(V)$, i.e. there is a unique $\overline{f} : D_2(V) \to M$ such that $\overline{f}(v_1 + v_2) = g(v_1, v_2)$ for the v_i in $D_1(V)$ $(i = 1, 2)$.

So to construct $\exp_x^{(2)}$ it suffices to construct a symmetric $f : D_1(T_xM) \times D_1(T_xM) \to M$, and for f we take

$$f(\tau_1, \tau_2) := \lambda(x, \exp(\tau_1), \exp(\tau_2)),$$

for $\tau_i \in D_1(T_xM)$ $(i = 1, 2)$, which is symmetric because of the assumed symmetry of λ.

We shall analyse $\exp^{(2)}$ in terms of the Christoffel symbols for λ. So assume M is an open subset of a finite-dimensional vector space V; then as in (2.3.12), we have (for $x \sim_1 y$ and $x \sim_1 z$)

$$\lambda(x, y, z) = y - x + z + \Gamma(x; y - x, z - x) \qquad (8.2.1)$$

with $\Gamma : M \times V \times V \to V$ bilinear in the last two arguments, and also symmetric, because of the assumed symmetry of λ. Then we claim that

$$\exp_x^{(2)}(u) = x + u + \tfrac{1}{2}\Gamma(x; u, u) \qquad (8.2.2)$$

(where we identify tangent vectors $\in D_2(T_xM)$ with their principal part $u \in D_2(V)$).

Under the identification of tangent vectors at $x \in M \subseteq V$ with their principal parts, it follows from Proposition 4.3.1 that $\exp_x(v) = x + v$ for $v \in D_1(V)$. Therefore the equation defining f reads, under this identification

$$f(v_1, v_2) = \lambda(x, x + v_1, x + v_2)$$

for $v_i \in D(V)$ $(i = 1, 2)$. Expressing λ in terms of Christoffel symbols, $\lambda(x, y, z) = y - x + z + \Gamma(x; y - x, z - x)$, the defining equation for f, as in (8.2.1), is therefore the first equality sign in

$$f(v_1, v_2) = (x + v_1) - x + (x + v_2) + \Gamma(x; v_1, v_2) = x + v_1 + v_2 + \Gamma(x; v_1, v_2).$$

Using symmetry of λ, and hence of $\Gamma(x;-,-)$, this can be rewritten

$$f(v_1,v_2) = x+v_1+v_2+\tfrac{1}{2}\Gamma(x;v_1+v_2,v_1+v_2),$$

so now we have identified the unique factorization $\overline{f}$ of f over the addition map;

$$\overline{f}(u) = x+u+\tfrac{1}{2}\Gamma(x;u,u),$$

for $u \in D_2(V)$. Since $\exp^{(2)}$ was defined as this $\overline{f}$, we have therefore proved (8.2.2).

The map $\exp_x^{(2)} : D_2(T_xM) \to \mathfrak{M}_2(x)$ thus described is invertible; its inverse $\log_x^{(2)}$ is given in the coordinatized situation in terms of Γ as

$$\log_x^{(2)}(x+u) = u - \tfrac{1}{2}\Gamma(x;u,u) \tag{8.2.3}$$

for $u \in D_2(V)$ (identifying a tangent vector at x with its principal part $\in V$). The fact that the map $\log_x^{(2)}$ thus described is indeed inverse for $\exp_x^{(2)}$ is a simple calculation using bilinearity of $\Gamma(x;-,-)$, together with the fact that $\Gamma(x;u,\Gamma(x;u,u)) = 0$ and $\Gamma(x;\Gamma(x;u,u),\Gamma(x;u,u)) = 0$, due to $u \in D_2(V)$.

Point reflection, and midpoint formation

One can use the $\exp^{(2)}$ and $\log^{(2)}$ maps to construct affine combinations of pairs of 2-neighbours, $x \sim_2 z$, extending the canonical affine combinations of 1-neighbours considered in Section 2.1. We shall only consider the combination $2x - z$, "reflection of z in the point x", and the combination $\frac{1}{2}x + \frac{1}{2}z$, "midpoint of x and z". We give the point reflection a special notation, $\rho_x(z)$. (Later on, when x is understood from the context, we shall write z' instead of $\rho_x(z)$.) In terms of $\exp^{(2)}$ and $\log^{(2)}$, point reflection is defined by

$$\rho_x(z) := \exp_x^{(2)}(-\log_x^{(2)}(z))$$

for $z \sim_2 x$. Equivalently

$$\log_x^{(2)}(\rho_x(z)) = -\log_x^{(2)}(z). \tag{8.2.4}$$

To see that for $z \sim_1 x$, this is actually $2x - z$, we utilize that $\exp^{(2)}$ and $\log^{(2)}$ extend the first-order exp and log. It suffices to see that for $z \sim_1 x$, we have $-\log_x(z) = \log_x(2x-z)$ in T_xM. So for $d \in D$, we should prove $(-\log_x(z))(d) = (\log_x(2x-z))(d)$, and this is a simple calculation using the very definition of $\log_x$ in terms of affine combinations (cf. Section 4.3):

$$(-\log_x(z))(d) = (\log_x(z))(-d) = (1+d)\cdot x - d\cdot z,$$

whereas

$$(\log_x(2x-z))(d) = (1-d)\cdot x + d\cdot(2x-z),$$

and these expressions are equal by plain algebra.

One may alternatively prove the result by using the expressions for $\log^{(2)}$ and $\exp^{(2)}$ in terms of Christoffel symbols Γ; the terms involving Γ vanish for first-order neighbours:

Exercise 8.2.1 Prove that the reflection ρ_x, in terms of Christoffel symbols, may be expressed (for $z = x+u$ with $u \in D_2(V)$)

$$\rho_x(x+u) = x - u + \Gamma(x;u,u).$$

In a similar vein, we may describe the midpoint of $x \sim_2 z$ as $\exp_x^{(2)}(\frac{1}{2}\log_x^{(2)}(z))$. It extends the affine combination $\frac{1}{2}x + \frac{1}{2}z$ in case $x \sim_1 z$. The apparent asymmetric role of x and z in the formula is easily seen to be spurious.

Exercise 8.2.2 Prove that the midpoint formation, in terms of Christoffel symbols Γ, may be expressed

$$\text{midpoint }(x, x+u) = x + \tfrac{1}{2}u - \tfrac{1}{8}\Gamma(x;u,u)$$

for $u \in D_2(V)$. Prove also that, for u and v in $D_1(V)$

$$\text{midpoint }(x+u, x+v) = x + \tfrac{1}{2}u + \tfrac{1}{2}v + \tfrac{1}{4}\Gamma(x;u,v).$$

The following exercises may be solved using Christoffel symbols.

Exercise 8.2.3 Prove that x is the midpoint of z and $\rho_x(z)$, and that z is the reflection of $\rho_x(z)$ in this midpoint.

Exercise 8.2.4 Let $y \sim_1 x$ and $z \sim_1 x$, so that $\lambda(x,y,z)$ may be formed. Prove the "parallelogram law" that the midpoint of y and z is the same as the midpoint of x and $\lambda(x,y,z)$.

Exercise 8.2.5 Prove that λ may be reconstructed from point reflection and midpoint formation: $\lambda(x,y,z)$ is the reflection of x in the midpoint of y and z.

Exercise 8.2.6 Prove that λ may be reconstructed from midpoint formation alone: $\lambda(x,y,z)$ is the midpoint of $2y-x$ and $2z-x$ (these two point reflections are affine combinations of first-order neighbour points,so do not depend on any further structure on the manifold). The relevant picture is here:

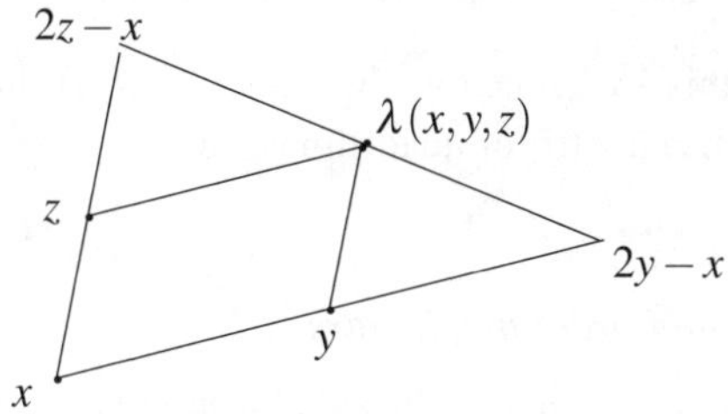

Exercise 8.2.7 Assume that g is a quadratic differential form on M (not necessarily related to the connection λ). Prove that $g(x,z) = g(x,\rho_x(z))$, and that $g(z,\rho_x(z)) = 4\cdot g(x,z)$. (*Hint:* It suffices to consider the coordinatized situation $M \subseteq V$, and to express g in terms of $G(-;-,-) : M\times V\times V \to R$, as in (8.1.1). Also, express λ in terms of Christoffel symbols.)

Remark 8.2.8 In case g is a (pseudo-) Riemannian metric, midpoint formation for the associated Levi–Civita connection may be characterized purely "metrically" as follows: the midpoint y_0 of $x \sim_2 z$ is the unique stationary point $\sim_2 x$ for the function $\mathfrak{M}_3(x) \to R$ given by $y \mapsto \overline{g}(x,y) + \overline{g}(y,z)$ where $\overline{g} : M_{(3)} \to R$ is the unique symmetric extension of g, cf. Proposition 8.1.2; see Kock (1998, Theorem 3.6).

We consider now a manifold equipped with a symmetric affine connection λ; it gives rise to $\log^{(2)}$ and $\exp^2$, as described. Assume further that M is provided with a quadratic differential form $g : M_{(2)} \to R$; it gives rise to bilinear forms $T_xM \times T_xM$, also described above. We do not assume that λ is the Levi–Civita connection for g (in fact, we do not even assume that g is non-degenerate).

We then have an "isometry" property of the $\log^{(2)}$–$\exp^{(2)}$ bijections.

Proposition 8.2.9 *For $z \sim_2 x$, $g(x,z) = \langle \log^{(2)}_x z, \log^{(2)}_x z\rangle$.*

Proof. We work in a coordinatized situation $M \subseteq V$, so that g is encoded by $G : M \times V \times V \to R$, and the connection is encoded by $\Gamma : M \times V \times V \to V$, with both G and Γ symmetric bilinear in the two last arguments. Let $z \sim_2 x$, so z is of the form $x + u$ with $u \in D_2(V)$. Then on the one hand

$$g(x,z) = G(x;u,u),$$

and on the other hand, by (8.2.3), $\log_x^{(2)}(z) = u - \frac{1}{2}\Gamma(x;u,u)$ so that

$$\langle \log_x^{(2)} z, \log_x^{(2)} z\rangle = G(x; u - \tfrac{1}{2}\Gamma(x;u,u), u - \tfrac{1}{2}\Gamma(x;u,u)),$$

and expanding this by bilinearity, we get $G(x;u,u)$ plus some terms which vanish because they are tri- or quadrilinear in u.

Transport of second-order neighbours

An affine connection λ on M induces a linear connection on the tangent bundle $TM \to M$, cf. Proposition 4.3.7. In particular, for $x \sim y$, we get a linear isomorphism $T_xM \to T_yM$. It restricts to a bijection $D_2(T_xM) \to D_2(T_yM)$. If λ is symmetric (as we assume throughout the present section), combining this bijection with the $\log^{(2)}$–$\exp^{(2)}$ bijections, we obtain a bijection $\overline{\lambda}(x,y,-) : \mathfrak{M}_2(x) \to \mathfrak{M}_2(y)$, extending the first-order transport $\lambda(x,y,-)$. From the very construction, it is clear that these transports are compatible with the $\log^{(2)}$–$\exp^{(2)}$ bijections.

8.3 Laplacian (or isotropic) neighbours

For any pseudo-Riemannian manifold M of dimension $n \geq 2$, we introduce (cf. Kock, 2001, 2004) the "L-neighbourhood $M_{(L)}$ of the diagonal" with

$$M_{(1)} \subseteq M_{(L)} \subseteq M_{(2)}.$$

(The letter "L" stands for "Laplacian", and this term was chosen because $\sim_L$ provides us with a geometric description of the Laplace operator Δ, see Section 8.4.) We write $x \sim_L y$ for $(x,y) \in M_{(L)}$. In the case where the pseudo-Riemannian metric is actually Riemannian, the structure $\sim_L$ will allow us to formulate the notion of *divergence* of a vector field, and the notion of *harmonic* map $M \to R$, in a geometric way. Also the notion of *conformal* map $M \to M'$ can be formulated in terms of $\sim_L$; in fact, $\sim_L$ is a conformal invariant.

The $D_L(V)$-construction

We begin by the pure linear algebra underlying the L-neighbour relation. Let V be an n-dimensional vector space ($n \geq 2$), equipped with a symmetric bilinear map $\langle -,-\rangle : V \times V \to R$. Then the subset $D_L(V) \subseteq V$ is defined as the set of $a \in V$ such that for all u and v in V,

$$\langle a,a\rangle \cdot \langle u,v\rangle = n \cdot \langle a,u\rangle \cdot \langle a,v\rangle. \tag{8.3.1}$$

In particular, putting $u = v = a$, one gets that $\langle a,a\rangle \cdot \langle a,a\rangle = n \cdot \langle a,a\rangle \cdot \langle a,a\rangle$, so $\langle a,a\rangle^2 = 0$. It does not follow, however, that $\langle a,a\rangle$ itself is 0, see Remark 8.3.6 below. (For V 1-dimensional, we put $D_L(V) = D_2(V)$.)

It is clear that if $f : V \to V'$ is a linear isomorphism between n-dimensional vector spaces, and f preserves given symmetric bilinear forms $\langle -,-\rangle$ on V and V', then f takes $D_L(V)$ to $D_L(V')$. It even suffices that f preserves the bilinear forms *up to a scalar factor* Λ,

$$\langle f(u), f(v)\rangle = \Lambda \cdot \langle a,b\rangle$$

for all u and v in V. In this sense $D_L(V)$ is not only invariant under isometries, but under *conformal* linear isomorphisms; we return to this point.

Let V and V' be n-dimensional vector spaces, and assume a symmetric bilinear form $\langle -,-\rangle$ on V', as above.

Proposition 8.3.1 *Consider maps f and g from $D_2(V)$ to V' which agree on $D_1(V)$ and take 0 to 0. Let $a \in D_2(V)$. Then $f(a) \in D_L(V')$ iff $g(a) \in D_L(V')$.*

Proof. From Taylor expansion (Theorem 1.4.2) and the assumptions on f and g it follows that there is a symmetric bilinear $B : V \times V \to V'$ such that for all $x \in D_2(V)$, $g(x) = f(x) + B(x,x)$. Also by Taylor expansion and $f(0) = 0$, we may, for $x \in D_2(V)$ write $f(x) = f_1(x) + f_2(x,x)$ with f_1 linear and f_2 bilinear. Assume $f(a) \in D_L(V')$. To prove that $g(a) \in D_L(V')$, we must prove for arbitrary u, v in V' that

$$\begin{aligned}&\langle f(a) + B(a,a), f(a) + B(a,a)\rangle \cdot \langle u,v\rangle \\ &\quad - n \cdot \langle f(a) + B(a,a), u\rangle \cdot \langle f(a) + B(a,a), v\rangle = 0.\end{aligned}$$

Expanding by bilinearity, one gets $\langle f(a), f(a)\rangle \cdot \langle u,v\rangle - n \cdot \langle f(a),u\rangle \cdot \langle f(a),v\rangle$ which is 0 since $f(a) \in D_L(V')$, plus some terms like $\langle f_1(a), B(a,a)\rangle \cdot \langle u,v\rangle$ or $-n \cdot \langle f_1(a),u\rangle \cdot \langle B(a,a),v\rangle$, and they vanish since $a \in D_2(V)$.

We proceed to give a purely equational description of $D_L(V)$, in case $V = R^n$ with the standard bilinear form (dot-product) $\langle \underline{x}, \underline{y}\rangle = \sum x_i y_i$. We write $D_L(n)$ for $D_L(R^n)$.

Proposition 8.3.2 *For $n \geq 2$, the set $D_L(n)$ consists of the vectors $\underline{a} = (a_1, \ldots, a_n)$ which satisfy the equations*

$$a_i \cdot a_j = 0 \textit{ for } i \neq j \quad \textit{and} \quad a_1^2 = a_2^2 = \ldots = a_n^2.$$

Proof. Assume $\underline{a} = (a_1, \ldots, a_n) \in D_L(n)$. Let $\underline{e}_i$ be the ith standard basis vector in R^n, so $a_i = \langle \underline{a}, \underline{e}_i \rangle$. Then for $i \neq j$, we have $\langle \underline{e}_i, \underline{e}_j \rangle = 0$, so putting $\underline{u} = \underline{e}_i$, $\underline{v} = \underline{e}_j$ in the defining condition for $D_L(R^n)$, we get

$$0 = \langle \underline{a}, \underline{a} \rangle \cdot \langle \underline{e}_i, \underline{e}_j \rangle = n \cdot \langle \underline{a}, \underline{e}_i \rangle \cdot \langle \underline{a}, \underline{e}_j \rangle = n \cdot a_i a_j,$$

whence $a_i a_j = 0$. On the other hand, putting $\underline{u} = \underline{v} = \underline{e}_i$ in the defining condition for $D_L(R^n)$, we get

$$\langle \underline{a}, \underline{a} \rangle \cdot 1 = \langle \underline{a}, \underline{a} \rangle \cdot \langle \underline{e}_i, \underline{e}_i \rangle = n \cdot \langle \underline{a}, \underline{e}_i \rangle \cdot \langle \underline{a}, \underline{e}_i \rangle = n \cdot a_i^2,$$

and since the left-hand side here is independent of i, then so is the right-hand side, and hence a_i^2 is independent of i.

Conversely, assume $\underline{a}$ satisfies the equations. Let $\underline{u} = (u_1, \ldots, u_n)$ and $\underline{v} = (v_1, \ldots, v_n)$. Then

$$\begin{aligned} \langle \underline{a}, \underline{a} \rangle \cdot \langle \underline{u}, \underline{v} \rangle &= (\textstyle\sum a_i^2) \cdot (\sum u_j v_j) \\ &= n \cdot a_1^2 \cdot (\textstyle\sum u_j v_j) \end{aligned}$$

since all the a_i^2 equal a_1^2; on the other hand

$$n \cdot \langle \underline{a}, \underline{u} \rangle \cdot \langle \underline{a}, \underline{v} \rangle = n \cdot (\textstyle\sum a_i u_i) \cdot (\sum a_j v_j);$$

when we multiply this out, we get n^2 terms, but due to $a_i a_j = 0$ for $i \neq j$, only the n terms with $i = j$ survive, so we get

$$\begin{aligned} &= n \cdot \sum_i a_i^2 u_i v_i \\ &= n \cdot \sum_i a_1^2 u_i v_i \end{aligned}$$

(since all the a_i^2 are equal) and taking the factor a_1^2 outside the sum now gives the desired result. This proves the proposition.

By definition, $D_2(1) = D_L(1)$. Now assume $n \geq 2$. Clearly $D(n) \subseteq D_L(n)$; for, if $\underline{a} \in D(n)$, we have that all the a_i^2 are equal: they are all 0. On the other hand, $D_L(n) \subseteq D_2(n)$; for if $\underline{a} \in D_L(n)$, $a_i a_j a_k = 0$ unless $i = j = k$; and $a_i^3 = a_i^2 a_i = a_j^2 a_i$ for any $j \neq i$ (and such j exists, since $n \geq 2$), but this equals $a_j(a_j a_i)$, and the parenthesis here is again 0.

We consider now a general finite-dimensional vector space V equipped with a *positive definite* inner product. So in particular, we assume that the basic ring R is Pythagorean; in particular, invertible square sums have square roots; a *positive* element of R is by definition an invertible element having a square root.

Proposition 8.3.3 *Let $a \in D_L(V)$ have $\langle a,a\rangle = 0$. Then $a \in D_1(V)$.*

Proof. It suffices to consider $V = R^n$ with standard inner product. If $\underline{a} = (a_1,\ldots,a_n)$, the assumption $\langle \underline{a},\underline{a}\rangle = 0$ says $\sum a_i^2 = 0$, and since $\underline{a} \in D_L(n)$, all the terms a_i^2 here are equal, hence are 0. Also, for $i \neq j$, $a_i \cdot a_j = 0$, by the definition of $D_L(n)$. So $a_i \cdot a_j = 0$ for all i, j. These are the defining equations for $D_1(n)$.

Proposition 8.3.4 *If $W \subseteq V$ is a linear subspace of lower dimension than V, then $W \cap D_L(V) = D_1(W)$.*

Proof. Take a unit vector u orthogonal to W. If $a \in W \cap D_L(V)$,

$$\langle a,a\rangle \cdot \langle u,u\rangle = n \cdot \langle a,u\rangle \cdot \langle a,u\rangle;$$

the right-hand side is 0, by the orthogonality of u to W. Since $\langle u,u\rangle = 1$, we deduce $\langle a,a\rangle = 0$. Since $a \in D_L(V)$, we conclude by Proposition 8.3.3 that $a \in D_1(V)$.

Remark 8.3.5 An unprecise heuristics following from the proposition is that elements $z \in D_L(V)$ "point in no particular direction W"; this is why it is reasonable to call them *isotropic* neighbours of 0 ("isotropic" = "identical in all directions", according to the dictionary). Or: "if an elementary particle in the atom $D_L(n)$ can be located at all, it must belong to the nucleus $D_1(n)$".

The following axiom is an instance of the KL axiom scheme as discussed in Section 1.3:

• KL axiom for $D_L(n)$: Every map $h : D_L(n) \to R$ which vanishes on $D_1(n)$ is of the form

$$(x_1,\ldots,x_n) \mapsto c \cdot \sum_{i=1}^{n} x_i^2$$

for a unique $c \in R$.

Equivalently, in light of the KL axiom for $D(n) = D_1(n)$: for every $f : D_L(n) \to R$, there exists a unique affine map $\overline{f} : R^n \to R$ and a unique constant $c \in R$ so that

$$f(x_1,\ldots,x_n) = \overline{f}(x_1,\ldots,x_n) + c \cdot \sum_{i=1}^{n} x_i^2.$$

Let us note the following cancellation principle which is an immediate consequence of the KL axiom for $D_L(n)$:

• if $c \in R^n$ has $c \cdot \sum_i z_i^2$ for all $(z_1,\ldots,z_n) \in D_L(n)$, then $c = 0$.

Remark 8.3.6 This cancellation principle implies that $\sum_i z_i^2$ cannot be 0 for all $(z_1,\dots,z_n)$ in $D_L(n)$ (but $(\sum_i z_i^2)^2 = 0$, as we noticed above). Equivalently, $D_1(n) \subseteq D_L(n)$ is a *proper* inclusion.

We henceforth assume the KL axiom for $D_L(n)$. By the "square norm" of a vector $\underline{a} \in R^n$, we understand of course the number $\langle \underline{a}, \underline{a} \rangle = \sum a_i^2$.

Proposition 8.3.7 *Let $[a_{ij}]$ be an invertible $n \times n$ matrix, and let $A : R^n \to R^n$ be the linear automorphism that it defines. Then the following conditions are equivalent (and define the notion of* conformal matrix*):*

(1) A maps $D_L(n)$ into $D_L(n)$;

(2) A^{-1} maps $D_L(n)$ into $D_L(n)$;

(3) the rows of $[a_{ij}]$ are mutually orthogonal and have the same square norm;

(4) the columns of $[a_{ij}]$ are mutually orthogonal and have the same square norm.

Proof. This is purely equational, using the equational description of $D_L(n)$ given in Proposition 8.3.2, and the above cancellation principle: Assume (1). For $\underline{z} = (z_1,\dots,z_n) \in D_L(n)$, the ith entry of $A \cdot \underline{z}$ is $\sum_j a_{ij} z_j$. Since $A \cdot \underline{z} \in D_L(n)$, $(\sum_j a_{ij} z_j)^2$ is independent of i. We calculate this square:

$$(\sum_j a_{ij} z_j) \cdot (\sum_j a_{ij} z_j) = \sum_{j,j'} a_{ij} a_{ij'} z_j z_{j'} = \sum_j a_{ij}^2 z_j^2$$

(since $z_j z_{j'} = 0$ if $j \neq j'$)

$$= z_1^2 \sum_j a_{ij}^2 = \frac{1}{n} (\sum_k z_k^2)(\sum_j a_{ij}^2),$$

both the last equality signs because $z_1^2 = \ldots = z_n^2$. Since for all $\underline{z} \in D_L(n)$ this is independent of i, we conclude by the above cancellation principle that $\sum_j a_{ij}^2$ is independent of i, whence the rows of A have the same square norm. The proof that the rows of A are orthogonal is similar. This proves (3). The implication from (3) to (1) is similar calculationally, using the equations in Proposition 8.3.2.

Also, it is standard matrix algebra to see that if (3) holds, then the matrix for A^{-1} is the transpose of the matrix for A modulo a positive scalar factor; so the remaining bi-implications are now clear.

A linear isomorphism $f : V \to V'$ between n-dimensional vector spaces with positive definite inner product is therefore called *conformal* if it takes $D_L(V)$ into $D_L(V')$, or, equivalently, if in some, or any, pair of orthonormal bases for

V and V', the matrix of f is a conformal matrix. There exists then a positive scalar Λ such that $\langle f(u), f(v)\rangle = \Lambda\langle u, v\rangle$ for any $u, v \in V$.

L-neighbours in *M*

Let M be a Riemannian manifold. Assume $x \sim_2 z$ in M. Then we say that $z \sim_L x$ if $\log_x^{(2)}(z) \in D_L(T_x M)$. We express this verbally by saying that z is a *Laplacian* neighbour (written $z \sim_L x$) or an *isotropic neighbour* of x.

For $x \in M$, we write $\mathfrak{M}_L(x) \subseteq \mathfrak{M}_2(x)$ for the set of $z \in M$ with $z \sim_L x$.

Proposition 8.3.8 *The relation $\sim_L$ is reflexive and symmetric.*

Proof. Reflexivity is obvious. The symmetry is essentially an exercise in degree calculus: it suffices to consider the case where M is an open subset $M \subseteq V$ of an n-dimensional vector space, and with the metric g given by $g(x,z) = G(x; z-x, z-x)$ for $x \sim_2 z$, and with Levi–Civita connection λ given by its Christoffel symbols $\Gamma(-;-,-)$. We can express that $z \sim_L x$ in terms of G and Γ as follows. Let us write $z = x + a$ with $a \in D_2(V)$. We identify $T_x M$ with V via principal-part formation. We use the expression for $\log_x^{(2)}$ provided by (8.2.3). For brevity, we write $c(a)$ for $\frac{1}{2}\Gamma(x; a, a)$; it depends in a quadratic way on a. Recall (Proposition 8.1.3) that the inner product on $T_x M$ corresponds to the inner product $G(x; -, -)$ on V under the principal-part identification $T_x M \cong V$. So the condition that $z \sim_L x$ is that for all u and v in V,

$$G(x; a + c(a), a + c(a)) \cdot G(x; u, v) = n \cdot G(x; a + c(a), u) \cdot G(x; a + c(a), v). \tag{8.3.2}$$

Similarly, the condition that $x \sim_L z$ is that for all u and v in V,

$$\begin{aligned} &G(x+a; -a + c'(a), -a + c'(a)) \cdot G(x+a; u, v) \\ &\qquad = n \cdot G(x+a; -a + c'(a), u) \cdot G(x+a; -a + c'(a), v) \end{aligned} \tag{8.3.3}$$

where $c'(a)$ denotes $\Gamma(x+a; -a, -a)$, again quadratic in a. When expanding (8.3.2) by bilinearity of $G(x; -, -)$, all the terms containing $c(a)$ vanish, by degree calculus, so that (8.3.2) is equivalent to

$$G(x; a, a) \cdot G(x; u, v) = n \cdot G(x; a, u) \cdot G(x; a, v). \tag{8.3.4}$$

Similarly, (8.3.3) is equivalent to

$$G(x+a; -a, -a) \cdot G(x+a; u, v) = n \cdot G(x+a; -a, u) \cdot G(x+a; -a, v); \tag{8.3.5}$$

now a Taylor expansion from x in the direction a gives several terms that vanish for degree reasons, and we are left with

$$G(x;-a,-a)\cdot G(x;u,v) = n\cdot G(x;-a,u)\cdot G(x;-a,v),$$

which clearly is the same condition as (8.3.4) since the minus signs cancel. This proves the proposition.

In the same spirit, we leave it to the reader to prove the following. The set-up $M \subseteq V$ is the same as in the proof of the previous proposition.

Proposition 8.3.9 *For $x \sim_2 z$ in M, we have $z \sim_L x$ iff $z - x \in D_L(V)$, where V is provided with the inner product $G(x;-,-)$.*

8.4 The Laplace operator

In this section, we consider a fixed Riemannian manifold (M,g).

Proposition 8.4.1 *Let g be a Riemannian metric on an n-dimensional manifold M, and let f be some quadratic differential form on M. Then there exists a unique function $c: M \to R$ so that for all $x \sim_L z$,*

$$f(x,z) = c(x)\cdot g(x,z). \tag{8.4.1}$$

Proof. It suffices to consider the standard coordinatized case $M \subseteq V$ with $g(x,z) = G(x;z-x,z-x)$ with $G(x;-,-): V\times V \to R$ symmetric positive definite for each $x \in M$. By positive definiteness, there exists a linear isomorphism $V \cong R^n$ taking $G(x;-,-)$ to the standard inner product on R^n; it identifies $D_L(V)$ with $D_L(n)$. Consider for fixed $x \in M$ the composite

$$D_L(n) \cong D_L(V) \xrightarrow[\cong]{\exp_x^{(2)}} \mathfrak{M}_L(x) \xrightarrow{\ f\ } R.$$

It maps $D_1(n)$ to $\mathfrak{M}_1(x)$, so by the assumption on f, it takes $D_1(n)$ to 0, hence by the KL axiom for $D_L(n)$, it is of the form $c(x)\cdot\langle a,a\rangle$ ($a \in D_L(n)$) for a unique constant $c(x)$,

$$f(\exp_x^{(2)}(a)) = c(x)\cdot\langle a,a\rangle,$$

or equivalently (put $a = \log_x^{(2)}(z)$)

$$f(z) = c(x)\cdot\langle \log_x^{(2)}(z), \log_x^{(2)}(z)\rangle.$$

But $\langle \log_x^{(2)}(z), \log_x^{(2)}(z)\rangle = g(x,z)$ by Proposition 8.2.9. This proves the proposition.

Given any function $f : M \to R$, we can manufacture a function $\hat{f} : M_{(2)} \to R$ by the following recipe: for $x \sim_2 z$, we put

$$\hat{f}(x,z) := f(z) + f(\rho_x(z)) - 2 \cdot f(x).$$

This $\hat{f}$ is in fact a quadratic differential form, i.e. it vanishes if $x \sim_1 z$. For, in this case $\rho_x(z)$ is the affine combination $2x - z$ of 1-neighbours, as we saw in Section 8.2, and any function f preserves affine combinations of 1-neighbours, by Theorem 2.1.4. This clearly implies that the right-hand side of the defining equation for $\hat{f}$ vanishes on $M_{(1)}$.

By Proposition 8.4.1, there is a unique function $c(x)$ such that for all $z \sim_L x$ $\hat{f}(x,z) = c(x) \cdot g(x,z)$. We write Δf for this function c. Thus Δf is characterized by: for all $z \sim_L x$,

$$(\Delta f)(x) \cdot g(x,z) = f(z) + f(\rho_x(z)) - 2 \cdot f(x). \tag{8.4.2}$$

Definition 8.4.2 *The* Laplace (–Beltrami) operator $\Delta : C^\infty(M) \to C^\infty(M)$ *is the operator which to $f \in C^\infty(M)$ associates the unique function Δf satisfying (8.4.2) for all $z \sim_L x$.*

We shall see below that it agrees with the operator div $\circ$ grad, which is the classical way to describe the Laplace–Beltrami operator in terms of divergence of vector fields, and gradient vector fields of functions.

We shall calculate $(\Delta f)(x)$ in a coordinatized situation. Assume that (M,g) (locally around the given $x \in M$) is openly embedded in an n-dimensional vector space V, in a way which is *geodesic* at x in the sense that the metric tensor $G : M \times V \times V \to R$ is constant on $\mathfrak{M}_1(x)$, and hence the Christoffel symbols Γ for the Levi–Civita connection for g vanish at x. Then for $z \sim_2 x$, $\rho_x(z) = 2x - z$, with this affine combination formed in V (see Exercise 8.2.1), and so the characterizing equation for $(\Delta f)(x)$ reads

$$(\Delta f)(x) \cdot g(x,z) = f(z) + f(2x - z) - 2 \cdot f(x). \tag{8.4.3}$$

We Taylor expand f from x, and have partly

$$f(z) = f(x) + df(x; z - x) + \tfrac{1}{2} d^2 f(x; z - x, z - x)$$

and

$$f(2x - z) = f(x + (x - z)) = f(x) + df(x; x - z) + \tfrac{1}{2} d^2 f(x; x - z, x - z).$$

Substituting these expressions in (8.4.3), we get that the characterizing equation may be written (still assuming that $x \in M \subset V$ is a geodesic point, so the

Christoffel symbols vanish at x):

$$(\Delta f)(x) \cdot g(x,z) = d^2 f(x; z-x, z-x). \tag{8.4.4}$$

Divergence of a vector field

The relation $\sim_L$ allows us to give a geometric construction of the divergence of a vector field (a picture is drawn below), which does not involve either integration or volume form.

Consider a vector field X on an n-dimensional Riemannian manifold (M, g). The infinitesimal transformations X_d ($d \in D$) of X preserve the relations $\sim_1$ and $\sim_2$ (but not necessarily $\sim_L$). As a measure of how much the vector field *diverges*, one may consider the difference of square-distance g before and after applying X_d: so for $x \sim_2 z$, we may consider $g(X_d(z), X_d(x)) - g(z,x)$ which clearly is 0 if $d = 0$ so is of the form $d \cdot h(x,z)$ for some function $h : M_{(2)} \to R$:

$$g(X_d(z), X_d(x)) - g(z,x) = d \cdot h(x,z).$$

If $x \sim_1 z$, both terms in this expression for h vanish, so h likewise vanishes on $M_{(1)}$, so h is, like g, a quadratic differential form on M. By Proposition 8.4.1, there is a unique function $c : M \to X$ such that equation (8.4.1) holds (with $f = h$). This function, multiplied by $\frac{n}{2}$, we call the *divergence* of the vector field, denoted $\operatorname{div} X$. (The factor $\frac{n}{2}$ is included for comparison with a classical formula for divergence, cf. (8.4.5) below.) So for all $z \sim_L x$,

$$d \cdot (\operatorname{div} X)(x) \cdot g(x,z) = \frac{n}{2}\big(g(X_d(z), X_d(x)) - g(z,x)\big).$$

Some imprecise heuristics: the difference $H(d) - H(0) := g(X_d(z), X_d(x)) - g(z,x)$ depends in general not only on $g(z,x)$, but also on the "direction" from x to z; but if $x \sim_L z$, there is no particular *direction* in which one may locate z, by Remark 8.3.5; L-neighbours of x *average* over all directions from x simultaneously. In standard formulation, this average-over-directions is expressed in terms of a (flux) integral over the surface of a sphere.

Here is a picture:

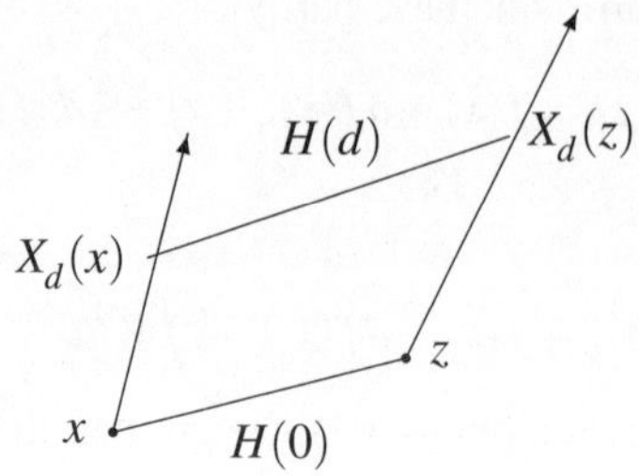

We shall give a coordinate-dependent description of div$X$. As in the proof of (8.4.4), we assume a coordinatized situation $M \subseteq V$, with $x \in M$ a geodesic point, i.e. the bilinear form $G(y;-,-) : V \times V \to R$ equals $G(x;-,-)$ for $y \sim_1 x$.

We write $\langle \underline{u}, \underline{v} \rangle$ for $G(x;\underline{u},\underline{v})$, $(= G(y;\underline{u},\underline{v})$ for $y \sim_1 x)$.

The vector field X is given by a function $\xi : M \to V$, such that for any $y \in M$, $X(y)$ is the tangent vector at y with principal part $\xi(y)$. We then calculate, for any $z \sim_2 x$,

$$\begin{aligned} d \cdot \operatorname{div} X(x) \cdot g(x,z) &= \frac{n}{2} \cdot \left[g\big(X_d(z), X_d(x)\big) - g\big(z,x\big) \right] \\ &= \frac{n}{2} \cdot \left[\big(\langle z + d \cdot \xi(z) - (x + d \cdot \xi(x)), \ldots \rangle - \langle z - x, \ldots \rangle \big) \right] \end{aligned}$$

(where "…" indicates that the argument after the comma is the same as that before the comma)

$$= d \cdot n \cdot \langle z - x, \xi(z) - \xi(x) \rangle$$

(using bilinearity and symmetry of $\langle -,- \rangle$, and using $d^2 = 0$)

$$= d \cdot n \cdot \langle z - x, d\xi(x; z - x) \rangle$$

plus a term which is of third degree in $z - x$ and therefore vanishes. We can further reduce this expression in case $z \sim_L x$. Let us pick orthonormal coordinates for $T_x M \cong V$. Then $T_x M \cong V$ gets identified with R^n with its standard inner product, and $z = x + \underline{u}$ with $\underline{u} = (u_1, \ldots, u_n) \in D_L(n)$. We now express $\langle z - x, d\xi(x; z - x) \rangle = \langle \underline{u}, d\xi(x; \underline{u}) \rangle$ in terms of these coordinates. With $\xi = (\xi_1, \ldots, \xi_n)$, the equation continues

$$\begin{aligned} &= d \cdot n \cdot \sum_i u_i \cdot d\xi_i(x; \underline{u}) \\ &= d \cdot n \cdot \sum_i u_i \big(\sum_k \frac{\partial \xi_i(x)}{\partial x_k} u_k \big) \\ &= d \cdot n \cdot \sum_{ik} u_i u_k \cdot \frac{\partial \xi_i(x)}{\partial x_k}; \end{aligned}$$

but now $\underline{u} \in D_L(n)$, so by Proposition 8.3.2, $u_i u_k = 0$ if $k \neq i$, so the double sum reduces to a single sum, and the equation continues

$$= d \cdot n \cdot \sum_i u_i u_i \cdot \frac{\partial \xi_i(x)}{\partial x_i},$$

and because u_i^2 is independent of i by Proposition 8.3.2, we may bring $u_i^2 = u_1^2$ outside the sum, so that we get

$$= d \cdot n \cdot u_1^2 \cdot \sum_i \frac{\partial \xi_i(x)}{\partial x_i}$$
$$= d \cdot (\sum_k u_k^2) \cdot \sum_i \frac{\partial \xi_i(x)}{\partial x_i}$$

(again by Proposition 8.3.2)

$$= d \cdot g(x,z) \cdot \sum_i \frac{\partial \xi_i(x)}{\partial x_i}.$$

Cancelling d and $g(z,x)$ (because the equation holds for all $d \in D$ and all $z \in \mathfrak{M}_L(x)$), we obtain

$$\operatorname{div} X(x) = \sum_i \frac{\partial \xi_i(x)}{\partial x_i}, \tag{8.4.5}$$

the standard expression, cf. e.g. Helgason, 1962, X.2 (3) (note that the two occurrences of $\sqrt{g}$ in the formula loc. cit. cancel, since G is constant on $\mathfrak{M}_1(x)$, and therefore $\sqrt{g}$ goes outside the differentiation).

Laplacian in terms of divergence and gradient

The treatment of this issue here is now completely standard. Consider a function $f : M \to R$ on a Riemannian manifold (M,g). Its differential df is a cross-section of the cotangent bundle $T^*M \to M$, as described in Section 4.4. Now g gives rise to an isomorphism (8.1.4) $\check{g} : T^*M \to TM$ of vector bundles. The *gradient vector field* of f is then the composite

$$M \xrightarrow{df} T^*M \xrightarrow{\check{g}} TM.$$

It is denoted $\operatorname{grad} f$. Thus for $x \in M$, $d \in D$, $(\operatorname{grad} f)(x,d) \in M$, and for $x \in M$, $(\operatorname{grad} f)(x)$ is a tangent vector at x. Thus

$$\langle (\operatorname{grad} f)(x), \tau \rangle = df(x;\tau),$$

where the left-hand side utilizes the inner product on T_xM derived from g, and where $\tau \in T_x(M)$.

Therefore using the definition of $df(x;\tau)$,

$$d \cdot \langle (\operatorname{grad} f)(x), \tau \rangle = f(\tau(d)) - f(x). \tag{8.4.6}$$

It is easy to see that if $M \subseteq V$ is an open subset of a finite-dimensional vector space, then we can calculate $\operatorname{grad} f$ at x by picking a basis $e_1, \ldots, e_n$

for V, orthonormal with respect to the inner product $G(x;-,-)$ (notation as in the proof of Proposition 8.3.8); then the principal part of the tangent vector $\mathrm{grad}(f)$ is

$$\sum_{i=1}^{n} df(x;e_i)\cdot e_i. \tag{8.4.7}$$

The following is not just a definition, since we did not define the Laplacian Δ in the standard way:

Proposition 8.4.3 *Let $f : M \to R$ be a function (M a Riemannian manifold). Then*

$$\Delta f = \mathrm{div}\,\mathrm{grad}\, f.$$

Proof. Let $x \in M$; we may pick a coordinate chart around $x \in M \subseteq V$ with x a geodesic point, as in the proof of (8.4.5). Pick an orthonormal basis for V as in the proof of (8.4.7). Combining this with the standard formula (8.4.5) for divergence, we arrive at the standard expression for $(\mathrm{div}\,\mathrm{grad}\, f)(x)$, namely

$$(\mathrm{div}\,\mathrm{grad}\, f)(x) = \sum_{i=1}^{n} \frac{\partial^2 f(x)}{(\partial x_i)^2},$$

and comparison with (8.4.4) then gives the result.

Note that to define $(\mathrm{div}\,\mathrm{grad}\, f)(x)$, one only needs to know the restriction of the function f to $\mathfrak{M}_2(x)$; but for $\Delta f(x)$, as we have defined it, we need even less, namely the restriction of f to $\mathfrak{M}_L(x) \subseteq \mathfrak{M}_2(x)$.

Exercise 8.4.4 The vector field $\mathrm{grad}\, f$ on M has the property that for all $x \sim_1 y$ in M and all $d \in D$,

$$f(d \cdot_x y) - f(x) = -\tfrac{1}{2} g((\mathrm{grad}\, f)(x,d), y).$$

Harmonic functions

For a Riemannian manifold (M, g), a function $f : M \to R$ is called *harmonic* if $\Delta f \equiv 0$. More generally, we say that f is harmonic *at* $x \in M$ if $(\Delta f)(x) = 0$. The coordinate-free description that we have been giving of the Δ operator gives as a corollary a geometric description of harmonicity:

Theorem 8.4.5 *A function $f : M \to R$ is harmonic at $x \in M$ iff for all $z \sim_L x$, $f(x)$ is the average of $f(z)$ and $f(z')$.*

(Here, z' denotes the reflection $\rho_x(z)$ of z in x.)

Proof. For, to say that $f(x)$ is the average of $f(z)$ and $f(z')$ is to say $2 \cdot f(x) = f(z) + f(z')$, and the difference of the two sides here is the one that enters in the definition (8.4.2) of $(\Delta f)(x)$.

One may say that this is the synthetic counterpart of the classical description of harmonicity at x (for R^n with standard metric); this classical description is that f has the *average value property* at x: $f(x)$ is the *average* of the values $f(z)$ as z ranges over any sphere centred at x. In the synthetic description, the spheres are replaced by pairs z, z' of antipodal isotropic neighbour points as seen from x.

Conformal diffeomorphisms

Let (M,g) and (M',g') be two Riemannian manifolds. A diffeomorphism (= bijection = invertible map) $f : M \to M'$ is an *isometry* if for all $x \sim_2 z$ in M, $g'(f(x), f(z)) = g(x,z)$. There is a weaker notion, namely that of *conformal* diffeomorphism: one says that f is *conformal* if there is a function $\Lambda : M \to R$ such that for all $x \sim_2 z$,

$$g'(f(x), f(z)) = \Lambda(x) \cdot g(x,z).$$

Theorem 8.4.6 *A diffeomorphism $f : M \to M'$ is conformal iff it preserves the relation $\sim_L$, in the sense*

$$x \sim_L z \quad \textit{iff} \quad f(x) \sim_L f(z).$$

Proof. This depends on Proposition 8.3.7. Consider the diagram

$$\begin{array}{ccc}
\mathfrak{M}_2(x) & \xrightarrow{\;f\;} & \mathfrak{M}_2(f(x)) \\
{\scriptstyle \log^{(2)}_x}\Big\downarrow & & \Big\downarrow{\scriptstyle \log^{(2)}_{f(x)}} \\
D_2(T_x M) & \underset{T_x f}{\overset{\tilde{f}}{\rightrightarrows}} & D_2(T_{f(x)} M')
\end{array}$$

where $\tilde{f}$ is the unique map making the diagram commutative. The map $T_x f$ does not make the diagram commutative, but when restricted to $\mathfrak{M}_1(x)$, it does, by naturality of log, (4.3.4). So $\tilde{f}$ and $T_x f$ agree on $\mathfrak{M}_1(x)$. It then follows from Proposition 8.3.1 that $\tilde{f}$ maps $D_L(T_x M)$ into $D_L(T_{f(x)} M')$ iff $T_x f$ does. By definition, $\mathfrak{M}_L(x)$ comes about from $D_L(T_x M)$ by transport along the $\log^{(2)}$–$\exp^{(2)}$ bijection, so f preserves $\mathfrak{M}_L$ iff $\tilde{f}$ preserves D_L. On the other hand, by Proposition 8.3.7, conformality of $T_x f$ is equivalent to $T_x f$ preserving D_L.

Appendix

This Appendix is a mixed bag; it contains things which are either foundational or technical. Of foundational nature is in particular Sections A.2 and A.3 – for those who need to lift the hood of the car to get an idea† how the engine works, before getting into the car and driving it. Section A.4 deals with microlinearity, a technical issue specific to SDG; and the remaining sections are completely standard mathematical technique, with due care taken of the constructive character of the reasoning, but otherwise probably found in many standard texts.

A.1 Category theory

Basic to the development of SDG is category theory, with emphasis on what the *maps* between the mathematical objects are.

In particular, SDG depends on the notion of *cartesian closed category* $\mathscr{E}$; recall that in such $\mathscr{E}$, one not only has a *set* of maps $X \to Y$, but an *object* (or "*space*") $Y^X \in \mathscr{E}$ of maps, with a well-known universal property, cf. e.g. Section A.3 below for the relevant diagram.

The category of sets is cartesian closed, in fact, any topos is so.

The category Mf of smooth manifolds is not cartesian closed, and this failure has historically caused difficulties for subjects like calculus of variations, path integrals, continuum mechanics, ..., and has led to the invention of more comprehensive "smooth categories" by Chen, Frölicher, Kriegl, Michor, and many others (see e.g. Frölicher and Kriegl, 1988; Kriegl and Michor, 1997 and the references therein), and also to the invention of toposes containing Mf like Grothendieck's "smooth topos" (terminology of Kock, 1981/2006) and the "well-adapted toposes" for SDG, as sketched below. See Lawvere (2000) for historical comments, and Kock (1986), Kock and Reyes (1979, 2006) for some concrete comparisons.

† For a more complete account, see Parts II and III of Kock (1981/2006).

Another basic categorical notion is that of a *slice* category $\mathscr{E}/X$ (a special case of a comma-category): if X is an object of $\mathscr{E}$, then $\mathscr{E}/X$ is the category whose objects are arrows in $\mathscr{E}$ with codomain X, and whose arrows are the obvious commutative triangles. One also says that an object in $\mathscr{E}/X$ is a *bundle* over X.

If the category $\mathscr{E}$ has pull backs, any map $\xi : N \to M$ in $\mathscr{E}$ gives rise to a functor

$$\xi^* : \mathscr{E}/M \to \mathscr{E}/N,$$

"pulling back along ξ".†

We henceforth assume that $\mathscr{E}$ has finite inverse limits, or equivalently, that it has pull backs and a terminal object 1. In this case, $\mathscr{E}$ itself appears as a slice, namely $\mathscr{E}$ is canonically isomorphic to $\mathscr{E}/1$. If N is an object in $\mathscr{E}$, and $\xi : N \to 1$ the unique such map, the functor $\xi^* : \mathscr{E}/1 \to \mathscr{E}/N$ may be identified with the functor $- \times N$ (note that for any $Z \in \mathscr{E}$, $Z \times N$ comes equipped with a canonical map to N, namely the projection).

It is easy to prove that $\mathscr{E}/M$ has finite inverse limits. The terminal object in $\mathscr{E}/M$ is the identity map $M \to M$.

An important result is the following (cf. e.g. Johnstone, 2002, A.1.5):

Theorem A.1.1 *The following two conditions are equivalent:*

(1) For any $\xi : N \to M$, the functor $\xi^* : \mathscr{E}/M \to \mathscr{E}/N$ has a right adjoint (denoted ξ_*).

(2) Each of the slice categories $\mathscr{E}/N$ is cartesian closed.

A category $\mathscr{E}$ (with finite inverse limits) satisfying these conditions is called *locally cartesian closed* (in Kock, 1981/2006, the term "stably cartesian closed" was used). If $\mathscr{E}$ is locally cartesian closed, then so is any slice $\mathscr{E}/N$. Any topos is locally cartesian closed.

Functors of the form ξ^* preserve all inverse limits (in fact, they are themselves right adjoint functors, the corresponding left adjoint $\xi_!$ is just the functor "composing with ξ"). For locally cartesian closed $\mathscr{E}$, the functors ξ^* also preserve cartesian closed structure, in the sense that the canonical comparison $\xi^*(B^A) \to \xi^*(B)^{\xi^*(A)}$ is an isomorphism.

We think of objects $\varepsilon : E \to M$ in $\mathscr{E}/M$ as *bundles* over M. When we have the notion of, say, a group object in a category, a group object in $\mathscr{E}/M$ may be thought of as a group *bundle* over M. Similarly for vector space objects vs. vector bundles.

† This involves *choosing* pull backs; this non-constructive aspect is not essential, but it simplifies some formulations. On the other hand, it makes certain other things complicated: it forces spurious coherence questions.

Example A.1.2 Let $\mathscr{E}$ be the category of sets. It is locally cartesian closed; let us for a given $\xi : N \to M$ describe the functor $\xi_* : \mathscr{E}/N \to \mathscr{E}/M$. We may view an object $\varepsilon : E \to N$ in $\mathscr{E}/N$ as an N-indexed family of sets $\{E_y \mid y \in N\}$ where $E_y := \varepsilon^{-1}(y)$, the "fibre over y". Similarly for objects in $\mathscr{E}/M$. Thus $\xi_*(\varepsilon)$ is to be an M-indexed family of sets; it is given (for $x \in M$) by

$$(\xi_*(\varepsilon))_x = \text{ the set of maps } s : \xi^{-1}(x) \to E \text{ with } \varepsilon(s(y)) = y \text{ for all } y \in \xi^{-1}(x).$$

Also, let us describe the cartesian closed structure of $\mathscr{E}/N$: if $\varepsilon : E \to N$ and $\phi : F \to N$ are objects in $\mathscr{E}/N$, then for $y \in N$,

$$(\varepsilon^{\phi})_y = E_y^{F_y},$$

where, as before, $E_y = \varepsilon^{-1}(y)$ and similarly $F_y = \phi^{-1}(y)$.

Example A.1.3 We consider, likewise in the category of sets, the situation of two parallel arrows $N \rightrightarrows M$, say α and β. We then have an endofunctor on $\mathscr{E}/M$, namely $\beta_* \circ \alpha^*$. The set-theoretic description of this bundle over M is given in Remark 7.3.1.

A *topos* (more precisely, an elementary topos) is a category $\mathscr{E}$ with finite inverse limits, which is cartesian closed, and has a subobject classifier Ω, in the sense explained in the textbooks, e.g. in Mac Lane and Moerdijk (1992, §I.3) and McLarty (1995, p. 13). Any topos $\mathscr{E}$ is automatically locally cartesian closed; in fact, each slice $\mathscr{E}/M$ is again a topos (cf. e.g. Johnstone, 1977, Theorem 1.42). Furthermore, the pull-back functor $\xi^* : \mathscr{E}/M \to \mathscr{E}/N$ (for $\xi : N \to M$) preserves not only the cartesian closed structure, but also the subobject classifier (cf. loc. cit.). We are not in the present text exploiting the subobject classifier very much; in fact, little is at present known about it in the topos models of SDG. But the existence of a subobject classifier in a locally cartesian closed $\mathscr{E}$ implies that $\mathscr{E}$ has finite colimits and is *exact* (Barr, 1971), i.e. it has many exactness properties: epimorphisms are coequalizers of their kernel pairs, and are stable under pull-back; and every epimorphism has a certain *descent* property (exploited in (A.5.1)).

A.2 Models; sheaf semantics

Synthetic differential geometry is a (hopefully) consistent body of notions, constructions, and assertions whose intended interpretation is geometric aspects of the real world. In this respect, it does not differ from, say, Euclid's books. This is also the reason for the adjective "synthetic".

But just as for Euclidean geometry, mathematicians today want to have a mathematical *semantics* for the theory; typically in terms of an analytic model, ultimately built on the field $\mathbb{R}$ of real numbers (which often is thought to be even more real than the real world itself).

This section aims to sketch in which sense the statements and constructions employed (those of naive set theory) make sense in any topos. It is another, more subtle, issue to construct toposes where the axioms are satisfied. But since the statements and constructions make sense in any topos, it is not necessary to choose one particular specific model prior to developing the theory. In fact, *no* model is needed.

There also exist models which do not depend on $\mathbb{R}$; differential calculus is known to exist in the context of algebraic geometry, without reference to the limit processes which require $\mathbb{R}$. One such model will be described in more detail in Section A.3 below. Models that contain the ordinary category Mf of smooth manifolds, in a way such that assertions proved in the model imply assertions about ordinary manifolds, also exist, the *well-adapted* models, cf. Dubuc (1979) and the textbooks Kock (1981/2006) and Moerdijk and Reyes (1991).

But what is meant by "a model"? It is a topos $\mathscr{E}$, with a ring object $R \in \mathscr{E}$ (and possibly with an openness or étaleness notion), such that the notions of the theory *make sense*, the constructions can be *performed*, and the axioms are *satisfied*.

For the well-adapted models $\mathscr{E}$, there is a full and faithful inclusion $i : \mathrm{Mf} \to \mathscr{E}$, and $i(\mathbb{R}) = R$.

The very notion of "ring object" makes sense in any category with finite products, and this is well understood in modern mathematics; similarly for several other of the primitive notions, like module, groupoid, ...

The crux is that there exists a method by which constructions and the satisfaction relation can be described in a simple set-theoretic language, in terms of "elements" of the objects of the category.

In the early days of category theory, a point was made of the fact that the objects of a category "have no elements", and arguments therefore had to be diagrammatic. This diagrammatic method allowed one to talk about, say, group *objects* in any category with finite products, cf. e.g. the Introduction to Mac Lane (1971).

However, the diagrammatic language is often cumbersome, and lacks expressivity. A way of using a language that talks about elements, even in purely diagrammatic situations, has its origins with Grothendieck and Yoneda; they exploit the fact that if, for instance, G is a group object in a category $\mathscr{E}$ with finite products, then each hom-set $\mathrm{Hom}_{\mathscr{E}}(X, G)$ (for any $X \in \mathscr{E}$) carries the structure of an "ordinary" group, i.e. is a *set* with a group structure. To say that the group object G is, say, *commutative* (in the diagrammatic sense) can then easily be proved to be equivalent to the assertion that each $\mathrm{Hom}_{\mathscr{E}}(X,G)$ is a commutative group. The same applies to any other purely algebraic notion, like ring object, module object, etc.

For any object G of $\mathscr{E}$, we call an element in $\mathrm{Hom}_{\mathscr{E}}(X,G)$ a *generalized element* of G defined at *stage* X. An element defined at stage 1 (= the terminal object of $\mathscr{E}$) is called a *global* element .

Here is an example of how to apply generalized elements. Let A and B be objects of $\mathscr{E}$. To construct an arrow $A \to B$ in $\mathscr{E}$ is, by the Yoneda Lemma, equivalent to constructing, for each $X \in \mathscr{E}$, a set-theoretic map, mapping elements of $\mathrm{Hom}_{\mathscr{E}}(X,A)$ to elements of $\mathrm{Hom}_{\mathscr{E}}(X,B)$ (in a way which is natural in X). For instance, if R is a ring object, we have for each X a map $\mathrm{Hom}_{\mathscr{E}}(X,R) \to \mathrm{Hom}_{\mathscr{E}}(X,R)$ given by

$$x \mapsto x^2 + 1, \tag{A.2.1}$$

say (using that $\mathrm{Hom}_{\mathscr{E}}(X,R)$ is a ring), and since these maps are natural in X, it follows by the Yoneda Lemma that there is a unique arrow $R \to R$ in $\mathscr{E}$ giving rise to the maps, and we say that this arrow $R \to R$ is given by the *description* $x \mapsto x^2 + 1$, a naive description in terms of (generalized) elements.

Sheaf semantics† is the method of interpreting naive assertions, like "the group object G is commutative", or naive constructions, like "the map $R \to R$ given by $x \mapsto x^2 + 1$" (for R a ring object, as above), "the centre $Z(G)$ of G", or "the group $\mathrm{Aut}(G)$ of automorphisms of G" (for G a group object), into assertions about generalized elements, respectively constructions, on the objects R and G, etc.

The possibility of sheaf semantics, in this comprehensive sense, depends on good categorical properties of $\mathscr{E}$. For instance, if G is a group object, we may form $\mathrm{Aut}(G)$: first, we utilize cartesian closedness to form G^G, and then construct $\mathrm{Aut}(G)$ as a subobject carved out by finite inverse limits. It is the *extension*, in the sense described below, of the following formula $\phi(t)$ (where t is a variable ranging over G^G):

$$\phi(t) = \text{“}t \text{ is a group homomorphism, and has an inverse”}.$$

Sheaf semantics is (with variation in detail) described in Chapter II of Kock (1981/2006), and in Mac Lane and Moerdijk (1992, §VI.6 and §VI.7); see also Lambek and Scott (1986, §II.8), McLarty (1983, chapter 18), and Moerdijk and Reyes (1991, §III.1). We refer to these treatises for the specifics. The crucial

† The terms "Kripke–Joyal sematics", "topos semantics", or "external semantics" have also been used.

point is that reasoning and performing constructions in naive (intuitionistic!) set theory is *sound* for this semantics; which means that one does not need to know the semantics in order to do the reasoning; this is the viewpoint taken in Part I of Kock (1981/2006), in Bunge and Dubuc (1987), in Lavendhomme (1996), and in several texts by Nishimura, Reyes, and by the present author.

Example A.2.1 Consider an arrow $f : A \to B$ in a topos $\mathscr{E}$. The semantics of the naive statement

for all $b \in B$, *there exists a unique* $a \in A$ *such that* $f(a) = b$

can be proved to be satisfied (in the sense of sheaf semantics) precisely when f is invertible (this is not quite trivial). Similarly, the semantics of the naive statement about f,

for all a_1 *and* a_2 *in* A, $f(a_1) = f(a_2)$ *implies* $a_1 = a_2$

can be proved to be satisfied precisely when f is monic. For the second example, note that in the naive statement, the universal quantifiers ("for all a_1 and a_2") seemingly range only over the object A, so are in some sense bounded, whereas to say "f is monic" involves a quantifier ranging over the whole universe $\mathscr{E}$: "for all $X \in \mathscr{E}$ and for all $X \rightrightarrows A \ldots$". This "unbounded" quantifier is in sheaf semantics hidden in the clause for interpreting naive universal quantifiers. See Kock (1981/2006, §II.2) Similarly, naive existential quantifiers involve hidden existential quantifiers ranging over the whole universe. See Kock (1981/2006, §II.8).

Remark A.2.2 A naive statement like "for all $b : C \to A, \ldots$" has to be read "for all $b \in A^C, \ldots$"; for, quantifiers used in the naive language range over objects of $\mathscr{E}$, like A^C, not over external sets like $\mathrm{Hom}_{\mathscr{E}}(C,A)$.

It can be proved that the validity notion for assertions about generalized elements is stable under change of stage: this means that if $a : X \to M$ is a generalized element satisfying a formula $\phi(x)$, then for any $\alpha : Y \to X$, the element $a \circ \alpha : Y \to M$ at stage Y likewise satisfies ϕ.† This makes it meaningful to ask for a *generic* element of M, satisfying ϕ. It can be proved that such generic element always exists, and is given as a subobject of M, the *extension* of the formula ϕ; we write $\{x \in M \mid \phi(x)\}$ for this extension.

For instance, if R is a ring object in $\mathscr{E}$, the extension of the formula $x^2 = 0$ is the equalizer of two maps $R \to R$, namely the squaring map and the constant-zero map. (This extension is what we elsewhere have denoted D.)

Because pull-back functors $\xi^* : \mathscr{E}/M \to \mathscr{E}/N$ (for $\xi : N \to M$) are *logical* functors (preserve limits and colimits, as well as exponential objects and

† In a notation which is standard in this connection: $\vdash_X \phi(a)$ implies $\vdash_Y \phi(a \circ \alpha)$.

subobject classifier), validity of *assertions* interpretable by sheaf semantics is preserved by ξ^*; thus if R is a local ring object in $\mathcal{E}/M$, $\xi^*(R)$ is a local ring object in $\mathcal{E}/N$; and if R satisfies the KL axioms, then so does $\xi^*(R)$. Similarly, the naive *constructions* expressed in set-theoretic language are preserved by ξ^*.

The data in $\mathcal{E}$ of an arrow $E \to M$ may, from the viewpoint of naive set theory, be seen as a *bundle* over M, or equivalently, as a *family* of sets *parametrized* by the set M, $\{E_x \mid x \in M\}$, where E_x is the *fibre* over $x \in M$. For the sheaf semantics, x is a generalized element $x : X \to M$, and the "fibre over x" is the left-hand arrow in the pull back

$$\begin{array}{ccc} E_x & \longrightarrow & E \\ \downarrow & & \downarrow \\ X & \underset{x}{\longrightarrow} & M \end{array}$$

(in French: produit fibré). Generalized elements b of E_x are then generalized elements of the object $E_x \to X$ in the slice topos $\mathcal{E}/X$; so the stage of definition of such element is some $\alpha : Y \to X$ (one sometimes says that b is an element defined at a stage Y *later than* X).

To say that $E \to M$ is, say, a vector *bundle* (= R-module bundle, where R is a ring object in $\mathcal{E}$) is to say that the object $E \to M$ in $\mathcal{E}/M$ is a module over the ring object $M \times R \to M$ in $\mathcal{E}/M$. On the other hand, to say that $E \to M$ is a vector bundle is to say the fibres E_x are R-modules, for any element of M. Externally, $x : X \to M$ is a generalized element, and E_x is an object in $\mathcal{E}/X$ (namely $x^*(E \to M)$). Among the generalized elements of M is the *generic* one, namely the identity $i : M \to M$. It is easy to prove that "E_x is an R-module for every (generalized) element of M" is equivalent to "E_x is an R-module for the generic element x of M"; but clearly, for the generic x, $x^*(E \to M) = (E \to M)$. So the notion of vector bundle comes out the same in the naive conception, and in the external one.

(Note that the generic element of M may be seen as the extension of (for instance) the formula $x = x$, where x is a variable ranging over M.)

A comment on the notion of "finite-dimensional vector space over R". There is a global and a sheaf-semantical sense to this. The global sense is that this is an R-module V, for which there exists a linear isomorphism $\phi : R^n \to V$ in $\mathcal{E}$. The sheaf-semantical sense is validity of the assertion "$\exists \phi \in \mathrm{Iso}(R^n, V)$" (where $\mathrm{Iso}(R^n, V)$ is the object in $\mathcal{E}$ of isorphisms, easily constructed using the cartesian closedness of $\mathscr{E}$, and finite inverse limits). The existential quantifier here has to be interpreted according to sheaf semantics, where existence means "localized along a jointly epic family" (cf. Kock, 1981/2006, §II.8). Here, concretely, it comes out to mean that there exists an epic $X \to 1$, and an isomorphism ϕ over X

$$R^n \times X \cong V \times X,$$

fibrewise linear. (If X admits a global element, then such ϕ implies the existence of a global isomorphism $R^n \to V$.)

For the case where the topos is $\mathscr{E}/M$, a finite-dimensional vector space $V \to M$ (in the sheaf-semantical sense) is a vector bundle which trivializes along an epic $X \to M$. This epic is not, in so far as the semantics goes, required to be étale.† When we in some places in the text talk about vector bundles $E \to M$ which are *locally* trivial, then this means that a trivializing epic exists which is in fact étale, in fact, the trivialization is usually along an *atlas* for a *manifold*.

Since we assume that $\mathscr{E}$ is a topos and thus has a subobject classifier Ω, there is for each space M in $\mathscr{E}$ also a "space Ω^M of subspaces of M", and this is relevant for the sheaf semantics of, say, the conclusion in the Frobenius Theorem: "for every $x \in M$, there exists a *leaf* $Q(x)$ through x", since here we have an existential quantifier ranging over the space of subspaces of M. The assertion "$Q(x)$ is itself a *manifold*" is not meaningful, because to say that a space N is a manifold involves a quantifier ("there exists an atlas") ranging over the *open* subspaces of N, and we have not assumed any such "space of *open* subspaces of M". (For any naively described notion of open, like formal open, or Penon open (Penon, 1985), however, the assertion *is* meaningful.)

It does, however, make sense to say that such and such explicitly constructed object is a manifold; for instance, we prove in Section A.6 that the Grassmann spaces, as constructed in Section A.5, are manifolds.

Constructions vs. choices

We deal with entries that are *constructed*; in this sense, we are doing constructive mathematics. One has not made a construction just by making a choice. Constructions are not in general allowed to depend on choices; however, a fundamental principle in mathematics, and hence in naive set theory, is that *if a "construction" is described in terms of choices, but can be proved to be independent of the choices made, then it is a real construction.* This is the crux of

† The notion of étale map (= local homeomorphism) may be derived from the notion of open inclusion in the standard way.

the definition of a *sheaf*, in contrast to a general presheaf: local data can be glued, provided they *match* on overlaps ("matching" being a consequence of "independence of the choice").

Examples of the use of this principle abound in differential geometry; typically, a (local) construction in a manifold may be made by choosing a coordinate chart around each point, and making the construction in terms of this chart, *and* proving that the constructed entry does not depend on the charts chosen.

In sheaf semantics, the justification of this principle hinges on the exactness property of toposes: epimorphism are coequalizers of their kernel pairs.

In our text, we have attempted to minimize the use of this principle, by having constructions which a priori are coordinate free. One notable exception is the description of affine combinations of mutual neighbour points in a manifold, Section 2.1. Another exception is e.g. in the proof of Lemma 7.1.5.

In terms of semantics in sites (which generalizes sheaf semantics), the justification depends on "representable functors are sheaves" for the Grothendieck topology that define the semantics of "existence"; see Kock (1981/2006, §II.8.3) for an exact formulation.

A.3 A simple topos model

We shall here present one of the simplest topos models $(\mathscr{E}, R)$ for the KL axiomatics.† It is a presheaf topos, and actually $\mathscr{E}$ can be proved to be the classifying topos for the theory of commutative rings, with $R \in \mathscr{E}$ the generic commutative ring (cf. e.g. Mac Lane and Moerdijk, 1992, §VIII.5 or Johnstone, 2002, §D.3). We shall not use or exploit the notion of "classifying topos" or "generic ring", but describe the topos $\mathscr{E}$ and the ring object $R \in \mathscr{E}$ directly. To simplify the arguments a little, we shall, instead of commutative rings consider commutative k-algebras, where k is a field, and consider the classifying topos $\mathscr{E}$ for commutative k-algebras, and the generic k-algebra $R \in E$. (The results are also proved in Kock, 1981/2006, but with greater generality, and in a less elementary way.)

Let k be a field. We consider the category k-alg of commutative k-algebras. If $W \in k$-alg is finite dimensional as a vector space, then any choice of a basis $\underline{\varepsilon} = (\varepsilon_1, \dots, \varepsilon_n)$ for W as a vector space defines a linear isomorphism $W \cong k^n$; more generally, for any $A \in k$-alg, we have a linear isomorphism (natural in A)

$$A \otimes W \to A^n; \tag{A.3.1}$$

† It does not satisfy all the other assumptions on R, for instance, it is not a *local* ring.

explicitly, an $F \in A \otimes W$ may uniquely be written $F = \sum_{i=1}^{n} a_i \otimes \varepsilon_i$, and the isomorphism (A.3.1) sends F to $(a_1, \ldots, a_n)$. Conversely, an n-tuple $f = (a_1, \ldots, a_n) \in A^n$ defines an element F in $A \otimes W$, namely $\sum a_i \otimes \varepsilon_i$. We may identify elements of A with algebra homomorphisms $k[T] \to A$, and n-tuples of elements in A with algebra homomorphisms $k[S_1, \ldots, S_n] \to A$. There is a canonical algebra homomorphism $\tilde{\varepsilon} : k[T] \to k[S_1, \ldots, S_n] \otimes W$, corresponding to the element $\sum S_i \otimes \varepsilon_i \in k[S_1, \ldots, S_n] \otimes W$.

Under the identification of elements, or n-tuples of elements, with algebra homomorphisms, the bijective correspondence between Fs and fs may be expressed diagrammatically as follows: for every $F : k[T] \to A \otimes W$, there is a unique $f : k[S_1, \ldots, S_n] \to A$ making the following diagram commutative:

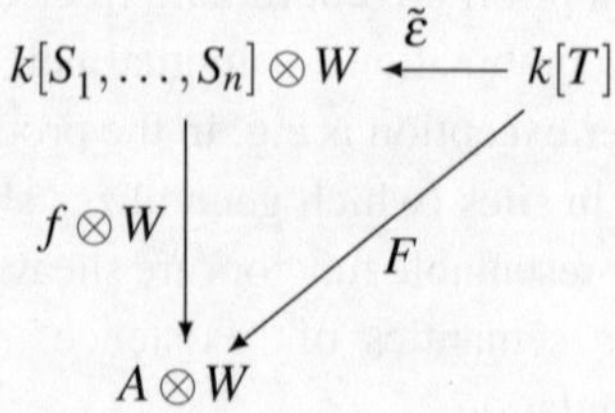

We now express this in the dual category, k-alg^{op}, the category of affine schemes over k. This just means turning the arrows around; $\otimes$, which is the coproduct in k-alg, gets replaced by $\times$ (product); we write $\overline{A}$ for A when considered in this dual category, and similarly for $\overline{W}$. We write R for $\overline{k[T]}$, and since $k[S_1, \ldots, S_n]$ is the coproduct of n copies of $k[T]$, we have $\overline{k[S_1, \ldots, S_n]} = R^n$.

With this change of notation, and of the direction of the arrows, the diagram above reads, in the category of affine schemes,

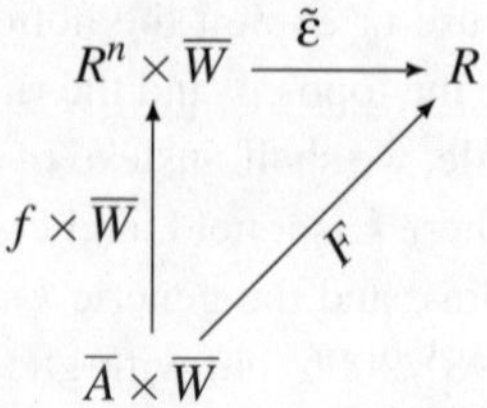

Thus, in the category of affine schemes, the map $\tilde{\varepsilon} : R^n \times \overline{W} \to R$ mediates a bijective correspondence between maps $F : \overline{A} \times \overline{W} \to R$, and maps $f : \overline{A} \to R^n$. This is precisely to say that R^n, by virtue of the map $\tilde{\varepsilon}$, qualifies as exponential object $R^{\overline{W}}$.

Let us record this:

Theorem A.3.1 *In the category of affine schemes over k, every affine scheme* $\overline{W}$ *corresponding to a finite-dimensional k-algebra W is exponentiable; more*

explicitly, every linear basis $\underline{\varepsilon}$ for W provides an isomorphism of affine schemes $R^n \to R^{\overline{W}}$.

We now consider some *small* category of k-algebras, stable under the constructions used above; say, the category of finitely presentable k-algebras. (It contains all finite-dimensional W, and also $k[S_1, \ldots, S_n]$.) Let us denote it k-alg$_0$. Then we have the topos $\mathscr{E}$ of covariant functors from k-alg$_0$ to the category of sets, or equivalently, contravariant functors from the dual category $\mathbf{A}$ (the category of those affine schemes that correspond to algebras in k-alg$_0$). So $\mathscr{E} = \hat{\mathbf{A}}$, the presheaves on $\mathbf{A}$, and we have the Yoneda embedding $y : \mathbf{A} \to \mathscr{E}$. It is well known that a Yoneda embedding preserves those limits that happen to exist, and also preserves those exponential objects that happen to exist. Therefore, the above theorem has as a consequence that any linear basis $\underline{\varepsilon}$ for W provides an isomorphism in $\tilde{\varepsilon}$ in $\hat{\mathbf{A}}$, namely $(y(R))^n \cong y(R)^{y(\overline{W})}$, or, omitting y from the notation,

$$R^n \xrightarrow[\cong]{\tilde{\varepsilon}} R^{\overline{W}}. \tag{A.3.2}$$

Recall from Example A.2.1 that to say "the arrow $f : A \to B$ in $\mathscr{E}$ is invertible" in the naive language is rendered:

$$\forall b \in B \, \exists! a \in A \text{ such that } f(a) = b.$$

Therefore, the invertibility of the arrow $\tilde{\varepsilon}$ exhibited in (A.3.2) may be expressed naively in terms of such an "$\forall \, \exists!$"-assertion:

$$\forall f \in R^{\overline{W}} \, \exists! (a_1, \ldots, a_n) \in R^n \text{ such that } \tilde{\varepsilon}(a_1, \ldots, a_n) = f;$$

this is (see Remark A.2.2) the same naive assertion as

$$\forall f : \overline{W} \to R \, \exists! (a_1, \ldots, a_n) \in R^n \text{ such that } \tilde{\varepsilon}(a_1, \ldots, a_n) = f.$$

This naive assertion can be made more elementary by replacing the map $\tilde{\varepsilon}$ by a naive *description* of it in terms of elements. Such a description is easy to give, but utilizes the ring structure on $R \in \mathscr{E}$. We have not here described this ring structure explicitly; it is of course quite tautological (after all, R is the *generic* k-algebra!). In fact R may be seen as the functor associating to a k-algebra $A \in k$-alg$_0$ the underlying *set* of A, and this set of course has a natural ring structure, since A is a ring. We shall not go into this tautological detail, but note another one: elements ϕ of a k-algebra $W \in k$-alg$_0$ may be identified with arrows $\overline{W} \to R$ in $\mathscr{E}$ (namely, with arrows $k[T] \to W$ in k-alg$_0$). In particular, the assumed basis $(\varepsilon_1, \ldots, \varepsilon_n)$ for W as a vector space over k may be identified with an n-tuple of maps $\varepsilon_i : \overline{W} \to R$. The description of the map $\tilde{\varepsilon} : R^n \to R^{\overline{W}}$

can now be given completely in naive terms as the map with description

$$(a_1,\ldots,a_n) \mapsto [d \in \overline{W} \mapsto \sum_{i=1}^{n} a_i \cdot \varepsilon_i(d)]$$

where d ranges over $\overline{W}$. Thus the fact that the map (A.3.2) is invertible implies the following result:

Theorem A.3.2 *Let $W \in k$-alg_0 be finite dimensional as a vector space over k. Then for any basis $\varepsilon_1,\ldots,\varepsilon_n$ for this vector space, the following naive assertion is valid: for all $f : \overline{W} \to R$, there exist unique $a_1,\ldots,a_n$ in R such that for all $d \in \overline{W}$,*

$$f(d) = \sum_{i=1}^{n} a_i \cdot \varepsilon_i(d).$$

The reader will recognize that all the KL axioms are special cases; for instance, with $W = k[T]/(T^2) = k[\varepsilon]$ (= the ring of dual numbers over k – which is 2-dimensional), one has $\overline{W} = D$, and (choosing the basis 1 and ε for $k[\varepsilon]$), the KL axiom for D follows. The complete KL axiom scheme refers not to all finite dimensional algebras W, but only to those where the nilpotent elements form an ideal of codimension 1; such algebras are often called *Weil algebras* in the SDG literature.

Remark A.3.3 For this topos $\mathscr{E}$, it is easy to describe $M_{(1)} \subseteq M \times M$, whenever M is an affine scheme (i.e. is of the form $\overline{A}$ for a k-algebra A in k-alg$_0$) – even without assuming that M is a manifold. For, then $M_{(1)}$ is itself an affine scheme, namely given by the k-algebra $(A \otimes A)/I^2$, where $I \subseteq A \otimes A$ is the kernel of the multiplication map $A \otimes A \to A$. The elements of $(A \otimes A)/I^2$ may be identified with functions $M_{(1)} \to R$, and the submodule I/I^2 gets identified with those functions $M_{(1)} \to R$ which vanish on the diagonal; i.e. with combinatorial 1-forms. In this sense, the combinatorial R-valued 1-forms on M exactly make up the Kähler differentials on M; see Kock (2003) for an exposition.

The viewpoint of $M_{(1)}$ as the scheme $A \otimes A/I^2$ is the starting point for the theory developed in Breen and Messing (2001).

A.4 Microlinearity

In its full form, to say that a space M is microlinear, cf. Bergeron (1980), is to say that it satisfies a certain axiom *scheme*; if R satisfies the full KL axiom (i.e. R satisfies the axiom with respect to all Weil algebras), then not only R itself, but also any manifold, is automatically microlinear in the full sense. For a formulation of the full axiom scheme, see Definition V.1.1 in Moerdijk and Reyes (1991), or Section 2.3 in Lavendhomme (1996).

Here, we shall only describe those instances of this axiom scheme which we have explicitly used. They *hold* for any manifold M.

Recall the object $D(n) \subseteq R^n$. We have n canonical inclusions $\text{incl}_i : D \to D(n)$, with $\text{incl}_i(d) = (0,\ldots,d,\ldots,0)$ (the d placed in the ith position).

ML 1. *If $t_i : D \to M$ ($i = 1,\ldots,n$) have $t_1(0) = \ldots = t_n(0)$, then there exists a unique $\tau : D(n) \to M$ with $\tau \circ \text{incl}_i = t_i$ for all i.*

The case $n = 2$ is the one that gives addition of tangent vectors on M $(t_1 + t_2)(d) := \tau(d,d)$. Associativity of this addition comes from ML 1 for the case $n = 3$; see Kock (1981/2006, §I.7) (in loc. cit., this property is called *infinitesimal linearity*).

The formulas for inclusions $\text{incl}_i : D \to D(n)$ likewise describe inclusions $\text{incl}_i : D \to D^n$. For the case $n = 2$, we have

ML 2. *If $\tau : D^2 \to M$ has $\tau \circ \text{incl}_1 = \tau \circ \text{incl}_2 = \tau(0,0)$, then there exists a unique $t : D \to M$ such that $\tau(d_1,d_2) = t(d_1 \cdot d_2)$ for all $(d_1,d_2) \in D^2$.*

The construction in (4.9.1) of Lie bracket of vector fields on M depends on ML 2; see Kock (1981/2006, §I.9) where ML 2 is called "Property W" ("W" for "Wraith").

ML 3. *If $f : D(n) \times D(n) \to M$ is symmetric, i.e. $f(u,v) = f(v,u)$, there is a unique $F : D_2(n) \to M$ such that $f(u,v) = F(u+v)$ for all $(u,v) \in D(n) \times D(n)$.*

There is a natural generalization to more than two factors. It is called the "symmetric functions property" in Kock (1981/2006, §I.4).

The strong difference construction considered for manifolds in Remark 2.1.9, in Proposition 4.9.1, and in Section 5.5, is available in any space M which satisfies the following instance of the microlinearity scheme. To formulate it, we need to consider a new infinitesimal object, classically denoted $D^2 \vee D \subseteq D^3$ (Kock and Lavendhomme, 1984; Lavendhomme, 1996, Chapter 3): it is described by

$$\{(d_1,d_2,e) \in D^3 \mid d_1 \cdot e = d_2 \cdot e = 0\}.$$

There are two inclusion maps ϕ and ψ from D^2 into $D^2 \vee D$: $\phi(d_1,d_2) = (d_1,d_2,0)$ and $\psi(d_1,d_2) = (d_1,d_2,d_1 \cdot d_2)$. There is also the inclusion $\varepsilon : D \to D^2 \vee D$ given by $\varepsilon(d) = (0,0,d)$.

ML 4. *If $\tau_i : D^2 \to M$ ($i = 1,2$) agree on $D(2) \subseteq D^2$, there exists a unique $T : D^2 \vee D \to M$ such that*

$$T \circ \phi = \tau_1 \text{ and } T \circ \psi = \tau_2.$$

ML 4'. *If $\tau : D^2 \to M$ and $t : D \to M$ have $\tau(0,0) = t(0)$, there exists a unique $T : D^2 \vee D \to M$ such that*

$$T \circ \phi = \tau \text{ and } T \circ \varepsilon = t.$$

If τ_1 and τ_2 are maps $D \times D \to M$, and if they agree on $D(2) \subseteq D \times D$, the strong difference $\tau_2 \dot{-} \tau_1$ is then constructed as $T \circ \varepsilon$, with T as in ML 4.

A.5 Linear algebra over local rings; Grassmannians

Since the naive way of reasoning only is sound for interpretation in toposes if it proceeds according to intuitionistic/constructive logic (e.g. proceeds without using the law of excluded middle, or proof by contradiction), some standard issues, say of linear algebra, have to be dealt with carefully. Also, the basic ring of the axiomatics of SDG cannot be assumed to be a field in the sense that "for all $x \in R$, either x is 0 or x is invertible".† We develop here the needed fragment of linear algebra over a local ring R, proceeding constructively.

Definition A.5.1 *A commutative ring R is called* local *if whenever $x_1, \ldots, x_n \in R$ are elements in R with $\sum x_i$ invertible, then at least one of the x_is is invertible. (Also $\neg(0 = 1)$ is required.)*

In the following, R is assumed to be a local ring.

Definition A.5.2 *A vector $\underline{x} = (x_1, \ldots, x_n) \in R^n$ is called* proper *if at least one of its entries x_i is invertible.*

Proposition A.5.3 *Let $f : R^n \to R^m$ be a linear map. If $\underline{x} \in R^n$ has the property that $f(\underline{x})$ is proper, then $\underline{x}$ is proper.*

Proof. The linear map f is given by an $m \times n$ matrix $[a_{ij}]$. The ith coordinate of $f(\underline{x})$ is $\sum_j a_{ij}x_j$. For some i, this is invertible, since $f(\underline{x})$ is proper. But if $\sum_j a_{ij}x_j$ is invertible, one of the terms, say $a_{ij}x_j$ is invertible, by localness of R. But this implies that x_j is invertible, so we conclude that $\underline{x}$ is proper.

Corollary A.5.4 *If $g : R^n \to R^n$ is a linear isomorphism, and $\underline{x} \in R^n$ is proper, then so is $g(\underline{x})$.*

† However, R may be assumed to be a field in the sense that "for all $x \in R$, if $x \neq 0$, then x is invertible"; these two field notions are not equivalent in constructive logic.

Proof. Apply the proposition with $f = g^{-1}$.

It follows that the notion of properness of vectors is invariant under the group $GL(n,R)$, and so makes sense for any n-dimensional vector space (= R-module isomorphic to R^n).

An n-tuple of vectors $v_1,\ldots,v_n$ in a vector space (R-module) V gives rise to a linear map $R^n \to V$, namely $(t_1,\ldots,t_n) \mapsto \sum_i t_i \cdot v_i$. The image of this map is the *span* of the vectors $v_1,\ldots,v_n$, and the n-tuple is called *linearly independent* if the map is injective. The n-tuple is called a *basis* for V if the map is an isomorphism, i.e. if the vectors are independent and span V. Clearly, a vector space is finite dimensional iff it admits some basis.

Proposition A.5.5 *Assume $v \in V$ is a member of an (n-element) basis for V (so V is necessarily finite dimensional). Then v is a proper vector in V.*

Proof. It suffices to see this for the case where $V = R^n$. Then the assumed basis of which v is a member (say, the *first* member) defines a linear isomorphism $R^n \to R^n$, equivalently, an invertible matrix $[a_{ij}]$, and its first column is v. So it suffices to see that in an invertible $n \times n$ matrix, the first column is a proper vector in R^n (in fact, every column is proper). For since $[a_{ij}]$ is an invertible $n \times n$-matrix, its determinant is invertible, i.e. (with σ ranging over the $n!$ permutations of $\{1,\ldots n\}$) $\sum_\sigma \pm \Pi_{j=1}^n a_{\sigma(j)j}$ is invertible. By localness of R, one of the terms in this sum, say the one corresponding to the permutation σ, is invertible. But if the product $\Pi_{j=1}^n a_{\sigma(j)j}$ is invertible, then so are all its factors, in particular the factor corresponding to $j = 1$; so $a_{\sigma(1)1}$ is invertible, proving that the first column in $[a_{ij}]$ is a proper vector in R^n.

Let V be a vector space equipped with a bilinear form $\langle -,- \rangle : V \times V \to R$.

Definition A.5.6 *An n-tuple of vectors $v_1,\ldots,v_n$ in V is called* orthogonal *if $\langle v_i, v_j \rangle = 0$ for $i \neq j$, and is invertible for $i = j$. If furthermore, more particularly, $\langle v_i, v_i \rangle = 1$ for all i, the n-tuple is called* orthonormal.

It is clear that to say $v_1,\ldots,v_n$ in V is an orthonormal basis in V is tantamount to saying that the linear isomorphism $R^n \to V$ to which it gives rise is an *isometry*, i.e. that it takes the standard bilinear form $\sum x_i y_i$ on R^n to the given bilinear form on V.

Definition A.5.7 *A vector v such that $\langle v, v \rangle = 1$ is called a* unit vector. *To* normalize *a vector v means to find a scalar $\lambda \in R$ such that $\lambda \cdot v$ is a unit vector.*

Unit vectors, more generally, vectors that can be normalized, are necessarily proper. It does not follow, conversely, that all proper vectors can be normalized. It is easy to see that if v can be normalized, then $\langle v,v\rangle$ is invertible, in fact has an invertible square root. (A *square root* of an element $b \in R$ is an element c so that $c^2 = b$, of course.)

Proposition A.5.8 *If $v_1,\ldots,v_n$ is an orthogonal set in V, then it is linearly independent.*

Proof. Consider the linear map $R^n \to V$ given, as above, by $(t_1,\ldots,t_n) \mapsto \sum t_i \cdot v_i$. To see that it is injective, it is enough to see that if $\sum t_i v_i = 0$, then all the t_is are 0. To see for instance that $t_1 = 0$, we note

$$0 = \langle v_1, 0\rangle = \langle v_1, \sum t_i v_i\rangle$$

$$= \sum t_i \langle v_1, v_i\rangle = t_1\langle v_1, v_1\rangle,$$

the last using the orthogonality condition for the v_is. Since $\langle v_1, v_1\rangle$ is invertible, we conclude $t_1 = 0$.

A local ring R is called a *formally real local ring* if it satisfies: whenever one of $x_1,\ldots,x_n$ is invertible, then the square sum $\sum x_i^2$ is invertible.

It is easy to see that for a formally real local ring R, the subset R_+ of R, consisting of all invertible elements of the form $\sum x_i^2$ is stable under addition and multiplication in R. Also $1,2,3,\ldots$ all belong to R_+ and hence are invertible, so that a formally real local ring automatically is a $\mathbb{Q}$-algebra.

Exercise A.5.9 Let R be a formally real local ring. Let $C = R \times R$ be equipped with the multiplication $(a,b)\cdot(a',b') := (aa' - bb', ab' + a'b)$ (so "$C = R[i]$", the "*complex numbers* over R"). Then C is local, but not formally real.

Exercise A.5.10 Let $a \in R$ be an invertible element in a local ring R. If both a and $-a$ have square roots, then R is not formally real.

Exercise A.5.11 Let R be a local ring in which 2 is invertible. (1) If b is an invertible element in R, then either $b+c$ or $b-c$ is invertible. (2) Assume $a \in R$ is invertible, and that $b^2 = c^2 = a$. Then either $b = c$ or $b = -c$.

So an invertible element in a local ring (in which 2 is invertible) can have at most two square roots.

Definition A.5.12 *An* inner product space *is a finite-dimensional vector space V, equipped with a symmetric bilinear $\langle -,-\rangle : V \times V \to R$, such that for every proper vector $v \in V$, $\langle v,v\rangle$ is invertible and has a square root.*

The following terminology is from Kock (1981/2006, §III.2). The motivation for the name is that the length of the hypotenuse in a right triangle by Pythagoras Theorem is calculated as $\sqrt{a^2+b^2}$.

Definition A.5.13 *A commutative ring R is called* Pythagorean *if it is formally real local, and if all invertible square sums have square roots.*

In a Pythagorean ring, one says that an element $a \in R$ is *positive* if it is invertible and has a square root (which then necessarily is invertible). It is easy to prove that the set of positive elements in R is stable under addition, multiplication, and multiplicative inversion.

If R is Pythagorean, then R^n (for each n), with its standard bilinear form $\underline{x}, \underline{y} \mapsto \sum x_i y_i$, is an inner-product space.

Assuming that R is Pythagorean:

Theorem A.5.14 (Gram–Schmidt) *If V is an n-dimensional inner product space, then V is isometric to R^n (R^n with its standard bilinear form).*

Proof. An isometry $R^n \to V$ is tantamount to an orthonormal basis in V. Now any basis in V can by the Gram–Schmidt process be converted into an orthogonal basis; this only uses that for proper vectors $v \in V$, $\langle v,v\rangle$ is invertible. This orthogonal basis can then be normalized, using scalars of the form $\sqrt{\langle v,v\rangle}$, and the existence of such square roots is part of the assumption that V is an inner-product space.

Remark A.5.15 A symmetric bilinear form $V \times V \to R$ making V into an inner product space is often said to be *positive definite*. We have refrained from considering *positivity* as an extra structure on the ring R. In case there is a (strict) positivity notion "> 0" on R, with all positive elements invertible and possessing square roots, then one may replace the notion of inner-product space V, as given here, by the assumption that $\langle a,a\rangle > 0$ for all proper $a \in V$.

Such a positivity notion can be obtained from the algebraic structure of R, if R (assumed formally real local) satisfies: for every invertible $a \in R$, either a or $-a$ has a square root. (They cannot both have square roots, since R is formally real.) The theory of formally real local rings with this property is a geometric and ε-stable theory, in the sense of Kock (1981/2006, §III.2).

Some classical manifolds

We consider here just the Stiefel and Grassmann manifolds. Until we have indicated in which sense they are manifolds, we call them just Stiefel and Grassmann *spaces*.

Let R be a local ring. Let $1 \leq k \leq n$ be integers, and let p denote the binomial coefficient (n,k). There is a canonical map $g : R^{k \cdot n} \to R^p$; it associates to $\underline{x} \in R^{k \cdot n}$ the p-tuple of $k \times k$ subdeterminants of $\underline{x}$ (where we view elements in $R^{k \cdot n}$ as $k \times n$ matrices). The *Stiefel space* $V(k,n)$ is the set of those $\underline{x} \in R^{k \cdot n}$ such that $g(\underline{x})$ is a proper vector. To say that $g(\underline{x})$ is a proper vector is equivalent to saying that at least one of the $k \times k$ subdeterminants of $\underline{x}$ is invertible. For H a k-element subset of $\{1,\ldots,n\}$, we let $Q_H \subseteq V(k,n)$ be the subset consisting of those $\underline{x}$ such that the subdeterminant corresponding to H is invertible.

The set $V(k,k)$ equals $GL(k)$, the group of invertible $k \times k$ matrices over R. This group acts on the left on $V(k,n)$ by matrix multiplication. The action restricts to an action on each Q_H. These actions are free. From exactness properties of the category of sets (or in fact of any exact category, and in particular, of any topos), it follows that we, for each H, have the following diagram with exact rows (kernel pair/coequalizer) and with the two left-hand squares both being pull backs:

$$\begin{array}{ccccc} GL(k) \times Q_H & \rightrightarrows & Q_H & \longrightarrow & Q_H/GL(k) \\ \downarrow & & \downarrow & & \downarrow \\ GL(k) \times V(k,n) & \rightrightarrows & V(k,n) & \longrightarrow & V(k,n)/GL(k) \end{array} \qquad \text{(A.5.1)}$$

where the parallel arrows in the top row, as well as in the bottom row, are projection and $GL(k)$-action, respectively.

Because of one of the more sophisticated exactness properties of exact categories (see e.g. Barr, 1971, p. 73), it follows that the right-hand square is a pull back as well; and because the middle vertical map is monic, then so is the right-hand vertical map.

The set $V(k,n)/GL(k)$ is the *Grassmann space* of k-planes in R^n. It is easy to see that the sets $Q_H/GL(k)$ may in fact be identified with $R^{k \cdot (n-k)}$; if for instance $H = \{1,\ldots,k\}$, a $k \times (n-k)$ matrix $\underline{y}$ gives rise to a matrix $\underline{x}$ in Q_H, namely by concatenating the identity $k \times k$ matrix and $\underline{y}$, and modulo the $GL(k)$-action, every matrix in Q_H comes about this way.

A.6 Topology

The very definition of what a manifold is requires a notion of *open* inclusion map. We do not want to describe when an individual inclusion map is open, but rather describe how the *class* of open maps should behave inside the category (topos) $\mathcal{E}$ in which we work. This behaviour may be described axiomatically, by requiring suitable stability properties of the class. One such axiomatization, due to Joyal, axiomatizes rather a class of (not necessarily monic) maps called *étale* maps; the *open inclusions* are then defined to be the monic étale maps. The Joyal axiomatization is reproduced in Kock (1981/2006, §I.19 (i)–(vii)); axiom (ii) says for instance that the pull back of an étale map along any map is étale.

A class satisfying these axioms is called an *étaleness notion* or an *openness notion*.

Under the assumption that all infinitesimal spaces D' are atoms, one can in naive terms define a notion (cf. Kock, 1981/2006, §I.17) of *formally étale map*, which satisfies the Joyal axioms, cf. Kock (1981/2006, §I.19). The *formally open* maps are then defined as the monic formally étale maps.

We shall not elaborate much on these issues, but refer to Kock (1981/2006, §I.17 and §I.19); we shall, however, show that the Grassmannian spaces, as described in Section A.5, are manifolds, for the Zariski notion to be described. We shall need the following properties which any étaleness notion has. Let $\mathcal{R}$ denote the class of étale maps. Then

(i) $\mathcal{R}$ is closed under composition and contains all isomorphisms.

(ii) $\mathcal{R}$ is stable under pull backs.

(iii) $\mathcal{R}$ is stable under descent along epics, meaning that if we have a pull-back

$$\begin{array}{ccc} \cdot & \longrightarrow & \cdot \\ {\scriptstyle v}\downarrow & & \downarrow{\scriptstyle u} \\ \cdot & \underset{g}{\longrightarrow} & \cdot \end{array}$$

and g is epic, then $v \in \mathcal{R}$ implies $u \in \mathcal{R}$.

(iv) If $U_i \subseteq V$ is a family of open inclusions, then their union $\bigcup U_i \subseteq V$ is open.

(The properties (i)–(iii) are part of the axioms for étaleness notions; (iv) is a consequence.)

There is a smallest étaleness notion $\mathcal{R}$ for which $Inv(R) \subseteq R$ is open (where $Inv(R) = R^* \subseteq R$ is the space of elements having a multiplicative inverse). It is called the *Zariski* notion. If $U \subseteq V$ is open in the Zariski sense, it is also formally open.

Then it follows that the following inclusions, used in the construction of the Grassmann spaces, are Zariski open: first, $Inv(R) \subseteq R$ is open. Second, for each $i = 1, \ldots, p$, the subspace of R^p, consisting of $(x_1, \ldots, x_p) \in R^p$ with x_i invertible, is open, since it comes about from $Inv(R)$ by pulling back along $\text{proj}_i : R^p \to R$. Third, the subset of R^p, consisting of proper vectors, is open, since it is a union of the p open subsets of R^p just considered. Fourth, the Stiefel space $V(k,n) \subseteq R^{k \cdot n}$ is open, since it comes about by pulling the set of proper vectors in R^p back along the map $g : R^{k \cdot n} \to R^p$ (where $p = (k,n)$).

Being an open subset of $R^{k \cdot n}$, the Stiefel space $V(k,n)$ is therefore a manifold.

For the Grassmann space, we first note that $Q_H \subseteq V(k,n)$ is open: it comes by pulling $Inv(R)$ back along the projection corresponding to $H \in (n,k)$. Since the right-hand square in (A.5.1) is a pull back and $V(k,n) \to V(k,n)/GL(k)$ is epic, it follows from the descent axiom (iii) that the monic $Q_H/GL(k) \to V(k,n)/GL(k)$ is open. Since the union of the Q_Hs is $V(k,n)$ it follows that the union of the $Q_H/GL(k)$ is $V(k,n)/GL(k)$. Finally, we have bijections $Q_H/GL(k) \cong R^{k \cdot (n-k)}$. So they provide a (finite) atlas showing that $V(k,n)/GL(k)$ is a manifold of dimension $k \cdot (n-k)$.

We need to have a notion of when a space is *connected*. For definiteness, we use this in the sense of path connected. To define this, we define the notion of a *piecewise path* in a space M.

Definition A.6.1 *A* piecewise path $\underline{\gamma}$ *in* M *is a finite sequence of paths* $\gamma_i : R \to M$ *(*$i = 1, \ldots, n$*), such that*

$$\gamma_1(1) = \gamma_2(0), \gamma_2(1) = \gamma_3(0), \ldots, \gamma_{n-1}(1) = \gamma_n(0).$$

We say that such $\underline{\gamma}$ *is a piecewise path* from *the point* $\gamma_1(0)$ to *the point* $\gamma_n(1)$, *or that* $\underline{\gamma}$ connects *these two points.*

Definition A.6.2 *A space is* path connected *if any two points in it can be connected by a piecewise path.*

It is clear that R itself is path connected.

Remark A.6.3 The notion of path connected is formulated in the naive language. To the extent that the notion of open inclusion is also naively formulated, one has a naive topological notion of connectedness of M: "any equivalence relation on M with *open* equivalence classes is trivial". For the (naive) notion of *formally* open, the line R is not connected in this topological sense: the relation "$x \equiv y$ if $x - y$ is nilpotent" has formally open equivalence classes, none of which exhaust R. (In fact, the equivalence classes for $\equiv$ are the cosets of the subgroup D_∞ of $(R, +)$; this subgroup is formally open in R.)

If the openness notion is such that R is connected in the topological sense, then any path connected space is connected in the topological sense.

From (path) connectedness of R, path connectedness of many other spaces easily follows. We list a few:

Proposition A.6.4 *The space R^n is connected. The space $D(n) \subseteq R^n$ is connected. If $x \in M$ (M a manifold), the monad $\mathfrak{M}(x)$ is connected. Any linear subset of a monad is connected.*

(Note that any linear subset of any monad in a manifold is isomorphic to some $D(n)$.)

A map $f : M \to N$ between manifolds is a *submersion* at $x \in M$ if f maps $\mathfrak{M}(x) \subseteq M$ surjectively to $\mathfrak{M}(f(x)) \subseteq N$. And f is a submersion if it is so at every $x \in M$.

The following is a non-trivial "infinitesimal-to-local" result, whose validity depends on the notion of "open", and, of course, on the model under consideration.

Theorem A.6.5 (Open Image Theorem) *The image of a submersion $f : M \to N$ between manifolds is an open subset of N.*

(At a certain point, we need a slightly stronger notion: we say that $f : M \to N$ is a submersion at $x \in M$ in the *strong* sense if for any k and for any infinitesimal k-simplex in N, say $(y_0, \ldots, y_k)$ with $y_0 = f(x)$, there exists an infinitesimal k-simplex in M with first vertex x, and mapping by f to the given k-simplex in N. The case $k = 1$ is the case considered previously. Under the assumption of a suitable implicit function theorem, the two notions of submersion are equivalent.)

A *Lie group* is a manifold with a group structure. The following is now a standard argument (assuming that path connected implies topologically connected):

Proposition A.6.6 *Let $i : H \to G$ be a group homomorphism between Lie groups, and assume that G is path connected. If i is a submersion at the neutral element $e \in H$, then i is surjective.*

Proof. From the algebraic properties of i, it is easy to see that i is in fact a submersion at every $x \in H$. Thus by the open image theorem, its image is an open subset of G, and from the algebraic properties of i, this image is a subgroup of G. So $i(H) \subseteq G$ is an open subgroup. Now a subgroup of a group G defines an equivalence relation on G whose equivalence classes are the cosets of the subgroup. In the present case, the openness of $i(H)$ implies the openness of all its cosets. By the connectedness of G, one of these cosets is all of G, and this implies that $i(H)$ itself is all of G.

Corollary A.6.7 (Linear Sufficiency Principle) *Let $H \subseteq G$ be a Lie group, subgroup of a path connected Lie group G, and assume that $\mathfrak{M}_G(e) = \mathfrak{M}_H(e)$. Then $H = G$.*

Proof. The assumption about the monads implies that the inclusion map $H \to G$ is a submersion at e.

A.7 Polynomial maps

Let V and W be R-modules (with R a commutative ring). Then we have the notion of *linear* (= R-linear) map, *bilinear* map $V \times V \to W$ (= R-bilinear), ..., k-linear $V^k \to W$, etc.; 0-linear means “constant”. We also have the notion of k-linear *symmetric* map.

Let $Lin_k(V,W)$ denote the R-module of k-linear maps $V^k \to W$, and $SLin_k(V,W)$ the submodule of k-linear symmetric maps. If R contains the field $\mathbb{Q}$ of rational numbers, which we shall henceforth assume, this submodule is a retract, i.e. there is a canonical symmetrization procedure which to a k-linear $\phi : V^k \to W$ associates $\frac{1}{k!}\sum_\sigma \phi \circ \sigma$, where σ ranges over the set of permutations of $\{1,\ldots,k\}$, and also denotes the resulting $V^k \to V^k$.

A k-linear $\phi : V^k \to W$ may be diagonalized into a map $f : V \to W$,

$$f(v) := \phi(v,\ldots,v).$$

The diagonalization of a k-linear ϕ agrees with the diagonalization of its symmetrization.

We call a map $f : V \to W$ *homogeneous of degree k* if it appears as the diagonalization of some k-linear $V^k \to W$, or equivalently, if it appears as the diagonalization of a k-linear symmetric $V^k \to W$. Sometimes, we say *homogeneous in the strong sense*, to distinguish it from Euler homogeneity; clearly if f is k-homogeneous in the strong sense, then f has the Euler homogeneity property: $f(t \cdot x) = t^k \cdot f(x)$ for all $t \in R$ and $x \in V$; the converse is not true without some further assumptions. Theorem 1.4.1 is a converse, for $k = 1$.

It is clear that if $f : V \to W_1$ and $g : V \to W_2$ are homogeneous of degree p and q, respectively, and if $* : W_1 \times W_2 \to W_3$ is a bilinear map, then $f * g : V \to W_3$ is homogeneous of degree $p+q$. For if $F : V^p \to W_1$ is a p-linear map which witnesses p-homogeneity of f, and similarly $G : V^q \to W_2$ witnesses q-homogeneity of g, then the $p+q$-linear map $V^{p+q} \to W_3$ given by

$$(v_1, \ldots, v_{p+q}) \mapsto F(v_1, \ldots, v_p) * G(v_{p+1}, \ldots, g_{p+q})$$

diagonalizes to $f * g$.

Thus, we have surjective linear maps

$$Lin^k(V,W) \xrightarrow{\text{symm}} SymLin^k(V,W) \xrightarrow{\delta} Homg^k(V,W), \tag{A.7.1}$$

where $Lin^k(V,W)$ denotes the module of k-linear maps, $SymLin^k$ the module of k-linear symmetric maps, $Homg^k(V,W)$ the module of k-homogeneous maps, and where symm and δ denote symmetrization and diagonalization, respectively; the maps exhibited are linear.

We have in particular a surjective linear map

$$SymLin^k(V,R) \xrightarrow{\delta} Homg^k(V,R). \tag{A.7.2}$$

We say that the commutative ring R is *polynomially faithful* if for every $k = 1, 2, \ldots$, it is the case that if $a_0, a_1, \ldots, a_k \in R$ are so that the function $R \to R$ given by $t \mapsto a_0 + a_1 \cdot t + \ldots + a_k \cdot t^k$ is constant 0, then all the a_is are 0. In standard commutative algebra, a sufficient condition that R be polynomially faithful is that R is an infinite integral domain. The ring R studied in the present book is not an integral domain, since it has a rich supply of nilpotent elements. However, as we noted in Section 1.3, the KL axiom for the D_ks implies that R is polynomially faithful.

Theorem A.7.1 *Assume that R is polynomially faithful. Then the diagonalization map exhibited in (A.7.2) is bijective.*

Proof. By definition of $Homg^k$, δ is surjective. To see that it is injective, it suffices to see that its kernel is 0, i.e. we should prove that if a k-linear symmetric $\phi : V^k \to R$ diagonalizes to the zero map, then ϕ itself is 0. This comes by putting $p = k$ in the following lemma.†

Lemma A.7.2 *For every non-negative integer p, we have validity of the following assertion: for any p-linear symmetric $\phi : V^p \to R$, if ϕ diagonalizes to 0, then ϕ itself is 0.*

† Essentially from Greub (1978, §9.7).

Proof. This is by induction in p. The assertion is clearly valid if $p = 0$ or $p = 1$. Assume that it is valid for $p-1$ ($p \geq 2$). Let ϕ be p-linear and symmetric, and diagonalize to 0. Then for all u and v in V, and for all $t \in R$, we have

$$\phi(u+t\cdot v,\ldots,u+t\cdot v) = 0.$$

We may expand by the binomial formula, using that ϕ is p-linear and symmetric, and we get (with the (p,k) denoting binomial coefficients)

$$\sum_{k=0}^{p} t^k \cdot (p,k)\cdot \phi(u,\ldots,u,v,\ldots,v) = 0,$$

where u is written $p-k$ times and v is written k times. For fixed u and v, the left-hand side is a polynomial R-valued function of $t \in R$. But coefficients in polynomial functions $R \to R$ are unique, by polynomial faithfulness, so all the $p+1$ coefficients are 0; we conclude for each k that $\phi(u,\ldots,u,v,\ldots,v) = 0$, in particular, for $k = 1$:

$$\phi(u,\ldots,u,v) = 0, \tag{A.7.3}$$

with u written $p-1$ times. Now for fixed $v \in V$, the function $V^{p-1} \to R$ given by $\phi(-,\ldots,-,v)$ is symmetric $p-1$-linear, and diagonalizes to 0 by (A.7.3), hence by induction hypothesis,

$$\phi(u_1,\ldots,u_{p-1},v) = 0$$

for all $u_1,\ldots,u_{p-1}$. Since v was arbitrary, we conclude that ϕ itself is the zero map $V^p \to R$. This gives the induction step, and hence the lemma is proved.

An R-module V is called a *finite-dimensional vector space over R* if it is linearly isomorphic to some R^n. If V is finite dimensional, any linear $g : V \to W$ is of the form $g(x) = \sum_j \gamma_j(x)\cdot w_j$ with $\gamma_j : V \to R$ linear, and $w_j \in W$. For, V is free on some finite basis, and then the γ_j may be taken to be the elements in the dual basis. Succinctly,

$$Lin(V,W) \cong Lin(V,R) \otimes_R W,$$

where $Lin(V,W)$ denotes the R-module of R-linear maps from V to W, and similarly for $Lin(V,R)$.

More generally:

Proposition A.7.3 *Assume that V is a finite-dimensional vector space. Then a map $f : V \to W$ is homogeneous of degree k iff it can be written $f(x) = \sum_j \psi_j(x)\cdot w_j$ (finite sum) with each $\psi_j : V \to R$ homogeneous of degree k, and $w_j \in W$.*

Proof. This follows because k-linear maps out of V^k are given by linear maps out of $\otimes^k V$ (k-fold tensor product), and because $\otimes^k V$ is a finite-dimensional vector space, hence a free R-module on a finite set. Succinctly:

$$Homg^k(V,W) \cong Homg^k(V,R) \otimes_R W.$$

A polynomial map $V \to W$ (where V is finite dimensional) can be written uniquely as a sum of its homogeneous components. For polynomial maps $V \to R$, this follows by an explicit description of polynomial maps $R^n \to R$ in terms of formal polynomials. For polynomial maps $V \to W$, it follows from the case $W = R$, by the sequence of natural isomorphisms

$$\begin{aligned} Pol_{\leq k}(V,W) &\cong Pol_{\leq k}(V,R) \otimes_R W \\ &\cong \left(\oplus_i Homg_i(V,R)\right) \otimes_R W \\ &\cong \oplus_i \left(Homg_i(V,R) \otimes_R W\right) \\ &\cong \oplus_i Homg_i(V,W). \end{aligned}$$

A.8 The complex of singular cubes

In this section, we place ourselves in the context of smooth manifolds, to be specific (everything applies equally well in the context of topological spaces, say). So all maps mentioned are smooth. The content is probably not new; it is formulated in completely classical terms.

For a manifold M, we consider "singular k-cubes in M"; by this is usually meant maps $I^k \to M$, where $I = \{x \in R \mid 0 \leq x \leq 1\}$; however, to simplify things, we do not want to consider a partial order $\leq$ on the number line R (hence there is no such thing as the unit interval I); we prefer to define singular cubes as maps $R^k \to M$. The set of these is denoted $S_{[k]}(M)$.

As k ranges, the sets $S_{[k]}(M)$ form a cubical complex, which we denote $S_{[\bullet]}(M)$. It has face and degeneracy maps; it has a symmetry structure: the symmetric group in k letters acts on $S_{[k]}(M)$, by an action induced by permutation of the coordinates of R^k; it also has a "reversion" structure (Grandis and Mauri, 2003), which we shall not need here (although it is important in Kock, 2008). Finally, it has a subdivision structure. All the structures mentioned here are induced by certain affine maps $R^k \to R^l$, and this is crucial for the applications here.

The face maps $\partial_i^\alpha : S_{[k]}(M) \to S_{[k-1]}(M)$ ($\alpha = 0$ or 1, $i = 1,\ldots,k$) are induced by the affine maps $\delta_i^\alpha : R^{k-1} \to R^k$ given by

$$\delta_i^\alpha(t_1,\ldots,t_{k-1}) = (t_1,\ldots,\alpha,\ldots,t_{k-1})$$

with α inserted as ith coordinate. Then for a singular $\gamma : R^k \to M$, $\partial_i^\alpha(\gamma) := \gamma \circ \delta_i^\alpha$. The degeneracy operators are similarly precomposition by the projection maps $R^k \to R^{k-1}$; we don't need to be more specific here.

We do, however, need to be specific about the notion of subdivision. First, for any $a, b \in R$, we have an affine map $R \to R$, denoted $[a,b] : R \to R$; it is the unique affine map sending 0 to a and 1 to b, and is given by $t \mapsto (1-t)\cdot a + t \cdot b$. Precomposition with it, $\gamma \mapsto \gamma \circ [a,b]$ defines an operator $S_{[1]}(M) \to S_{[1]}(M)$ which we denote $\mid [a,b]$, and we let it operate on the right, thus

$$\gamma \mid [a,b] := \gamma \circ [a,b].$$

Heuristically, it represents the restriction of γ to the interval $[a,b]$. Note that $\gamma = \gamma \mid [0,1]$, since $[0,1] : R \to R$ is the identity map.

More generally, for $a_i, b_i \in R$ for $i = 1, \ldots, k$, we have the map

$$[a_1, b_1] \times \cdots \times [a_k, b_k] : R^k \to R^k.$$

It induces by precomposition an operator $S_{[k]}(M) \to S_{[k]}(M)$, which we similarly denote $\gamma \mapsto \gamma \mid [a_1, b_1] \times \ldots \times [a_k, b_k]$.

Given an index $i = 1, \ldots, k$, and $a, b \in R$, we use the abbreviated notation $[a,b]_i$ for the map $[0,1] \times \cdots \times [a,b] \times \cdots \times [0,1]$ with the $[a,b]$ appearing in the ith position; the corresponding operator is denoted $\gamma \mapsto \gamma \mid_i [a,b]$. Given a, b and $c \in R$, and an index $i = 1, \ldots, k$, then we say that the ordered pair

$$\gamma \mid_i [a,b], \quad \gamma \mid_i [b,c]$$

form a subdivision of $\gamma \mid_i [a,c]$ *in the ith direction.*

There are compatibilities between the subdivision relation and the face maps; we shall record some of these relations. First, let us note that the maps $\delta_i^\alpha : R^{k-1} \to R^k$ considered above for $\alpha = 0$ or $= 1$ may be similarly defined for any $\alpha = a \in R$; namely

$$\delta_i^a(t_1, \ldots, t_{k-1}) = (t_1, \ldots, t_{i-1}, a, t_i, \ldots, t_{k-1}).$$

The corresponding $S_{[k]}(M) \to S_{[k-1]}(M)$ we denote of course ∂_i^a. It is easy to see that we have $[a,b]_i \circ \delta_i^0 = \delta_i^a$, and similarly $[a,b]_i \circ \delta_i^1 = \delta_i^b$, from which follows, for any $\gamma \in S_{[k]}(M)$, that

$$\partial_i^0(\gamma \mid_i [a,b]) = \partial_i^a(\gamma) \text{ and } \partial_i^1(\gamma \mid_i [a,b]) = \partial_i^b(\gamma). \tag{A.8.1}$$

Also, for $\alpha = 0, 1$ (in fact for every $\alpha = a \in R$) $[a,b]_i \circ \delta_j^\alpha = \delta_j^\alpha \circ [a,b]_i$ if $i < j$ and $= \delta_j^\alpha \circ [a,b]_{i-1}$ if $i > j$, and from this follows, for any $\gamma \in S_{[k]}(M)$, that

$$\partial_j^\alpha(\gamma \mid_i [a,b]) = (\partial_j^\alpha(\gamma)) \mid_i [a,b] \text{ for } i < j \tag{A.8.2}$$

and $= (\partial_j^\alpha(\gamma)) \mid_{i-1} [a,b]$ for $i > j$.

Recall that an *affine combination* in a vector space is a linear combination where the sum of the coefficients is 1. An affine space is a set E where one may form affine combinations, and where these combinations satisfy the same equations as those that are valid for affine combinations in vector spaces. An affine map is a map preserving affine combinations. The vector space R^n is a free affine space on $n+1$ generators. More concretely, given an $n+1$-tuple of points $(x_0, x_1, \ldots, x_n)$ in an affine space E, there is a unique affine map $R^n \to E$, which we denote $[x_0, x_1, \ldots, x_n]$ with $0 \mapsto x_0$ and $e_i \mapsto x_i$ for $i > 0$, where e_j is the jth canonical basis vector $e_j \in R^n$.

This map $[x_0, x_1, \ldots, x_n]$ is given by

$$(t_1, \ldots, t_n) \mapsto (1 - \sum t_i)x_0 + t_1 x_1 + \ldots + t_n x_n. \tag{A.8.3}$$

(Recall from Section 2.1 that if the x_is are mutual neighbours in a manifold, then the affine combination used here likewise makes sense; and the map $R^n \to M$ defined by it is denoted the same way, and has similar properties.)

An affine map between vector spaces is of the form: a constant plus a linear map. This allows us to have a matrix calculus for affine maps between the coordinate vector spaces R^n. Recall that a linear map $f : R^n \to R^m$ is given by an $m \times n$ matrix, and that composition of maps corresponds to matrix multiplication. The jth column $a_j \in R^m$ of such a matrix A is $f(e_j)$ $(e_j \in R^n)$.

An affine map $R^n \to R^m$ may be given by an $m \times (1 \times n)$ matrix, where the first column $a_0 \in R^m$ denotes the constant, and the remaining $m \times n$ matrix A is the $m \times n$ matrix of the linear map. We display this "augmented" matrix in the form $\|a_0 \mid A\|$ or $\|a_0 \mid a_1, \ldots, a_n\|$ where as before a_j $(j = 1, \ldots, n)$ is the jth column of A.

With this notation, composition of affine maps corresponds to "semidirect matrix multiplication":

$$\|a_0 \mid A\| \cdot \|b_0 \mid B| = \|a_0 + A \cdot b_0 \mid A \cdot B\|.$$

For $E = R^m$, the affine map $[x_0, \ldots, x_n] : R^n \to R^m$ considered above has as augmented matrix the matrix $\|x_0 \mid x_1 - x_0, \ldots, x_n - x_0\|$, and conversely, the augmented matrix $\|x_0 \mid a_1, \ldots, a_n\|$ defines the affine map $[x_0, x_0 + a_1, \ldots, x_0 + a_n]$.

Let us also give the augmented matrices for the affine maps δ_i^α that were used for defining the cubical face maps $\delta_i^\alpha : R^{k-1} \to R^k$, $\alpha = 0$ or 1, $i = 1, \ldots, k$:

$$\delta_i^0 = \|0 \mid e_1, \ldots, \widehat{i}, \ldots, e_k\| \tag{A.8.4}$$

(the e_js here are the canonical basis vectors of R^k), and

$$\delta_i^1 = \|e_i \mid e_1, \dots, \widehat{i}, \dots, e_k\|. \tag{A.8.5}$$

With the matrix calculus for augmented matrices, we can calculate the cubical faces of a singular k-cube in R^n of the form $[x_0, x_1, \dots, x_k]$. We may verify the equations (2.8.1) and (2.8.2),

$$\partial_i^0([x_0, x_1, \dots, x_k]) = [x_0, x_1, \dots, \widehat{x}_i, \dots, x_k]$$

and

$$\partial_i^1([x_0, x_1, \dots, x_k]) = [x_i, x_1 - x_0 + x_i, \dots, \widehat{i}, \dots, x_k - x_0 + x_i].$$

(Note that the entries like $x_k - x_0 + x_i$ are affine combinations, so they also make sense for points in an affine space E, and, in fact, if the points x_i are mutual neighbours, they make sense in a manifold; $x_k - x_0 + x_i$ is the fourth vertex (opposite x_0) of a parallelogram whose three other vertices are x_0, x_i, and x_k.)

Among the affine maps $[x_0, x_1, \dots, x_k]$ from R^k to R^k are the "axis-parallel rectangular boxes", for short: *rectangles*; they are those where $x_i - x_0$ ("the ith side") is of the form $t_i \cdot e_i$ where e_i is the ith canonical basis vector. In matrix terms, these are matrices of the form $\|x_0 \mid T\|$ where T is a diagonal matrix. Let us spell out a subdivision for rectangles in matrix terms:

$$\left\| \begin{array}{c|ccccc} x_{01} & t_1 & & & & \\ \vdots & & \ddots & & & \\ x_{0i} & & & t_i + s_i & & \\ \vdots & & & & \ddots & \\ x_{0k} & & & & & t_k \end{array} \right\| \tag{A.8.6}$$

is subdivided in the ith direction into

$$\left\| \begin{array}{c|ccccc} x_{01} & t_1 & & & & \\ \vdots & & \ddots & & & \\ x_{0i} & & & t_i & & \\ \vdots & & & & \ddots & \\ x_{0k} & & & & & t_k \end{array} \right\| \quad \text{and} \quad \left\| \begin{array}{c|ccccc} x_{01} & t_1 & & & & \\ \vdots & & \ddots & & & \\ x_{0i} + t_i & & & s_i & & \\ \vdots & & & & \ddots & \\ x_{0k} & & & & & t_k \end{array} \right\|$$

We invite the reader to spell out subdivisions of $[x_0, x_1, \dots, x_k]$ explicitly, and in particular to prove that $[x_0, x_1, \dots, x_0, \dots, x_k]$ (with x_0 appearing again in the ith position) subdivides in the ith direction into two copies of itself.

The chain complexes of singular cubes

Out of the cubical complex $S_{[\bullet]}(M)$, we can manufacture a chain complex $C_\bullet(M)$ in the standard way. We let $C_k(M)$ be the free abelian group generated by $S_{[k]}(M)$. The boundary operator $\partial : C_k(M) \to C_{k-1}(M)$ is defined on the generators $\gamma \in S_{[k]}(M)$ by the standard formula (see e.g. Hilton and Wylie, 1960, §8.3) with $2k$ terms

$$\partial(\gamma) := \sum_{i=1}^{k}(-1)^i\big(\partial_i^0(\gamma) - \partial_i^1(\gamma)\big). \tag{A.8.7}$$

We let $N_k(M) \subseteq C_k(M)$ be the subgroup generated by

$$\gamma - \gamma' - \gamma''$$

for all γ which are subdivided in some direction into γ' and γ''.

Proposition A.8.1 *The boundary operator $\partial : C_k(M) \to C_{k-1}(M)$ maps $N_k(M)$ into $N_{k-1}(M)$.*

Proof. Assume γ is subdivided in the ith direction into γ' and γ''. By (A.8.2) we have that, for $j \neq i$, $\partial_j^\alpha(\gamma)$ is subdivided (in the ith direction, or in the $i-1$th direction, according to whether $j > i$ or $j < i$) into $\partial_j^\alpha(\gamma')$ and $\partial_j^\alpha(\gamma'')$; the difference of these terms is in N_{k-1}. In $\partial(\gamma - \gamma' - \gamma'')$ only remain the six ∂_i^α-terms. Omitting i from the notation, these six terms are (plus or minus)

$$[\partial^0(\gamma) - \partial^0(\gamma') - \partial^0(\gamma'')] - [\partial^1(\gamma) - \partial^1(\gamma') - \partial^1(\gamma'')].$$

The two first terms in the left-hand square bracket cancel by (A.8.1), and the two outer terms in the last square bracket cancel for the same reason. So we are left with $\partial^1(\gamma') - \partial^0(\gamma'')$. This is 0, likewise by (A.8.1). This proves the proposition.

We have the cochain complex of R-valued cochains on the cubical complex just described. A k-cochain on M is thus a map $\Phi : S_{[k]}(M) \to R$, or equivalently an additive map $\Phi : C_k(M) \to R$. Such cochains behave contravariantly, i.e. given a map $f : M' \to M$ between manifolds and a cochain Φ on M, we get a cochain $f^*(\Phi)$ on M'.

Also, the (cubical) boundary operator $\partial : C_{k+1}(M) \to C_k(M)$ gives rise to a coboundary operator d from k-cochains to $k+1$-cochains.

A k-cochain Φ is said to satisfy the *subdivision law* if

$$\Phi(\gamma) = \Phi(\gamma') + \Phi(\gamma'')$$

whenever a singular cube γ subdivides, in some direction, into γ' and γ''. This is equivalent to saying that $\Phi : C_k(M) \to R$ kills $N_k(M)$.

A k-cochain Φ is called *alternating* if $\Phi(\gamma \circ \sigma) = \text{sign}(\sigma) \cdot \Phi(\gamma)$, for any map $\sigma : R^k \to R^k$ given by some permutation σ of the coordinates.

Definition A.8.2 *Consider a k-cochain, i.e. a map* $\Phi : S_{[k]}(M) \to R$*. It is called an* observable *(of dimension k) if it satisfies the subdivision law and is alternating.*

(The term "observable" I picked up from Meloni and Rogora, 1988, who considered such functionals, for similar reasons as ours. The terminology can be motivated by the idea that a (combinatorial) differential form is microscopic (has infinitesimal values), whereas its integral exhibits its macroscopic, hence observable, effect. Similar notions appear in Felix and Lavendhomme 1990, reproduced also in Lavendhomme, 1996, §4.5.3.)

Proposition A.8.3 *If* Φ *is an observable, then so is* $d\Phi$*.*

Proof. Since $N_\bullet(M)$ is stable under the boundary operator, it follows that if Φ satisfies the subdivision law, then so does $d\Phi$. Next, for the alternating property: it suffices to consider those permutations σ_i which interchange the ith and $i+1$st coordinate. We must prove that $\Phi(\partial(\gamma \circ \sigma_i)) = -\Phi(\partial(\gamma))$. Writing $\partial_j^\alpha(\gamma)$ in terms of its definition by the affine maps $\delta_j^\alpha : R^k \to R^{k+1}$,

$$\partial(\gamma \circ \sigma) = \sum_{j=1}^{k+1} (-1)^j (\gamma \circ \sigma_i \circ \delta_j^0 - \gamma \circ \sigma_i \circ \delta_j^1).$$

Now we need some relations between the σ_i and δ_j^α. They can be found in formula (29) (middle line) of Grandis and Mauri (2003). Let us elaborate on the case $i = 1$, and leave the remaining cases to the reader. For $j \geq 3$, we have $\sigma_1 \circ \delta_j^\alpha = \delta_j^\alpha \circ \sigma_1$, so when applying Φ, we get the required sign change. There remain the terms $j = 2$ and $j = 1$. Here, the sign change occurs already at the level of the chain complex. The $j = 2$ term of the chain $\partial(\gamma \circ \sigma_1)$ is

$$(-1)^2 (\gamma \circ \sigma_1 \circ \delta_2^0 - \gamma \circ \sigma_1 \circ \delta_2^1);$$

now, by loc. cit. $\sigma_1 \circ \delta_2^\alpha = \delta_1^\alpha$, so the $j = 2$ term in $\partial(\gamma \circ \sigma_1)$ equals minus the $j = 1$ term in $\partial(\gamma)$. Similarly, the $j = 1$ term in $\partial(\gamma \circ \sigma_1)$ equals minus the $j = 2$ term in $\partial(\gamma)$, because $\sigma_1 \circ \delta_1^\alpha = \delta_2^\alpha$ by loc. cit.

This proves the proposition.

We thus have a cochain complex of observables; we don't give it a name, since it is isomorphic to the cochain complex of (cubical) differential forms, see Section 3.4.

A.9 "Nullstellensatz" in multilinear algebra

In classical terms (multilinear algebra over a field R of characteristic 0), this theorem says the following. Let V be an n-dimensional vector space, and let $\omega_i : V \to R$ be linearly independent linear maps ($i = 1, \ldots, q$); let U be the meet of the null spaces of the ω_is (so U in an $n-q$-dimensional linear subspace of V). If now $\theta : V^k \to R$ is a k-linear alternating map which vanishes on $U^k \subseteq V^k$, then θ belongs to the ideal generated by the ω_is in the exterior algebra of multilinear alternating functions on V.

In order not to get involved in how notions like "field" and "linear independence" ramify in the context of SDG (where we cannot assume that R is a field, and some logical laws, like the law of excluded middle, proof by contradiction, etc. have limited validity), we shall only prove a very special, totally coordinatized, case: it works for any commutative ring R. Also, we shall only consider the case of $k = 2$, for simplicity of notation. It is the only case we need.

If α and β are linear maps $V \to R$, we get a bilinear alternating map $\alpha \wedge \beta : V \times V \to R$,

$$(\alpha \wedge \beta)(v_1, v_2) := \alpha(v_1) \cdot \beta(v_2) - \alpha(v_2) \cdot \beta(v_1).$$

Consider $V = R^n$. Let $\text{proj}_i : R^n \to R$ be the projection onto the ith factor ($i = 1, \ldots, n$). A pair $\underline{x}_1, \underline{x}_2$ of vectors in R^n defines a $2 \times n$ matrix $\underline{\underline{x}}$ whose rows are $\underline{x}_1$ and $\underline{x}_2$. For $i < j$,

$$(\text{proj}_i \wedge \text{proj}_j)(\underline{x}_1, \underline{x}_2) = \det \underline{\underline{x}}_{i,j}$$

where $\underline{\underline{x}}_{i,j}$ is the 2×2 matrix obtained from $\underline{\underline{x}}$ by taking the ith and jth column.

Any bilinear alternating $\theta : R^n \times R^n \to R$ is a linear combination of such $\text{proj}_i \wedge \text{proj}_j$; more explicitly

$$\theta = \sum_{i<j} c_{ij} \cdot \text{proj}_i \wedge \text{proj}_j, \tag{A.9.1}$$

where $c_{ij} = \theta(\underline{e}_i, \underline{e}_j)$ ($\underline{e}_i$ the ith canonical basis vector).

We consider the "vector space" (R-module) $V = R^n$, and we let $\omega_i : R^n \to R$ be projection proj_i to the ith factor ($i = 1, \ldots, q$). The meet U of the null spaces of the ω_is is then the subspace of coordinate vectors whose k first coordinates are 0. With these notations:

Proposition A.9.1 *Let $\theta : R^n \times R^n \to V$ be bilinear alternating, and assume that $\theta(\underline{x}_1, \underline{x}_2) = 0$ whenever $\underline{x}_1$ and $\underline{x}_2$ are in U. Then θ may be written*

$$\theta = \sum_{i=1}^{q} \omega_i \wedge \alpha_i$$

for suitable linear $\alpha_i : R^n \to R$ (in other words, θ belongs to the ideal generated by the ω_is).

Proof. From the assumption, we see that $\theta(\underline{e}_i, \underline{e}_j)$ vanishes if both i and j are $> q$. So in (A.9.1), c_{ij} vanishes if $q < i$. The remaining terms are of the form $c_{ij}\,\mathrm{proj}_i \wedge \mathrm{proj}_j$ with $i \leq q$; since $\mathrm{proj}_i = \omega_i$ for $i \leq q$, we thus have

$$\theta = \sum_{i \leq q, i < j} \omega_i \wedge c_{ij}\,\mathrm{proj}_j .$$

So $\alpha_i := \sum_{i<j} c_{ij}\,\mathrm{proj}_j$ will do the job.

Bibliography

Ambrose, W. and I. M. Singer (1953) A theorem on holonomy, *Transactions of the American Mathematical Society* **75**, 428–443.

Baez, J. and U. Schreiber (2006) Higher Gauge Theory, arXiv:math/0511710v2 [math.DG].

Barr, M. (1971) Exact categories, in M. Barr, P. A. Grillet and D. H. van Osdol (eds), *Exact Categories and Categories of Sheaves*, Springer Lecture Notes in Mathematics, Vol. 236, pp. 1–120.

Bell, J. (1998) *A Primer of Infinitesimal Analysis*, Cambridge University Press.

Bergeron, F. (1980) Objet infinitésimal en géométrie différentielle synthétique, Exposé 10 in Rapport de Recherches du Dépt. de Math. et de Stat. 80-11 and 80-12, Université de Montréal.

Bishop, R. L. and R. J. Crittenden (1964) *Geometry of Manifolds*, Academic Press.

Breen, L. and W. Messing (2001) Combinatorial differential forms, *Advances in Math.* **164**, 203–282.

Brown, R. and P. J. Higgins (1981) On the algebra of cubes, *Journal of Pure and Applied Algebra* **21**, 233–260.

Brown, R. and C. Spencer (1976) Double groupoids and crossed modules, *Cahiers de Topologie et Géométrie Différentielle* **17**, 343–362.

Bunge, M. and E. Dubuc (1987) Local concepts in synthetic differential geometry and germ representability, in D. Kueker, E. G. K. Lopez-Escobar and C. H. Smith (eds), *Mathematical Logic and Theoretical Computer Science*, Marcel Dekker.

Burke, W. L. (1985) *Applied Differential Geometry*, Cambridge University Press.

Demazure, M. and P. Gabriel (1970) *Groupes Algébriques* Tome I, Masson and North Holland.

Dubuc, E. (1979) Sur les modèles de la géometrie différentielle synthétique, *Cahiers de Topologie et Géométrie Différentielle* **20**, 231–279.

Dubuc, E. (1990) Germ representability and local integration of vector fields in a well adapted model of SDG, *Journal of Pure and Applied Algebra* **64**, 131–144.

Dubuc, E. and A. Kock (1984) On 1-form classifiers, *Communications in Algebra* **12**, 1471–1531.

Ehresmann, C. (1954) Structures locales, *Annali di Matematica*, 133–142.

Ehresmann, C. (1950) Les connexions infinitésimales dans un espace fibré differentiable, Colloque de Topologie, Bruxelles, CBRM.

Faran, J. (1998) A synthetic Frobenius Theorem, *Journal of Pure and Applied Algebra* **128**, 11–32.

Felix, Y. and R. Lavendhomme (1990) On DeRham's theorem in synthetic differential geometry, *Journal of Pure and Applied Algebra* **65**, 21–31.

Frölicher, A. and A. Kriegl (1988) *Linear Spaces and Differentiation Theory*, Wiley-Interscience.

Grandis, M. (2001) Finite sets and symmetric simplicial sets, *Theory and Applications of Categories* **8**, 244–252.

Grandis, M. and L. Mauri (2003) Cubical sets and their site, *Theory and Applications of Categories* **11**, 186–211.

Greub, W. (1978) *Multilinear Algebra* (2nd edition), Springer Universitext.

Grothendieck, A. (1967) *Élements de Géométrie Algébrique* IV, Étude locale de schémas et des morphismes de schémas, part 4, Publ. Math. 32, Bures-sur-Yvette.

Helgason, S. (1962) *Differential Geometry and Symmetric Spaces*, Academic Press.

Hilton, P. J. and S. Wylie (1960) *Homology Theory*, Cambridge University Press.

Johnstone, P. T. (1977) *Topos Theory*, Academic Press.

Johnstone, P. T. (2002) *Sketches of an Elephant: A Topos Theory Compendium*. Oxford Logic Guides, Vols 43, 44, Oxford University Press.

Joyal, A. Structures Infinitésimales, Lecture, 30 March 1979 (handwritten notes by G. Reyes).

Joyal, A. and I. Moerdijk (1994) A completeness theorem for open maps, *Journal of Pure and Applied Logic* **70**, 51–58.

Klein, F. (1926) *Vorlesungen über höhere Geometrie*, Springer Verlag.

Kobayashi, S. and K. Nomizu (1963) *Foundations of Differential Geometry*, Wiley New York.

Kock, A. (1977) A simple axiomatics for differentiation, *Mathematica Scandinavica* **40**, 183–193.

Kock, A. (ed.) (1979) *Topos Theoretic Methods in Geometry*, Aarhus Mathematical Institute Various Publications Series No. 30.

Kock, A. (1980) Formal manifolds and synthetic theory of jet bundles, *Cahiers de Topologie et Géométrie Différentielle* **21**, 227–246.

Kock, A. (1981/2006) *Synthetic Differential Geometry*, LMS 51, Cambridge University Press (2nd edition, LMS 333, Cambridge University Press).

Kock, A. (1982a) Differential forms with values in groups, *Bulletin of the Australian Mathematical Society* **25**, 357–386.

Kock, A. (1982b) The algebraic theory of moving frames, *Cahiers de Topologie et Géométrie Différentielle* **23**, 347–362.

Kock, A. (ed.) (1983a) *Category Theoretic Methods in Geometry*, Proceedings Aarhus 1983, Aarhus Mathematical Institute Various Publications Series No. 35.

Kock, A. (1983b) Some problems and results in synthetic functional analysis, in A. Kock (ed.) *Category Theoretic Methods in Geometry*, Proceedings Aarhus 1983, Aarhus Mathematical Institute Various Publications Series No. 35, pp. 168–191.

Kock, A. (1984) A combinatorial theory of connections, in J. Gray (ed.) *Mathematical Applications of Category Theory*, Proceedings 1983, AMS Contemporary Mathematics Vol. 30, pp. 132–144.

Kock, A. (1985) Combinatorics of non-holonomous jets, *Czechoslovak Mathematical Journal* **35**, 419–428.

Kock, A. (1986a) Convenient vector spaces embed into the Cahiers topos, *Cahiers de Topologie et Géométrie Différentielle Catégoriques* **27**, 3–17. Corrections in Kock and Reyes (1987).

Kock, A. (1986b) Introduction to synthetic differential geometry, and a synthetic theory of dislocations, in F. W. Lawvere and S. Schanuel (eds), *Categories in Continuum Physics*, Proceedings Buffalo 1982, Springer Lecture Notes Vol. 1174.

Kock, A. (1989) On the integration theorem for Lie groupoids, *Czechoslovak Mathematical Journal* **39**, 423–431.

Kock, A. (1996) Combinatorics of curvature, and the Bianchi Identity, *Theory and Applications of Categories* **2**, 69–89.

Kock, A. (1998) Geometric construction of the Levi–Civita parallelism, *Theory and Applications of Categories* **4**, 195–207.

Kock, A. (2000) Differential forms as infinitesimal cochains, *Journal of Pure and Applied Algebra* **154**, 257–264.

Kock, A. (2001) Infinitesimal aspects of the Laplace operator, *Theory and Applications of Categories* **9**, 1–16.

Kock, A. (2003) First neighbourhood of the diagonal, and geometric distributions, *Universitatis Iagellonicae Acta Mathematica* **41**, 307–318.

Kock, A. (2004) A geometric theory of harmonic and semi-conformal maps, *Central European Journal of Mathematics* **2**, 708–724.

Kock, A. (2006) Connections and path connections in groupoids, Aarhus Mathematical Institute Preprint No. 10. http://www.imf.au.dk/publs?id=619.

Kock, A. (2007a) Envelopes – notion and definiteness, *Beiträge zur Algebra und Geometrie* **48**, 345–350.

Kock, A. (2007b) Principal bundles, groupoids, and connections, in J. Kubarski, J. Pradines, T. Rybicki and R. Wolak (eds), *Geometry and Topology of Manifolds* (The Mathematical Legacy of Charles Ehresmann), Banach Center Publications 76, pp. 185–200.

Kock, A. (2007c) Infinitesimal cubical structure, and higher connections, arXiv:0705.4406[math.CT]

Kock, A. (2008) Combinatorial differential forms – cubical formulation, *Applied Categorical Structures* DOI 10.1007/s10485-008-9143-6.

Kock, A. and R. Lavendhomme (1984) Strong infinitesimal linearity, with applications to strong difference and affine connections, *Cahiers de Topologie et Géometrie Differentielle Catégoriques* **25**, 311–324.

Kock, A. and G. E. Reyes (1979a) Manifolds in formal differential geometry, in *Applications of Sheaves*, Proceedings Durham 1977, Springer Lecture Notes in Mathematics Vol. 753.

Kock, A. and G. E. Reyes (1979b) Connections in formal differential geometry, in *Topos Theoretic Methods in Geometry*, Aarhus Mathematical Institute Various Publications Series No. 30.

Kock, A. and G. E. Reyes (1987) Corrigendum and addenda to "Convenient vector spaces embed", *Cahiers de Topologie et Géométrie Différentielle Catégoriques* **28**, 99–110.

Kock, A. and G. E. Reyes (2003) Some calculus with extensive quantities: wave equation, *Theory and Applications of Categories* **11**, 321–336.

Kock, A. and G. E. Reyes (2006) Distributions and heat equation in SDG, *Cahiers de Topologie et Géométrie Différentielle Catégoriques* **47**, 2–28.

Kock, A., G. E. Reyes and B. Veit (1979/80) Forms and integration in synthetic differential geometry, Aarhus Preprint Series No. 31.

Kolar, I. (1982) On the second tangent bundle and generalized Lie derivatives, *Tensor N.S.* **38**, 98–102.

Kriegl, A. and P. Michor (1997) *The Convenient Setting of Global Analysis*, American Mathematical Society.

Kumpera, A. and D. Spencer (1973) *Lie Equations.* Volume I: General Theory, Annals of Mathematics Studies Number 73, Princeton University Press.

Lambek, J. and P. Scott (1986) *Introduction to Higher Order Categorical Logic*, Cambridge Studies in Advanced Mathematics 7, Cambridge University Press.

Lavendhomme, R. (1987) Lecons de géométrie différentielle synthétique naïve, CIACO, Louvain-la-Neuve.

Lavendhomme, R. (1994) Algèbres de Lie et groupes microlinéaires, *Cahiers de Topologie et Géométrie Différentielle Catégoriques* **35**, 29–47.

Lavendhomme, R. (1996) *Basic Concepts of Synthetic Differential Geometry*, Kluwer Academic Publishers.

Lawvere, F. W. (1979) Categorical dynamics, in *Topos Theoretic Methods in Geometry*, Aarhus Mathematical Institute Various Publications Series No. 30, pp. 1–28.

Lawvere, F. W. (1998) Outline of Synthetic Differential Geometry, 14 pp. Unpublished manuscript, www.acsu.buffalo.edu.wlawvere/downloadlist.html.

Lawvere, F. W. (2000) Comments on the development of topos theory, in J.-P. Bier (ed.), *Development of Mathematics 1950–2000*, Birkhäuser Verlag, Basel, pp. 715–734.

Lawvere, F. W. (2002) Categorical algebra for continuum micro physics, *Journal of Pure and Applied Algebra* **175**, 267–287.

Lawvere, F. W., C. Maurer and G. C. Wraith (eds) (1975) *Model Theory and Topoi*, Springer Lecture Notes in Mathematics Vol. 445.

Libermann, P. (1971) Sur les prolongements des fibrés principaux et des groupoïdes différentiables banachiques, in *Analyse Globale*, Séminaire de Mathématiques Supérieures, Les Presses de l'Université de Montréal, pp. 7–108.

Lie, S. (1896/1977) *Geometrie der Berührungstransformationen*, Leipzig 1896 (reprinted by Chelsea Publishing Company, 1977).

Mac Lane, S. (1971) *Categories for the Working Mathematician*, Springer Graduate Texts in Mathematics No. 5, Springer Verlag.

Mac Lane, S. and I. Moerdijk (1992) *Sheaves in Geometry and Logic*, Springer Universitext.

Mackenzie, K. C. H. (1987) *Lie Groupoids and Lie Algebras in Differential Geometry*, LMS 124, Cambridge University Press.

Mackenzie, K. C. H. (1995) Lie algebroids and Lie pseudoalgebras, *Bulletin of the London Mathematical Society* **27**, 97–147.

Madsen, I. and J. Tornehave (1997) *From Calculus to Cohomology*, Cambridge University Press.

Malgrange, B. (1972) Equations de Lie, I, *Journal of Differential Geometry* **6**, 503–522.

McLarty, C. (1983) Local, and some global, results in synthetic differential geometry, in *Category Theoretic Methods in Geometry*, Proceedings Aarhus 1983, Aarhus Mathematical Institute Various Publications Series No. 35, pp. 226–256.

McLarty, C. (1995) *Elementary Categories, Elementary Toposes*, Oxford Logic Guides 21, Clarendon Press, Oxford.

Meloni, G.-C. and E. Rogora (1988) Global and infinitesimal observables, in F. Borceux (ed.), *Categorical Algebra and its Applications*, Proceedings Louvain-la-Neuve 1987, Springer Lecture Notes 1348, pp. 270–279.

Minguez Herrero, C. (1988) Wedge products of forms in synthetic differential geometry, *Cahiers de Topologie et Géométrie Différentielle Catégoriques* **29**, 59–66.

Moerdijk, I. and G. E. Reyes (1991) *Models for Smooth Infinitesimal Analysis*, Springer.

Mumford, D. (1965/1988) Introduction to algebraic geometry (Preliminary version of first three chapters), Harvard notes 1965, (reprinted in D. Mumford, *The Red Book of Varieties and Schemes*, Springer Lecture Notes in Mathematics 1358, 1988).

Nelson, E. (1967) *Tensor Analysis*, Princeton University Press.

Nishimura, H. (1997) Theory of microcubes, *International Journal of Theoretical Physics* **36**, 1099–1131.

Nishimura, H. (1998) Nonlinear connections in synthetic differential geometry, *Journal of Pure and Applied Algebra* **131**, 49–77.

Nishimura, H. (2004) Higher-order preconnections in synthetic differential geometry of jet bundles, *Beiträge zur Algebra und Geometrie* **45**, 677–696.

Nishimura, H. (2007) The Lie algebra of the group of bisections, *Far East Journal of Mathematical Sciences* **24**, 329–342.

Nishimura, H. (2008a) The Frölicher–Nijenhuis Calculus in Synthetic Differential Geometry, arXiv:0810.5492[math.DG]

Nishimura, H. (2008b) Curvature in synthetic differential geometry of groupoids, *Beiträge zur Algebra und Geometrie* **49**, 369–381.

Noll, W. (1967) Materially uniform simple bodies with inhomogeneities, *Archive for Rational Mechanics and Analysis* **27**, 1–32.

Palais, R. *et al.* (1965) Seminar on the Atiyah–Singer Index Theorem, *Annals of Mathematical Studies* **57**.

Penon, J. (1985) De l'infinitésimal au local, These de Doctorat d'Etat, Paris 7.

Pradines, J. (1967) Theorie de Lie pour les groupoides differentiables, *C.R. Academy Paris* **266**, 245–248.

Reyes, G. and G. C. Wraith (1978) A note on tangent bundles in a category with a ring object, *Mathematica Scandinavica* **42**, 53–63.

Saunders, D. J. (1989) *The Geometry of Jet Bundles*, LMS 142, Cambridge University Press.

Schreiber, U. and K. Waldorf (2008), Smooth Functors vs. Differential Forms, arXiv:0802.0663v2[mathDG]

Serre, J.-P. (1965) *Lie Algebras and Lie Groups*, Benjamin Publishing Company.

Spivak, M. (1979) *A Comprehensive Introduction to Differential Geometry* (Vols 1–5), Publish or Perish, Inc.

Virsik, J. (1971) On the holonomity of higher order connections, *Cahiers de Topologie et Géométrie Différentielle* **12**, 197–212.

White, J. E. (1982) *The Method of Iterated Tangents with Applications to Local Riemannian Geometry*, Pitman Press.

Index